ATLAS OF ISOSEISMAL MAPS OF NEW ZEALAND EARTHQUAKES

G.L. DOWNES

Institute of Geological & Nuclear Sciences monograph 11

Institute of Geological & Nuclear Sciences Limited
Lower Hutt, New Zealand
1995

BIBLIOGRAPHIC REFERENCE

Downes, G.L., 1995. Atlas of isoseismal maps of New Zealand earthquakes. *Institute of Geological & Nuclear Sciences monograph 11, 304* p., Lower Hutt, New Zealand: Institute of Geological & Nuclear Sciences Limited.

G.L. Downes
Institute of Geological & Nuclear Sciences Limited, Kelburn, New Zealand

Maps were draughted by C. Hume, Institute of Geological & Nuclear Sciences Limited
Prepared for publication by C. Hourihan and C. Eales, Institute of Geological & Nuclear Sciences Limited
Printed by GP Print Ltd, Wellington

ISSN 1172-028X
ISBN 0-478-09502-3

FRONT COVER

North Canterbury earthquake 1888: geologist Alexander McKay beside a fence offset 2.6 m dextrally by movement on the Hope Fault. (after McKay 1902)

CONTENTS

Abstract ii

Introduction 1

Earthquake data 2

Hypocentres 2

Magnitudes 6

Macroseismic data and the isoseismal maps 7

Acknowledgments 9

References 9

Intensity scales 12

Modified Mercalli (MM) intensity scale 12

Rossi-Forel (R-F) scale 19

R-F to MM conversion chart 19

Standard report form (pages 1 and 2) 20

Isoseismal maps with their descriptions, in chronological order 23

Map Symbols 23

Appendix

Selected bibliography (listed by earthquake)

Table 1. Index of isoseismal maps 4

Table 2. Definition of the velocity/depth relations for New Zealand. 6

Figure 1. Index map of earthquakes listed in Table 1 3

Figure 2. Map of New Zealand seismograph stations - December 1993 8

Figure 3. Map of districts and locations frequently referred to in text 11

ABSTRACT

This Atlas of 133 isoseismal maps of New Zealand earthquakes contains the felt intensity distributions of 123 earthquakes, 122 of which occurred within or close to the New Zealand mainland. A magnitude 6.7 event which originated south of the Kermadec Islands but was felt widely on the New Zealand mainland is also included. In five cases detailed maps of the epicentral area are given. Three other earthquakes are represented by two maps, these being either an alternative interpretation of the same data set or an interpretation based on a different data set.

The Atlas is arranged with the isoseismal maps presented in chronological order. Each map is accompanied by the most recently calculated hypocentral parameters and an indication of the method of their determination and placed opposite a brief description of the earthquake, the source of the map and the data on which it has been based. References are listed by earthquake at the end of the Atlas after the last isoseismal map. Sufficient references are given for the reader to be able to obtain a fuller description of the felt effects of the earthquake and an insight into what other aspects have been studied. The Atlas includes all known maps that are considered to have been sufficiently well-determined. Two maps have the isoseismal lines plotted in Rossi-Forel scale, no conversion being possible as the individual felt intensities have not been given and the original felt reports have not been located. The data, however, appear to be more extensive than that which is held in our files and for this reason the maps have been included. All other maps refer to the Modified Mercalli scale adapted for New Zealand conditions by Eiby (1966) or Study Group (1992).

The maps in the Atlas span the years 1848 to 1990, the earliest map being from the 1848 Marlborough earthquake and the most recent, the Weber earthquakes in February and May 1990. Since organised European colonisation began about 1840, New Zealand has experienced many large earthquakes, more than are represented in this collection. Few large earthquakes between 1855 and 1930 have been studied in detail and their inclusion in a collection similar to this awaits further research.

The largest earthquake to be included in the Atlas and the largest known in the recorded history of New Zealand is the Wairarapa earthquake on January 23 1855, estimated magnitude 8.0-8.2. The Wellington and the Wairarapa districts suffered severely, up to intensity MM10. The earthquake was also destructive as far north as Wanganui and along the northeast coast of the South Island and caused extensive landsliding and liquefaction. Regional uplift of the southern part of the North Island, including Wellington, occurred, accompanied possibly by subsidence in eastern Wairarapa and Marlborough as well as vertical and probably horizontal movement on the West Wairarapa Fault. Evaluation of the historic data on this event is continuing.

The Mangonui earthquake on Nov 16 1963, of magnitude 3.5 M_L, is the smallest earthquake included in the Atlas. It was one of several earthquakes which occurred in this area of Northland in November and December 1963, the largest event having a magnitude 4.9 M_L, for which there is also an isoseismal map. They were unusual as few earthquakes of damaging intensity have occurred north of Auckland.

INTRODUCTION

The purpose of this Atlas is to bring together in one volume all the available well determined isoseismal maps and to define the data on which they have been based. The distribution of the felt intensities from an earthquake is most commonly represented by an isoseismal map. Usually, observed intensities are indicated on the map and isoseismal lines drawn to contour areas which are judged to have experienced approximately equal intensity of shaking. The isoseismal map allows immediate visualisation of the extent of the effects of an earthquake, is invaluable in the estimation of magnitudes and epicentres of historic earthquakes and is useful in the formulation of attenuation models and hence, in the estimation of seismic hazard. However, it should be recognised that the quantity and quality of the data is often less than ideal, and researchers can differ in their interpretation of the intensity scale and technique of assessing intensities and drawing isoseismal lines, more in the latter than in the former. It has not been the purpose of this Atlas to question technique but sufficient background information on the data sources is supplied to enable users to make their own judgement. For the purposes of detailed intensity attenuation studies, accurate locations of the individual felt intensity data from the original data source are preferable to those approximated from the isoseismal maps.

Two intensity scales are used; the Modified Mercalli scale, adapted for New Zealand (Eiby (1966) and Study Group (1992)) and the Rossi-Forel scale. The scales are defined and the conversion from Rossi-Forel to Modified Mercalli scale is given after the references. The revision of the 1966 version of the Modified Mercalli scale has been proposed by a Study Group of the New Zealand National Society for Earthquake Engineering (Study Group, 1992). It has been used to determine intensities for the 1929 Buller earthquake only. Two maps have the isoseismal lines plotted using the Rossi-Forel scale.

The sources of the maps appearing in the Atlas are various. Many have been reproduced from the New Zealand Seismological Reports. Some have not been published previously. All maps of earthquakes occurring from 1943 to 1976 published in this collection have revised magnitudes and hypocentral parameters, using post-1976 analysis procedures. A number of maps published previously, primarily in the New Zealand Seismological Reports, have in this manner been superseded. Some revised epicentres differ significantly from previously calculated values, but few cases require revision of the isoseismal map. This is generally commented upon and both old and revised epicentres are indicated on the maps for comparison.

The New Zealand Seismological Reports, frequently quoted and referenced in the Atlas, first appeared in 1921 and Eiby (1988) writes: "...these have continued under the direction of what has now become the Seismological Observatory, Wellington, surviving a number of confusing changes in the Observatory's name and its administrative control". A predecessor of the Seismological Observatory referenced in the text is the Dominion Observatory, which became part of DSIR (Department of Scientific & Industrial Research) in 1926. In 1951 various units of the DSIR engaged in geophysical studies, including the Seismological Observatory, were combined to form one branch, Geophysics Division, which later, in 1990, merged with the Geological Survey DSIR, to be called DSIR Geology & Geophysics. In July 1992 the Seismological Observatory became part of the Institute of Geological and Nuclear Sciences Limited (GNS). Files that originated under previous organisational names are now part of GNS, and hence maps that have recently been revised or retrieved from these files are attributed to GNS in Table 1.

For some earthquakes the original felt reports have been reinterpreted. Where these have not been easily accessible or required more time to research than was feasible in the short term, the early summaries of felt intensity data, recorded as Rossi-Forel scale, have been converted to Modified Mercalli scale intensities (and marked as such). The latter category includes many of the events dating from 1900 until 1942 and maps of this period should only be used with the knowledge that there is not a one-to-one correspondence between the two scales. Complete revision and a new descriptive account using sources such as newspapers, other archival papers and felt information contained in GNS files and re-working of the available seismological data is required.

Not all large earthquakes that have occurred in New Zealand have isoseismal maps. The period 1950-1990 is represented by 96 maps, far in excess of the 37 maps for the years 1840, the beginning of organised European colonisation of New Zealand, to 1949. Yet the last forty or so years have been seismically quiet compared with the previous one hundred. Notable events for which there is little information and no isoseismal map are the 1863 Hawkes Bay earthquake which is thought to have caused surface faulting in southern Hawkes Bay, the 1897 Wanganui earthquake and the 1901 Cheviot earthquake, all estimated as having a magnitude of about 7.

To summarise, few significant events for the period 1950-1990 *lack* an isoseismal map. In contrast, for the period 1900-1949 few events *have* an isoseismal map, but it can be assumed that few earthquakes with a magnitude greater than 6.5 have escaped notice and that some felt information is readily available. For the earlier period, up to 1900, there are few maps *and* it is probable that some large and many moderate magnitude events have not been identified other than those close to the main centres of population. Further, there is little readily available felt information for events in this period, other than for those which occurred before 1855 and have been studied by Eiby (1968a, 1973) and the few selected events in the Atlas. A list of known larger earthquakes from 1460 to 1965 is given by Eiby (1968b).

It is hoped that this Atlas will be the first in a series as maps of all significant earthquakes in New Zealand become available or are revised with further research.

EARTHQUAKE DATA

Table 1 lists details of the earthquakes for which isoseismal maps are given in this volume. Figure 1 is an index map of the earthquakes listed in Table 1.

HYPOCENTRES

The first instruments for recording earthquakes in New Zealand were introduced about 1884 but of these little is known (Young 1984, Young et al. 1984). A Milne seismograph was installed in Wellington in 1900 and a second instrument in Christchurch in 1901. Data from these were published by the Seismological Committee of the British Association for the Advancement of Science and in the Transactions and Proceedings of the New Zealand Institute but the instruments were more effective in recording teleseisms than local shocks. Short period instruments were introduced in 1930-31 (in time to record the 1931 Hawke's Bay earthquake), and determinations of epicentres based upon instrumental readings first attempted. The network had developed sufficiently by the early 1940's for a reasonable coverage of shocks of magnitude 4 and above, between latitudes 38° and 42°S, provided that all stations were working.

Since 1987 a programme, one of upgrading the National Network, has replaced many of the analogue instruments with digitally recording seismographs, so that in 1993, a network of nearly seventy digital stations including several special purpose local networks cover the country (Figure 2). Many stations have three components and it seems reasonable to assume that all shallow earthquakes $M_L \geq 3.5$, and deep earthquakes $M_L \geq 3.8$ are detected on or near the New Zealand mainland. Frequently earthquakes of much lower magnitudes are analysed where close station spacing and good azimuthal coverage allow. Details of the growth and development of the New Zealand network may be found in the New Zealand Seismological Reports.

Prior to 1964, when computer analysis was first introduced, hypocentres were determined graphically (indicated by the word "GRAPHICAL" where epicentres were calculated from local data, and graded according to quality, A being the most reliable, after the epicentre location) or from the isoseismal pattern (indicated by a "I" after the epicentre location, also graded for reliability). Some early epicentres have Gutenberg & Richter (1949) as their source. Revision of the graphically determined hypocentres of all earthquakes from 1942, when reliable epicentres are considered possible, up to the end of 1963 is proceeding using post-1976 procedures. The years 1943-54 are completed, although not all published. For homogeneity of the data in this Atlas, hypocentres and magnitudes of events in the period 1955-63 which have isoseismal maps have been revised individually. Further, hypocentres of all events from 1964 to 1976 which had been calculated using an early velocity model have been recalculated using the New Zealand Standard Velocity Model, a laterally homogeneous model defined in Table 2. Use of this model is indicated by a "1" after the epicentre location. However in the data from 1964-1976 readings of crustal phases were often given equal weight with the first arrival data and no readings were excluded in the analysis because of distance from the epicentre. This is at variance with the procedure after 1976 and generally has not been modified. In 1987, when analysis of digitally recorded seismograms was first introduced, a new velocity model was introduced. This model allowed different, although still laterally homogeneous, velocity/depth structures within four regions where special studies had permitted a greater understanding. The New Zealand Standard Model, as defined in Table 2, is used outside these defined areas and the velocities smoothly merged at the interfaces. The four regions and their velocity/depth structure are also defined in Table 2. All hypocentres calculated with this model have a "2" appearing after the epicentre location. Refinement of the velocity model is continuing. Finally special studies of some earthquakes have used other techniques to derive hypocentres and these are appropriately noted.

A description of the technique of analysis of digitally recorded seismograms using the CUSP (Caltech-USGS

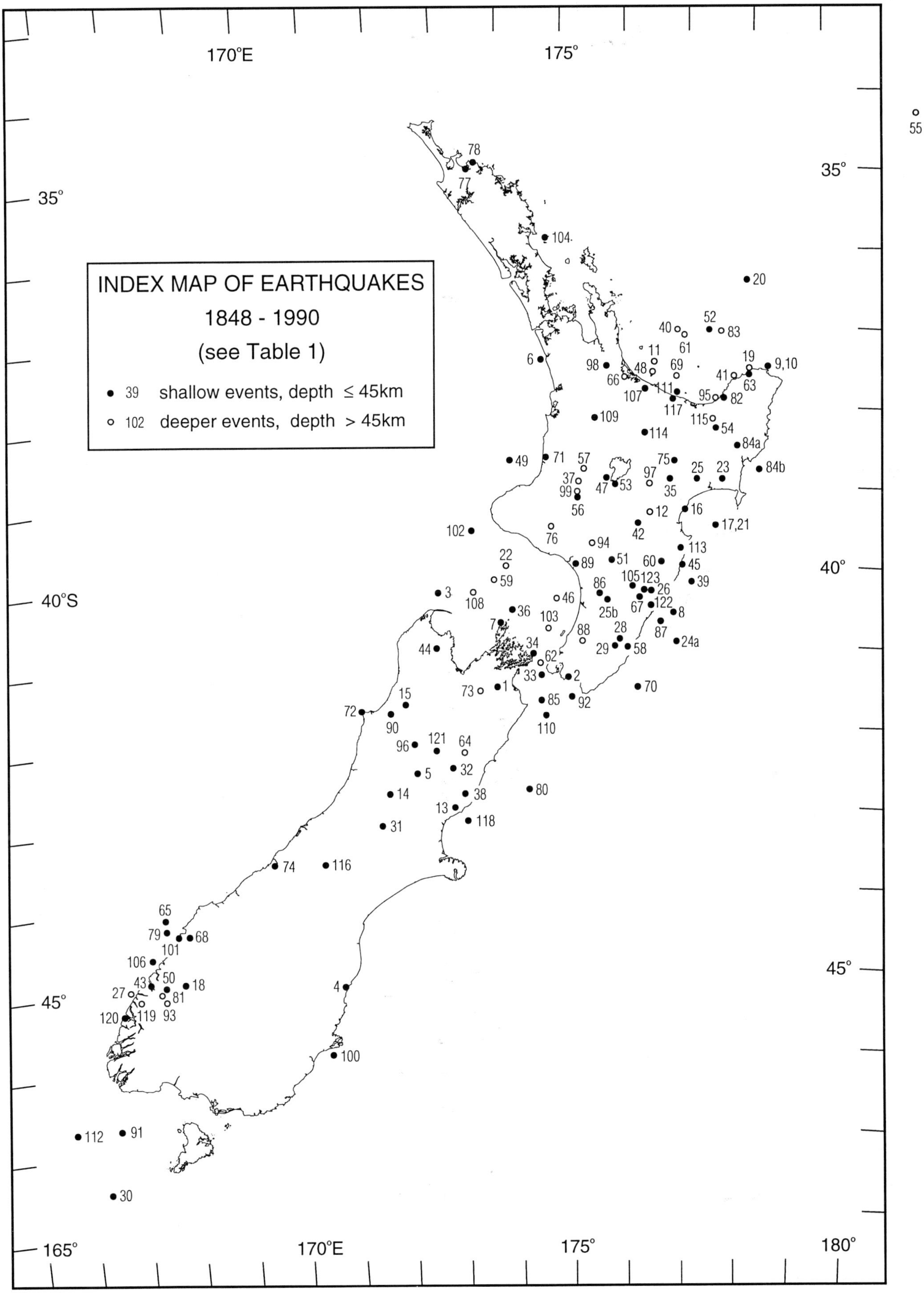

Figure 1. Index map of earthquakes listed in Table 1.

Table 1. Index of isoseismal maps

No.	Locality	Origin time Yr	M	D	UT H	M	S	Epicentre Lat. °S	Long. °E	Focal depth km	Magnitude M,M_L	M_s	m_b,M_w	Int. Scale	Map reference
1	* Marlborough	1848	Oct	15	14	10		41.5	173.8	S	7.1 I			MM	Eiby (1980)
2	* Wairarapa	1855	Jan	23	09	32		41.4	175.0	S	8.2 I			MM	Eiby (1989)
3	* Cape Farewell	1868	Oct	18	12	35		40.3	172.9	S	7-7.5 I			MM	Anderson et al. (1994)
4a	* Oamaru I	1876	Feb	25	15	50		45.2	170.9	S	5.75 I			MM	Reyners (pers. comm.)
4b	* Oamaru II	1876	Feb	25	21	15		45.2	170.9	S	5.75 I			MM	Reyners (pers. comm.)
4c	* Oamaru III	1876	Apr	11	00	10		45.2	170.9	S	5.75 I			MM	Reyners (pers. comm.)
5a	* North Canterbury	1888	Aug	31	16	45		42.6	172.4	S	7-7.3 I			MM	Cowan (1991)
5b	* North Canterbury	1888	Aug	31	16	45		42.6	172.4	S	7-7.3 I			MM	Cowan (1991)
6	* Waikato Heads	1891	Jun	23	23	25		37.4	174.6	C	5.5-6.0 I			MM	GNS
7	* Nelson	1893	Feb	11	20	30		40.7	173.9	LC	6.6-6.9 I			MM	Anderson et al. (1994)
8	* Cape Turnagain	1904	Aug	08	22	50		40.6	176.8	LC		6.7 D&S		MM	Downes (in prep.)
9	* East Cape I	1914	Oct	06	19	16		37.5	178.3	S		6.7 D&S		MM	Reyners (in prep)
10	* East Cape II	1914	Oct	28	00	16		37.5	178.3	S		6.5 D&S		MM	Reyners (in prep)
11	* Bay of Plenty	1914	Nov	22	08	14		37.5	176.5	300	7.2 I			MM	Reyners (in prep)
12	* Hawke's Bay	1921	Jun	28	13	58	54	39.3	176.4	80	7.0 G&R	6.4 D&S		RF-MM	GNS
13a	* Motunau	1922	Dec	25	03	33		43	173	C		6.4 D&S		RF-MM	GNS
13b	* Motunau	1922	Dec	25	03	33		43	173	C		6.4 D&S		RF	Skey (1925)
14	* Arthur's Pass	1929	Mar	09	10	50		42.8	171.9	<15		7.1 D&S		RF-MM	GNS
15	* Buller	1929	Jun	16	22	47	43	41.7	172.2	20		7.8 D&S		MM (rev)	Dowrick (1994)
16	* Hawke's Bay	1931	Feb	02	22	47		39.3	177	30		7.8 D&S		RF-MM	GNS
17	* Hawke's Bay	1931	Feb	13	01	27		39.5	177.5	30		7.3 D&S		RF-MM	GNS
18	* Southland	1931	Sep	15	21	09		45	168	<60		5.4 D&S		RF-MM	GNS
19	Bay of Plenty	1931	Sep	21	13	34		37.5	178	80	6.75 G&R	6.1 D&S		RF-MM	GNS
20	Bay of Plenty	1932	Mar	05	01	40		36.5	178	LC		6.0 D&S		RF-MM	GNS
21	Hawke's Bay	1932	May	05	08	23		39.5	177.5	C		5.9 D&S		RF-MM	GNS
22	Taranaki	1932	Jul	20	04	52		40	174	60	6.3 I	5.1 D&S		RF-MM	GNS
23	* Wairoa	1932	Sep	15	13	55		38.9	177.6	30		6.9 D&S		RF-MM	GNS
24a	* Pahiatua	1934	Mar	05	11	46		40.95	176.80	LC		7.6 D&S		RF	Hayes (1937)
24b	* Pahiatua	1934	Mar	05	11	46		40.45	175.56	LC		7.6 D&S		RF-MM	GNS
25	Hawke's Bay	1934	Mar	15	10	46		38.9	177.2	≤40		6.4 D&S		RF-MM	GNS
26	Hawke's Bay	1938	Dec	15	09	12		40.3	176.4	<45		5.6 D&S		RF-MM	GNS
27	* Charles Sound	1938	Dec	16	17	21		45	167	60		7.0 D&S		RF-MM	GNS
28	* Wairarapa I	1942	Jun	24	11	16	26	40.9	175.9	15		7.2 D&S		RF-MM	GNS
29	* Wairarapa II	1942	Aug	01	12	34	05	41.0	175.8	43		7.0 D&S		RF-MM	GNS
30	* Puysegur Bank	1945	Sep	01	22	44	08	47.47	166.12	12	6.5	7.0 D&S		RF-MM	GNS
31	* Lake Coleridge	1946	Jun	26	12	34	39	43.18	171.68	12	6.2	6.4 D		MM	Eiby (1990)
32	* Waiau	1948	May	22	19	21	27	42.50	173.00	12	6.4			MM	Eiby (1953)
33	* Cook Strait I	1950	Jan	07	14	21	03	41.14	174.56	33	5.6			MM	GNS
34	* Cook Strait II	1950	Jan	12	20	49	50	41.15	174.61	33	5.6			MM	GNS
35	Huiarau Range	1950	Feb	28	18	28	44	38.88	176.70	33	5.7			MM	GNS
36	Stephens Island	1950	Mar	13	09	38	28	40.55	174.04	33	5.7			MM	GNS
37	North Island	1950	Jun	17	15	56	33	38.92	175.25	185	5.9			MM	GNS
38	Cheviot	1951	Jan	10	19	15	18	42.79	173.18	12	5.5			MM	GNS
39	Aramoana	1951	Feb	10	03	27	57	40.21	177.04	33	6.1			MM	GNS
40	Bay of Plenty	1951	Mar	28	01	55	15	37.08	176.91	370	6.4			MM	GNS
41	Cape Runaway	1951	Apr	23	06	50	20	37.56	177.76	125	6.2	5.7 D&S		MM	GNS
42	Hawke's Bay	1951	Jun	24	04	41	51	39.46	176.20	33	6.3			MM	GNS
43	Fiordland	1951	Jul	07	10	15	24	44.87	167.35	33	5.6			MM	GNS
44	NW Nelson	1951	Oct	03	17	38	48	40.99	172.69	12	5.5			MM	GNS
45	Hawke's Bay	1952	Aug	28	10	40	15	39.99	176.96	12	5.8			MM	GNS
46	Foxton	1953	Apr	11	10	27	53	40.45	174.78	106	5.4			MM	GNS
47	Tokaanu	1953	Jul	04	02	07	28	38.86	175.68	12	5.5			MM	GNS
48	Bay of Plenty	1953	Sep	29	01	36	51	37.59	176.48	273	7.2			MM	GNS
49	N. Taranaki Bight	1953	Oct	18	03	38	06	38.64	174.03	12	5.3			MM	GNS
50	Fiordland	1954	May	04	02	41	58	44.92	167.60	12	5.4			MM	GNS
51	Waipuru	1955	Feb	08	10	38	57	39.90	175.80	33	5.1			MM	GNS
52	Bay of Plenty	1956	Jan	30	08	43	02	37.10	177.42	12	5.8	6.4 D&S		MM	GNS
53	Tokaanu	1956	Mar	02	22	43	51	38.90	175.80	12	5.4			MM	GNS
54	East Cape	1956	Dec	28	14	24	33	38.3	177.5	33	6.3			MM	GNS
55	S. Kermedec Ridge	1957	Feb	09	13	29	20	34.25	179.51W	314	6.7			MM	GNS
56	Tongariro	1957	Feb	22	00	30	10	39.19	175.14	5	5.6	4.9 D&S		MM	GNS
57	Lake Taupo	1957	Mar	13	09	11	30	38.68	175.33	273	6.0			MM	GNS
58	Castlepoint	1957	Aug	21	05	48	02	41.03	176.00	33	5.6			MM	GNS
59	S. Taranaki Bight	1957	Sep	26	12	03	06	40.21	173.80	145	5.7			MM	GNS
60a	* Ashley Clinton	1958	Jan	31	06	32	42	39.95	176.53	33	6.1	5.2 D&S		MM	GNS
60b	* Ashley Clinton	1958	Jan	31	06	32	42	39.95	176.53	33	6.1	5.2 D&S		MM	GNS
61	Bay of Plenty	1958	Dec	10	07	03	03	37.13	176.95	284	6.3			MM	GNS
62	Marlborough Sounds	1959	May	22	06	57	09	41.22	174.58	86	5.7	4.9 D&S		MM	GNS
63	East Cape	1960	Feb	03	02	21	08.0	37.56	177.98	33	6.0	5.1 D&S		MM	GNS
64	Acheron River	1960	Feb	21	00	46	54.8	42.28	173.10	50	5.8	5.4 D&S		MM	GNS
65	* Fiordland	1960	May	24	14	46	37.9	44.19	167.73	12	6.5	6.5 D&S		MM	GNS

No.	Locality	Origin time Yr	M	D	UT H	M	S	Epicentre Lat. °S	Long. °E	Focal depth km	Magnitude M, M_L	M_s	m_b, M_w	Int. Scale	Map reference
66	North Island	1961	Feb	03	12	33	29.5	37.66	175.97	308	5.9			MM	GNS
67	Dannevirke	1961	May	14	00	12	38.2	40.38	176.20	33	5.3			MM	GNS
68	Barrier Range	1961	Jul	04	08	23	31.2	44.45	168.08	12	5.1			MM	GNS
69	Bay of Plenty	1961	Jul	26	09	19	04.4	37.68	176.84	204	5.8			MM	GNS
70	* Hikurangi Trench	1961	Dec	27	23	47	53.2	41.51	176.11	12	6.3	6.5 D&S		MM	GNS
71	* Aria	1962	Jan	23	06	49	41.5	38.58	174.60	12	5.5	4.5 D&S		MM	GNS
72	* Westport	1962	May	10	00	27	14.7	41.67	171.44	12	5.7	5.9 D&S		MM	GNS
73	Marlborough	1962	Jul	29	18	19	49.7	41.45	173.47	109	5.7			MM	GNS
74	* Bruce Bay	1962	Oct	15	23	36	31.5	43.54	169.77	12	5.7	5.4 D&S		MM	GNS
75	Huiarau Range	1963	Apr	12	08	41	43.6	38.66	176.80	20	5.5	5.7 D&S		MM	GNS
76	Taranaki	1963	Jul	14	17	06	32.1	39.52	174.72	166	5.5			MM	GNS
77	* Mangonui	1963	Nov	16	15	17	02.3	35.01	173.54	10	3.5			MM	Eiby (1964)
78	* Peria	1963	Dec	22	13	35	34.4	34.93	173.67	10	4.9	4.4 D&S		MM	Eiby (1964)
79	* Big Bay	1964	Mar	08	01	35	47.2	44.26	167.72	12	5.8	5.8 D&S	5.9 M_w	MM	GNS
80	* Chatham Rise	1965	Apr	11	00	11	10.3	42.80	174.31	12	6.1	5.8 ISC	6.1 M_w	MM	GNS
81	Lake Te Anau	1965	May	20	20	37	41.0	45.05	167.51	105	5.6			MM	GNS
82	Bay of Plenty	1965	Jun	15	09	20	31.7	37.90	177.53	33	5.8	5.3 D&S		MM	GNS
83	Bay of Plenty	1965	Dec	08	18	05	20.7	37.11	177.66	234	6.5			MM	GNS
84a	* Gisborne	1966	Mar	04	23	58	56.8	38.52	177.85	25	6.0	5.8 D&S		MM	Hamilton et al. (1969)
84b	* Gisborne	1966	Mar	04	23	58	56.4	38.77	178.14	25	6.0	5.8 D&S		MM	GNS
85	* Seddon	1966	Apr	23	06	49	40.5	41.64	174.52	22	5.8	5.6 D&S	5.8 M_w	MM	GNS
86	Palmerston North	1967	Jan	16	11	40	59.1	40.34	175.52	12	4.9			MM	GNS
87	Owahanga	1967	Mar	24	19	09	17.4	40.70	176.51	12	5.3			MM	GNS
88	Te Horo	1967	Sep	21	17	45	46.0	40.93	175.37	64	4.7			MM	GNS
89	Whangaehu	1967	Dec	20	14	56	30.5	39.98	175.11	12	4.8			MM	GNS
90a	* Inangahua	1968	May	23	17	24	15.6	41.76	171.96	15	7-7.1	7.4 D&S	7.1 M_w	MM	GNS
90b	* Inangahua	1968	May	23	17	24	15.6	41.76	171.96	15	7-7.1	7.4 D&S	7.1 M_w	MM	GNS
91	* Solander	1968	Sep	25	07	02	45.5	46.67	166.50	12	5.9	6.2 D&S	6.3 M_w	MM	GNS
92	* Palliser Bay	1968	Nov	01	01	32	25.1	41.62	175.05	33	5.4	5.0 D&S		MM	GNS
93	Fiordland	1969	Jan	02	10	25	21.5	45.12	167.66	139	5.8			MM	GNS
94	North Island	1969	May	23	14	29	42.4	39.70	175.42	110	5.6			MM	GNS
95	Bay of Plenty	1970	Jul	27	12	31	18.9	37.91	177.53	138	5.7			MM	GNS
96	* Maruia Springs	1971	Aug	13	14	42	42.8	42.18	172.30	12	5.8	5.5 D&S	5.7 M_w	MM	GNS
97	Hawke's Bay	1971	Oct	31	12	10	18.3	38.96	176.37	141	5.9			MM	GNS
98	* Te Aroha	1972	Jan	08	21	33	00.1	37.57	175.69	12	5.3	5.0 D&S		MM	Adams et al. (1972)
99	North Island	1973	Jan	05	13	54	27.6	39.13	175.18	173	7.0	6.3 D&S		MM	GNS
100	* Dunedin	1974	Apr	09	07	49	46.1	45.97	170.52	12	4.9			MM	Adams & Kean (1974)
101	* Milford Sound	1974	Sep	20	19	48	39.5	44.40	167.99	12	5.5	5.3 D&S		MM	GNS
102	* Opunake	1974	Nov	05	10	38	38.9	39.54	173.46	12	6.0	5.4 D&S		MM	GNS
103	Kapiti Island	1975	Jan	04	20	37	17.5	40.77	174.67	72	5.9	4.6 D&S		MM	GNS
104	Hen & Chickens Is.	1975	Feb	11	16	45	20.4	35.93	174.76	12	4.4			MM	GNS
105	Oringi	1975	Jun	10	10	11	20.5	40.31	176.07	33	5.8	5.3 D&S		MM	GNS
106	* Milford Sound	1976	May	04	13	56	29.2	44.67	167.38	12	6.5	6.4 D&S	6.5 M_w	MM	GNS
107	* Te Puke	1976	Oct	27	20	57	09.5	37.83	176.34	12	5.0			MM	GNS
108	Golden Bay	1976	Dec	03	13	25	52.4	40.33	173.47	213	5.6			MM	GNS
109	* Korakonui	1976	Dec	05	04	57	16.7	38.17	175.51	1	5.1			MM	Eiby (1977)
110	* Cape Campbell	1977	Jan	18	05	41	48.9	41.84	174.58	33	6.0	5.9 D&S		MM	GNS
111	* Matata	1977	May	31	18	50	55.3	37.83	176.83	9	5.4			MM	Richardson (1989)
112	* Puysegur Bank	1979	Oct	12	10	25	19.5	46.67	165.73	12	6.5	7.2 D&S	7.3 M_w	MM	Webb & Lowry (1982)
113	* Hawke's Bay	1982	Sep	02	15	58	54.4	39.74	176.94	41	5.4		5.6 m_b	MM	GNS
114	* Waiotapu	1983	Dec	14	20	56	29.4	38.36	176.33	5	5.1	4.6 D&S		MM	GNS
115	* Motu River	1984	Mar	08	00	40	52.6	38.22	177.48	73	6.4			MM	GNS
116	* Macauley River	1984	Jun	24	13	29	39.9	43.60	170.64	5	5.9	6.1 D&S	6.1 M_w	MM	Reyners (unpubl.)
117a	* Edgecumbe	1987	Mar	02	01	42	35.0	37.89	176.80	10	6.1	6.6 D&S		MM	Lowry et al. (1989)
117b	* Edgecumbe	1987	Mar	02	01	42	35.0	37.89	176.80	10	6.1	6.6 D&S		MM	Lowry et al. (1989)
118	* Pegasus Bay	1987	Mar	08	19	17	59.0	43.22	173.20	30	5.2			MM	GNS
119	* Te Anau	1988	Jun	03	23	27	34.7	45.10	167.17	57	6.1	6.7 D&S	6.0 m_b 6.7 M_w	MM	Reyners et al. (1991)
120	* Doubtful Sound	1989	May	31	05	54	23.1	45.33	166.87	23	6.1	6.2	6.4 M_w	MM	GNS
121	* Lake Tennyson	1990	Feb	10	03	27	42.0	42.25	172.65	13	5.8	5.9	6.0 M_w	MM	GNS
122	* Weber I	1990	Feb	19	05	34	37.7	40.36	176.36	24	6.1	6.3	5.9 m_b 6.3 M_w	MM	GNS
123a	* Weber II	1990	May	13	04	23	09.3	40.28	176.30	12	6.2	6.3	6.0 m_b 6.4 M_w	MM	GNS
123b	* Weber II	1990	May	13	04	23	09.3	40.28	176.30	12	6.2	6.3	6.0 m_b 6.4 M_w	MM	GNS

Notes:
1 * - Recognised event name
2 G&R - Gutenberg & Richter (1949)
3 D&S - Dowrick & Smith (1990)
4 D - Dowrick (pers. comm.)
5 GNS - Institute of Geological & Nuclear Sciences Ltd
6 For references see individual maps.

Table 2. Definition of the velocity/depth relations for New Zealand. The New Zealand standard velocity model is used to calculate crustal velocities beneath all regions except within the three areas defined below.

MODEL	UPPER DEPTH BOUNDARY (km)	Vp (km/s)	Vs (km/s)	CORNERS OF REGION Lat.	Long.
				(in clockwise order)	
New Zealand Standard	0.0	5.5	3.3		
	12.0	6.5	3.7		
	33.0	8.1	4.6		
Wellington	0.0	4.40	2.54	41.0 S	178.0 E
	0.4	5.63	3.16	43.5 S	175.0 E
	5.0	5.77	3.49	42.0 S	173.0 E
	15.0	6.39	3.50	39.7 S	175.7 E
	25.0	6.79	3.92		
	35.0	8.07	4.80		
	45.0	8.77	4.86		
Taupo	0.0	3.00	1.70	35.6 S	180.0 E
	2.0	5.30	3.00	38.0 S	177.5 E
	5.0	6.00	3.50	39.7 S	175.7 E
	15.0	7.40	4.30	39.0 S	175.0 E
	33.0	7.78	4.39	37.0 S	176.0 E
	65.0	7.94	4.51	34.6 S	178.5 E
	96.4	8.08	4.52		
Clyde	0.0	4.4	2.6	45.5 S	172.0 E
	0.5	6.0	3.3	49.0 S	167.0 E
	12.0	6.5	3.7	44.5 S	168.0 E
	33.0	8.1	4.6	44.0 S	169.0 E

Seismic Processor) software system, which has been adapted for use in New Zealand, may be found in the 1987 New Zealand Seismological Report or subsequent Reports.

Generally the focal depth is calculated at the same time as the other parameters; origin time, latitude and longitude. However, the focal depths of some crustal earthquakes have been fixed at the conventional depths of 5 km, 12 km or 33 km. This is necessary when there is a lack of data close to the epicentre and there is no additional information on which to base a more accurate estimate of the focal depth. As the National Network is expanded the number and accuracy of acceptable free-depth solutions has increased. Further, the ability to deploy quickly a temporary array of seismographs to record aftershocks of major earthquakes has led to a better estimation of focal depths (and a better understanding of focal mechanism), particularly since 1987. This is reflected in the special studies referred to in the text.

For the deeper earthquakes it is recognised that the focal depths are likely to be in error because the New Zealand Standard Velocity Model does not take into account the subducting Pacific Plate. Discussion of the effect may be found in Adams and Ware (1977).

MAGNITUDES

For each earthquake, magnitudes as determined by different techniques are listed together with the source of the determination. Local magnitudes M_L are most commonly given, particularly for earthquakes after 1945. All have been calculated using the Haines (1981) formula which removes systematic errors for certain areas of New Zealand and for deep earthquakes. Since 1987 magnitudes from digital data have been used increasingly. The formulae used for derivation of M_L from analogue and digital data may be found in the more recent New Zealand Seismological Reports.

Magnitudes (M) of earthquakes prior to the instrumental period (about 1901) have necessarily been estimated from macroseismic information (designated by an I in Table 1) and must be considered as approximate and subject to revision as the result of further research. Several magnitudes are taken from Gutenberg & Richter (1949) and ISC (International Seismological Centre) listings, as indicated. Surface wave magnitudes (M_s) of a number of earthquakes have been determined by Dowrick and Smith (1990) (D&S in Table 1). The M_s of the Lake Coleridge earthquake has been determined by Dowrick (pers. comm.) (D in Table 1). For more recent earthquakes m_b and M_w have been included where available, most M_w resulting from source mechanism studies of large South Island earthquakes by Anderson, Webb and Jackson (1993)

MACROSEISMIC DATA AND THE ISOSEISMAL MAPS

Because the earliest years of the new colony in New Zealand were particularly active seismically, the early settlers developed a strong interest in documenting the effects of earthquakes, to the extent that even quite small events were noted in the press. In 1868, when government climatological observers were appointed, part of their duties was to record and report to Dr James Hector on earthquakes felt, thus forming the first "felt reporter" network. However the observers were not great in number and poorly distributed. Later, about 1888, George Hogben, a school master in Timaru, later in charge of the new seismograph in Wellington, enlisted the help of about fifty telegraph operators, providing them with report forms. Within the limits of the seismological knowledge of the time he attempted to find origins and velocities of propagation. He later increased the number of reporters to eighty and improved the layout of the report form. Some of these early report forms are to be found in the files of GNS.

While the "felt reporting" system had an early beginning it has not always been as comprehensive or homogeneous as might be desired. Depending on political circumstances, scientific priorities and leaders, the system has undergone a number of changes. With the writing of a New Zealand version of the Modified Mercalli scale (Eiby 1966), a new report form in 1963 and a system set up where at least two reporters were sought for each of a set of numbered localities bounded by half degrees of latitude and longitude, the felt reporting network was considered to have improved greatly. The 1963 form was modified in 1982 to remove ambiguities and improve the layout. At present there are about 700 nominal reporters of whom about 400 report regularly. With varying success, large numbers of questionnaires (the standard report form) have been distributed after major events in order to improve the coverage of felt information and on four occasions a simplified form has appeared in daily newspapers in Wellington, Wanganui, Gisborne and recently (1994) in Christchurch and many intensity assessments received by this means. The Wellington survey (after the Weber II earthquake, 1990 May 13) showed evidence of markedly higher intensities in areas of poorly consolidated deep sediments.

Unless otherwise indicated the source of the macroseismic data of maps which have been taken from the New Zealand Seismological Report has been the standard network of felt reporters.

The Modified Mercalli scale adapted for New Zealand by Eiby (1966) has been used in 130 out of 133 maps. Two maps (1922 Motunau, 1934 Pahiatua), in which the isoseismal lines refer to the Rossi-Forel scale, have been included as they are markedly different from the more recently drawn maps (also included) and the data on which they have been based appear to be more extensive. The individual intensities are not indicated on the maps and hence no conversion can be made. The necessity of including both versions of the felt distributions of these earthquakes emphasises the critical need for revision of the macroseismic and seismological data. The 1992 revision of the 1966 Modified Mercalli Scale for New Zealand (Study Group, 1992) has been used for one map.

In 19 maps (dating from 1921-1945) felt summaries of Rossi-Forel intensities have been converted to Modified Mercalli (both scales and the conversion from R-F to MM can be found after the references). These maps were conceived as an interim measure to be revised when time permitted full examination and extension of the original data. The end users in engineering seismology and hazard assessment should use these maps with an understanding of their limitations, i.e. the conversion from one scale to the other is not exact and unreliable as discussed by Dowrick (1994).

Maps which have been published elsewhere have generally been reproduced in this Atlas with no modification of the isoseismal pattern. Two maps have been replotted as the originals were misprinted (commented upon in the text).

Map of New Zealand seismograph stations - December 1993

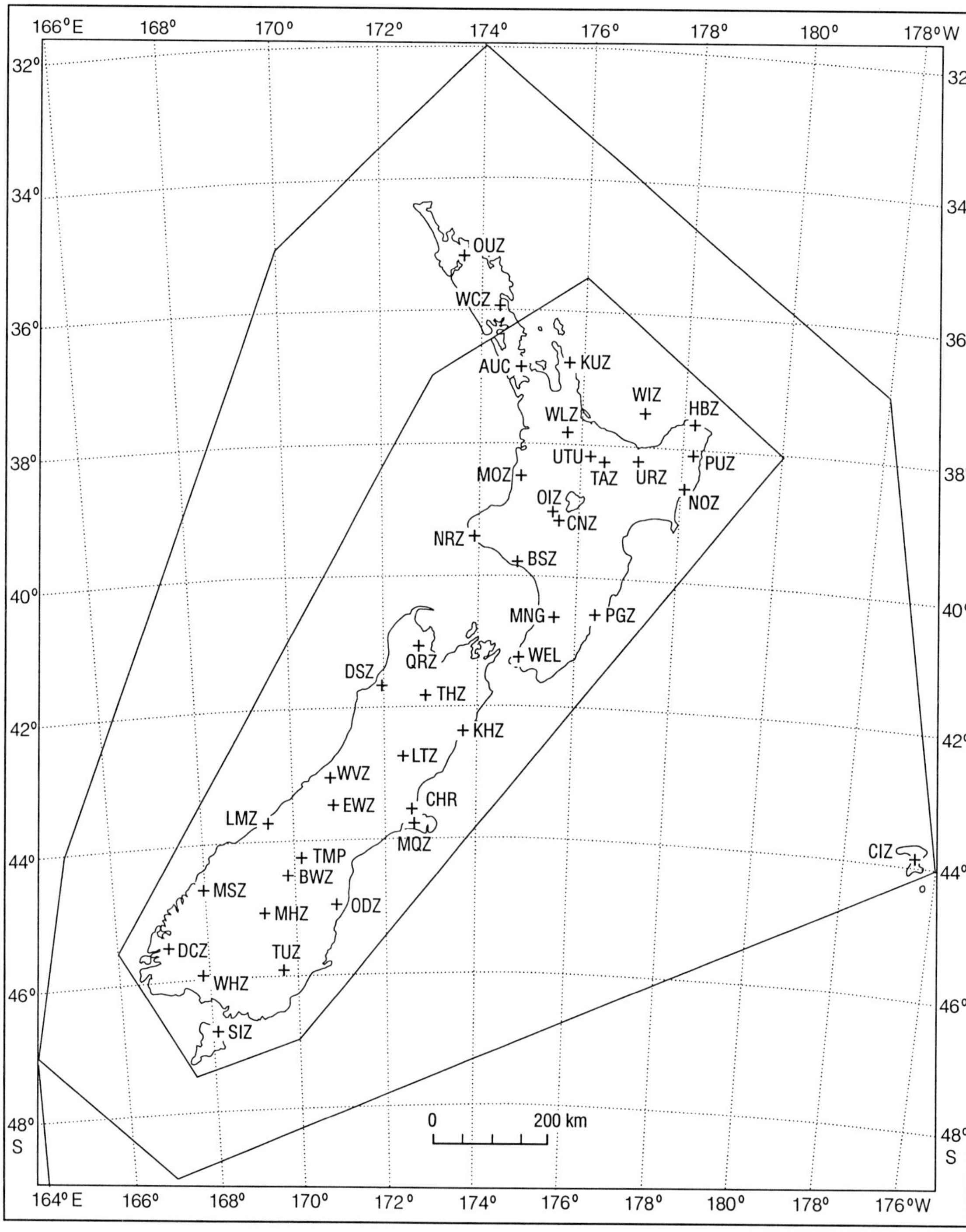

Figure 2. Local networks at Wellington, Hawke's Bay, Chateau, Rotorua and Clyde are too closely spaced to be shown on this map. The inner and outer polygons define three areas where accuracy of epicentre locations is considered reliable, less reliable and inadequate.

Where possible, maps have been redrawn using a computer-aided draughting package using the same projection (Lambert conformal conic; standard parallels, 37° and 45°) and the presentation standardised. The usual practice of indicating with a dashed line the uncertain parts of the isoseismal lines including that which is over the sea is adopted. Individual intensity assessments are indicated. In the text many small communities have been named and indicating these on the maps is not feasible, but the main cities and areas have been indicated in Fig 3. Some authors may omit the apostrophe when referring to the "Hawke's Bay" district.

Finally lists of references for each earthquake are given in the last section of the Atlas after the isoseismal maps. The lists are extensive but may not necessarily be complete and are designed to indicate the state of knowledge about each event.

ACKNOWLEDGMENTS

The production of this Atlas has been supported by a grant from the Earthquake Commission.

The project was initiated by the late George Eiby and he had revised and redrawn a large number of maps before his illness prevented him from continuing with the task. A Research Associate of the Seismological Observatory, since his retirement in 1979, he had put many hours of unpaid labour into the study of historic earthquakes and the author hopes that this Atlas is close to what he envisaged. His unpublished notes on the 1855 Wairarapa earthquake and the history of the felt reporter network and various other activities of the predecessors of the present Institute of Geological and Nuclear Sciences have proved very useful.

The assistance of a number of people is greatly appreciated: Warwick Smith for revising hypocentres and magnitudes of a number of events for inclusion in the Atlas and Martin Reyners for the text and data on the 1876 Oamaru and the 1914 East Cape earthquakes which he has recently studied. The careful reviewing of the text by Diane Maunder, David Dowrick and Warwick Smith is greatly appreciated. Finally I wish to thank Carolyn Hume for the enormous amount of time and careful attention she has put into draughting 131 new maps.

REFERENCES

Anderson, H., Webb, T., Jackson, J. 1993. Focal mechanisms of large earthquakes in the South Island of New Zealand: implications for the accommodation of Pacific-Australia plate motion. *Geophysical journal international 115*: 1032-1054.

Adams, R. D., Ware, D. E. 1977. Subcrustal earthquakes beneath New Zealand; locations determined with a laterally inhomogeneous velocity model. *New Zealand journal of geology and geophysics 20*: 59-83.

Dowrick, D. J. 1994. Damage and intensities in the Magnitude 7.8 1929 Murchison, New Zealand, earthquake. *Bulletin of the New Zealand National Society for Earthquake Engineering 27*: 190-204.

Dowrick, D. J., Smith, E. G. C. 1990. Surface wave magnitudes of some New Zealand earthquakes 1901-1988. *Bulletin of the New Zealand National Society for Earthquake Engineering 23*: 198-210.

Eiby, G. A. 1966. The Modified Mercalli Scale of earthquake intensity and its use in New Zealand. *New Zealand journal of geology and geophysics 9*: 122-129.

Eiby, G. A. 1968a. A descriptive catalogue of New Zealand earthquakes. Part 1 Shocks felt before the end of 1845. *New Zealand journal of geology and geophysics 11*: 16-40.

Eiby, G. A. 1968b. An annotated list of New Zealand earthquakes, 1460-1965. *New Zealand journal of geology and geophysics 11*: 630-647.

Eiby, G. A. 1973. A descriptive catalogue of New Zealand earthquakes. Part 2 Shocks felt from 1846 to 1854. *New Zealand journal of geology and geophysics 16*: 857-907.

Eiby, G. A. 1988. Documenting New Zealand earthquakes. *in* Historical Seismograms and Earthquakes of the

Gutenberg, B., Richter, C. F. 1949. Seismicity of the earth. 2nd Edition. Princeton University Press, Princeton, New Jersey.

Haines, A. J. 1981. A local magnitude scale for New Zealand earthquakes. *Bulletin of the Seismological Society of America 71*: 275-294.

McKay, A. 1902. Report on the recent seismic disturbances within Cheviot County in Northern Canterbury and the Amuri District of Nelson, New Zealand. Government Printer, Wellington.

Richter, C. F. 1958. Elementary Seismology. W. H. Freeman & Co.

Study Group of the New Zealand National Society for Earthquake Engineering. 1992. A revision of the Modified Mercalli seismic intensity scale. *Bulletin of the New Zealand National Society for Earthquake Engineering 25*: 345-356.

Young, R. M. 1984. Early New Zealand seismographs. "In search of New Zealand's Scientific Heritage." Hoare, M. E., Bell, L. G., (eds). *Royal Society of New Zealand bulletin 21*: 113-118.

Young, R. M., Ansell, J. H., Hurst, M. A. 1984. New Zealand's first seismograph: the Hector seismograph, 1884-1902. *Journal of the Royal Society of New Zealand 14*: 159-173.

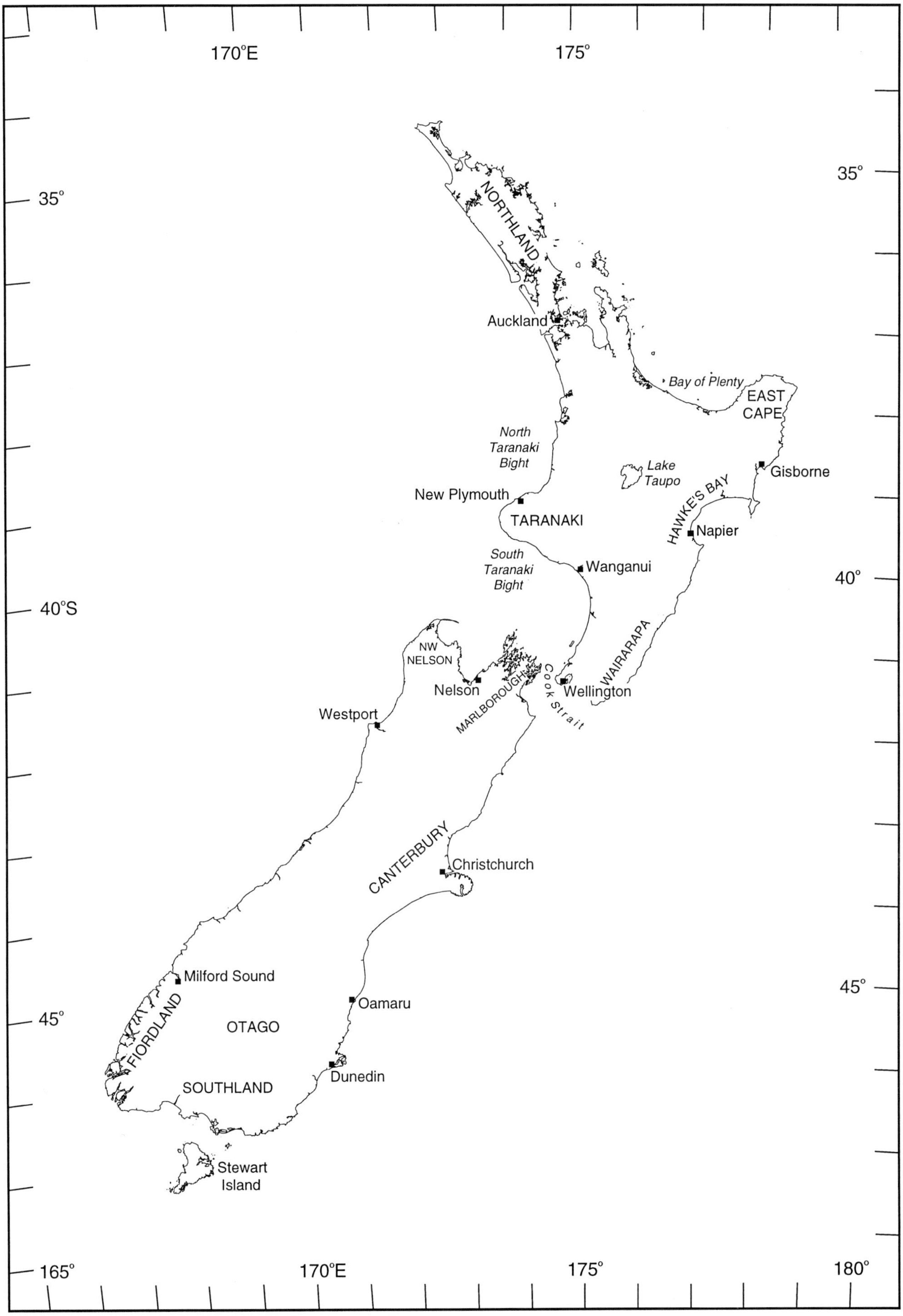

Figure 3. Map of districts and locations frequently referred to in the text.

INTENSITY SCALES

MODIFIED MERCALLI (MM) INTENSITY SCALE

After Eiby (1966)

MM 1 Not felt by humans, except in especially favourable circumstances, but birds and animals may be disturbed. Reported mainly from the upper floors of buildings more than 10 storeys high. Dizziness or nausea may be experienced.

Branches of trees, chandeliers, doors, and other suspended systems of long natural period may be seen to move slowly.

Water in ponds, lakes, reservoirs, etc., may be set into seiche oscillation.

MM 2 Felt by a few persons at rest indoors, especially by those on upper floors or otherwise favourably placed.

The long-period effects listed under MM1 may be more noticeable.

MM 3 Felt indoors, but not identified as an earthquake by everyone. Vibration may be likened to the passing of light traffic.

It may be possible to estimate the duration, but not the direction. Hanging objects may swing slightly. Standing motorcars may rock slightly.

After Study Group (1992)

MM 1 *People*

Not felt except by a very few people under exceptionally favourable circumstances.

MM 2 *People*

Felt by persons at rest, on upper floors or favourably placed.

MM 3 *People*

Felt indoors; hanging objects may swing, vibration similar to passing of light trucks, duration may be estimated, may not be recognised as an earthquake.

After Eiby (1966)		After Study Group (1992)	
MM 4	Generally noticed indoors, but not outside. Very light sleepers may be wakened. Vibration may be likened to the passing of heavy traffic, or to the jolt of a heavy object falling or striking the building. Walls and frame of buildings are heard to creak. Doors and windows rattle. Glassware and crockery rattle. Liquids in open vessels may be slightly disturbed. Standing motorcars may rock, and the shock can be felt by their occupants.	MM 4	*People* Generally noticed indoors but not outside. Light sleepers may be awakened. Vibration may be likened to the passing of heavy traffic or to the jolt of a heavy object falling or striking the building. *Fittings* Doors and windows rattle. Glassware and crockery rattle. Liquids in open vessels may be slightly disturbed. Standing motorcars may rock. *Structures* Walls and frame of buildings, and partitions and suspended ceilings in commercial buildings, may be heard to creak.
MM 5	Generally felt outside, and by almost everyone indoors. Most sleepers awakened. A few people frightened. Direction of motion can be estimated. Small unstable objects are displaced or upset. Some glassware and crockery may be broken. Some windows cracked. A few earthenware toilet fixtures cracked. Hanging pictures move. Doors and shutters may swing. Pendulum clocks stop, start, or change rate.	MM 5	*People* Generally felt outside, and by almost everyone indoors. Most sleepers awakened. A few people alarmed. Direction of motion can be estimated. *Fittings* Small unstable objects are displaced or upset. Some glassware and crockery may be broken. Hanging pictures knock against the wall. Open doors may swing. Cupboard doors secured by magnetic catches may open. Pendulum clocks stop, start, or change rate (H*). *Structures* Some Windows Type I* cracked. A few earthenware toilet fixtures cracked (H).

After Eiby (1966)

MM 6 Felt by all.
People and animals alarmed.
Many run outside.
Difficulty experienced in walking steadily.

Slight damage to Masonry D.
Some plaster cracks or falls.
Isolated cases of chimney damage.
Windows, glassware, and crockery broken.
Objects fall from shelves, and pictures from walls.
Heavy furniture moved.
Unstable furniture overturned.
Small church and school bells ring.

Trees and bushes shake, or are heard to rustle.
Loose material may be dislodged from existing slips, talus slopes, or shingle slides.

After Study Group (1992)

MM 6 *People*

Felt by all.
People and animals alarmed.
Many run outside.
Difficulty experienced in walking steadily.

Fittings

Objects fall from shelves.
Pictures fall from walls (H*).
Some furniture moved on smooth floors.
Some unsecured free-standing fireplaces moved.
Glassware and crockery broken.
Unstable furniture overturned.
Small church and school bells ring (H).
Appliances move on bench or table tops.
Filing cabinets or "easy glide" drawers may open (or shut).

Structures

Slight damage to Buildings Type I*.
Some stucco or cement plaster falls.
Suspended ceilings damaged.
Windows Type I* broken.
A few cases of chimney damage.

Environment

Trees and bushes shake, or are heard to rustle.
Loose material may be dislodged from sloping ground, e.g. existing slides, talus slopes, shingle slides.

After Eiby (1966)

MM 7 General alarm.
Difficulty experienced in standing.
Noticed by drivers of motorcars.

Trees and bushes strongly shaken.
Large bells ring.
Masonry D cracked and damaged.
A few instances of damage to Masonry C.
Loose brickwork and tiles dislodged.
Unbraced parapets and architectural ornaments may fall.
Stone walls cracked.
Weak chimneys broken, usually at the roof-line.
Domestic water tanks burst.
Concrete irrigation ditches damaged.

Waves seen on ponds and lakes.
Water made turbid by stirred-up mud.
Small slips, and caving-in of sand and gravel banks.

After Study Group (1992)

MM 7 *People*

General alarm.
Difficulty experienced in standing.
Noticed by motorcar drivers who may stop.

Fittings

Large bells ring.
Furniture moves on smooth floors, may move on carpeted floors.

Structures

Unreinforced stone and brick walls cracked.
Buildings Type I cracked and damaged.
A few instances of damage to Buildings Type II.
Unbraced parapets and architectural ornaments fall.
Roofing tiles, especially ridge tiles may be dislodged.
Many unreinforced domestic chimneys broken.
Water tanks Type I* burst.
A few instances of damage to brick veneers and plaster or cement-based linings.
Unrestrained water cylinders (Water Tanks Type II*) may move and leak.
Some Windows Type II* cracked.

Environment

Water made turbid by stirred up mud.
Small slides such as falls of sand and gravel banks.
Instances of differential settlement on poor or wet or unconsolidated ground.
Some fine cracks appear in sloping ground.
A few instances of liquefaction.

After Eiby (1966)

MM 8 Alarm may approach panic.

Steering of motorcars affected.

Masonry C damaged, with partial collapse.
Masonry B damaged in some cases.
Masonry A undamaged.

Chimneys, factory stacks, monuments, towers, and elevated tanks twisted or brought down.
Panel walls thrown out of frame structures.
Some brick veneers damaged.
Decayed wooden piles broken.
Frame houses not secured to the foundation may move.
Cracks appear on steep slopes and in wet ground.
Landslips in roadside cuttings and unsupported excavations.
Some tree branches may be broken off.
Changes in the flow or temperature of springs and wells may occur.
Small earthquake fountains.

After Study Group (1992)

MM 8 *People*

Alarm may approach panic.
Steering of motorcars greatly affected.

Structures

Buildings Type II damaged, some seriously.
Buildings Type III damaged in some cases.
Monuments and elevated tanks twisted or brought down.
Some pre-1965 infill masonry panels damaged.
A few post-1980 brick veneers damaged.
Weak piles damaged.
Houses not secured to foundations may move.

Environment

Cracks appear on steep slopes and in wet ground.
Slides in roadside cuttings and unsupported excavations.
Small earthquake fountains and other manifestations of liquefaction.

After Eiby (1966)

MM 9 General panic.

Masonry D destroyed.
Masonry C heavily damaged, sometimes collapsing completely.
Masonry B seriously damaged.

Frame structures racked and distorted.
Damage to foundations general.
Frame houses not secured to the foundations shifted off.
Brick veneers fall and expose frames.
Cracking of the ground conspicuous.
Minor damage to paths and roadways.
Sand and mud ejected in alluviated areas, with the formation of earthquake fountains and sand craters.
Underground pipes broken.
Serious damage to reservoirs.

After Study Group (1992)

MM 9 *Structures*

Very poor quality unreinforced masonry destroyed.
Buildings Type II heavily damaged, some collapsing.
Buildings Type III damaged, some seriously.
Damage or permanent distortion to some buildings and bridges Type IV.
Houses not secured to foundations shifted off.
Brick veneers fall and expose frames.

Environment

Cracking of ground conspicuous.
Landsliding general on steep slopes.
Liquefaction effects intensified, with large earthquake fountains and sand craters.

After Eiby (1966)	After Study Group (1992)
MM 10 Most masonry structures destroyed, together with their foundations. Some well built wooden buildings and bridges seriously damaged. Dams, dykes, and embankments seriously damaged. Railway lines slightly bent. Cement and asphalt roads and pavements badly cracked or thrown into waves. Large landslides on river banks and steep coasts. Sand and mud on beaches and flat land moved horizontally. Large and spectacular sand and mud fountains. Water from rivers, lakes and canals thrown up on the banks.	MM 10 *Structures* Most unreinforced masonry structures destroyed. Many Buildings Type II destroyed. Many Buildings Type III (and bridges of equivalent design) seriously damaged. Many Buildings and Bridges Type IV have moderate damage or permanent distortion.
MM 11 Wooden frame structures destroyed. Great damage to railway lines and underground pipes.	
MM 12 Damage virtually total. Practically all works of construction destroyed or greatly damaged. Large rock masses displaced. Lines of sight and level distorted. Visible wave-motion of the ground surface reported. Objects thrown upwards into the air.	

After Eiby (1966)
Categories of non-Wooden Construction

Masonry A

Structure designed to resist lateral forces of about 0.1 g, such as those satisfying the New Zealand Model Building Bylaw, 1955. Typical buildings of this kind are well reinforced by means of steel or ferro-concrete bands, or are wholly of ferro-concrete construction. All mortar is of good quality and the design and workmanship is good. Few buildings erected prior to 1935 can be regarded as in category A.

Masonry B

Reinforced buildings of good workmanship and with sound mortar, but not designed in detail to resist lateral forces.

Masonry C

Buildings of ordinary workmanship, with mortar of average quality. No extreme weakness, such as inadequate bonding of the corners, but neither designed nor reinforced to resist lateral forces.

Masonry D

Buildings with low standard of workmanship, poor mortar, or constructed of weak materials like mud brick and rammed earth. Weak horizontally.

Windows

Window breakage depends greatly upon the nature of the frame and its orientation with respect to the earthquake source. Windows cracked at MM5 are usually either large display windows, or windows tightly fitted to metal frames.

Water Tanks

The "domestic water tanks" listed under MM7 are of the cylindrical corrugated-iron type common in New Zealand rural areas. If these are only partly full, movement of the water may burst soldered and riveted seams.

Hot-water cylinders constrained only by supply and delivery pipes may move sufficiently to break the pipes at about the same intensity.

After Study Group (1992)
Categories of Construction

Buildings Type I

Weak materials such as mud brick and rammed earth; poor mortar; low standards of workmanship (Masonry D in other MM scales).

Buildings Type II

Average to good workmanship and materials, some including reinforcement, but not designed to resist earthquakes (Masonry B and C in other MM scales).

Buildings Type III

Buildings designed and built to resist earthquakes to normal use standards, i.e. no special damage limiting measures taken (mid-1930's to c. 1970 for concrete and to c. 1980 for other materials).

Buildings and Bridges Type IV

Since c.1970 for concrete and c.1980 for other materials, the loadings and materials codes have combined to ensure fewer collapses and less damage than in earlier structures. This arises from features such as: (i) "capacity design" procedure, (ii) use of elements (such as improved bracing or structural walls) which reduce racking (i.e. drift), (iii) high ductility, (iv) higher strength.

Windows

Type I - Large display windows, especially shop windows.
Type II - Ordinary sash or casement windows.

Water Tanks

Type I - External, stand mounted, corrugated iron water tanks.
Type II - Domestic hot-water cylinders unrestrained except by supply and delivery pipes.

H - (Historical)

Important for historical events. Current application only to older houses, etc.

General Comment

"Some" or "a few" indicates that the threshold of a particular effect has just been reached at that intensity.

Rossi-Forel (R-F) Scale

I. Microseismic shock: recorded by a single seismograph or by seismographs of the same model, but not by several seismographs of different kinds; the shock felt by an experienced observer.

II. Extremely feeble shock: recorded by several seismographs of different kinds; felt by a small number of persons at rest.

III. Very feeble shock: felt by several persons at rest; strong enough for the direction or duration to be appreciable.

IV. Feeble shock: felt by persons in motion; disturbances of movable objects, doors, windows; creaking of ceilings.

V. Shock of moderate intensity: felt generally by everyone; disturbance of furniture, beds, etc., ringing of swinging bells.

VI. Fairly strong shock: general awakening of those asleep; general ringing of house bells; oscillation of chandeliers; stopping of pendulum clocks; visible agitation of trees and shrubs; some startled persons leave their dwellings.

VII. Strong shock: overthrow of movable objects; fall of plaster; ringing of church bells; general panic, without damage to buildings.

VIII. Very strong shock: fall of chimneys, cracks in walls of buildings.

IX. Extremely strong shock: partial or total destruction of some buildings.

X. Shock of extreme intensity: great disaster, buildings ruined, disturbance of the strata, fissures in the ground, rock-falls from mountains.

Note: This version is reproduced from the 1921 *Bulletin of the Seismological Society of America 11*, 94, as an accepted translation of the Swiss version of the scale, which appeared in 1884, *Archives des Sciences Physiques et Naturelles 11*, 148-149.

R-F to MM Conversion Chart

Felt Intensities - Conversion RF - MM

RF		MM	
RF	1	MM	1
	1-2		2
	2)		2-3
	2-3)		
	3		3
	3-4)		3-4
	4)		
	4-5		4
	5		4-5
	5-6		5
	6		5-6
	6-7		6
	7)		6-7
	7-8)		
	8		7
	8-9		8
	9		9
	9-10		9-10
	10		10, 11, 12

STANDARD REPORT FORM (pages 1 and 2)

S.I.R. 229

SEISMOLOGICAL OBSERVATORY, WELLINGTON

EARTHQUAKE REPORT from

Name: . . . .

Postal Address: . . . .

. . . .

For Office Use Only	
U.T.:	. . . .
Locality:	. . . .
M.M:	. . . .

Please describe the shock by marking the statements which apply (X). Most shocks can be described by filling in Section 1 only. Section 2 applies only to moderate and Section 3 to damaging shocks. Include only effects you have noticed yourself. People in other parts of the district should be encouraged to fill in separate reports. Use a separate form for all shocks more than one minute apart.

---

1. FOR ALL EARTHQUAKES

Felt at (place): .... **Date:** .... **Day of week:** ....

Clock Time: .... hr. .... min. a.m. ( ) p.m. ( )   **Check also:** Night ( ) Day ( )   Early morning ( ) Evening ( )

The time is accurate to – half a minute or better ( ), One or two minutes ( ), 5 minutes ( ), Uncertain ( )

Observer's Position:

Walking ( ) on the .... storey of a building constructed of wood ( )
Standing ( ) brick ( )
Sitting ( ) concrete ( )
In bed ( )

Awake ( )   Out of doors ( )   Passenger in ( )   stationary ( )   car ( )
Awakened ( )   Driving ( )   moving ( )   lorry ( )
bus ( )
train ( )

Observer's Reactions:

Felt by everyone present ( )   By .... persons out of .... Only by those at rest ( )
Some not awakened ( )   No alarm ( ) Little alarm ( ) Some alarm ( ) Panic ( )

Nature of shock:

Light ( )   Jolt ( )   Double, with .... sec. between tremors
Moderate ( )   Vibration ( )
Heavy ( )   Swaying ( )   Estimated duration .... secs.

Sounds:

Before shock: boom ( ) crack ( ) rumble ( )   During shock: boom ( ) crack ( ) rumble ( )   After shock: boom ( ) crack ( ) rumble ( )

Crockery and windows rattle ( ). Walls or roof strain or creak ( ).

---

2. FOR MODERATE SHOCKS

Household effects, etc.:

Lights or other suspended objects swing – slightly ( ) violently ( )   Fire or burglar alarms triggered ( )

Crockery ( ), goods on shelves ( ), window displays ( ) – disarranged ( ), broken ( ) overturned ( ), fell off ( )

Heavy furniture – moved ( ), overturned ( ).

Liquids slopped from – jugs, etc. ( ), baths ( ), tanks ( ), ornamental pools ( ).

Damage to earthenware sanitary fittings ( ).

IF DAMAGE TO BUILDINGS TOOK PLACE, PLEASE COMPLETE SECTION 3 ON BACK OF FORM

Further Comment: . . . .

. . . .

. . . .

. . . .

. . . .

(Fold 1)

3. DAMAGING SHOCKS

Buildings:

Your own building damaged ( ). Nearby buildings damaged ( ).
Plaster: slightly cracked ( ), badly cracked ( ), fell ( ).
Domestic chimneys: cracked ( ), pots thrown down ( ), destroyed ( ).

Poorly constructed buildings:	slightly damaged	( )
	moderately damaged	( )
	severely damaged	( )
	destroyed	( )
Most ordinary buildings:	slightly damaged	( )
	moderately damaged	( )
	seriously damaged	( )
	destroyed	( )

Countryside, etc.:

Cracks and fissures in: roads ( ), embankments ( ), river banks ( ), bridge approaches ( ).
Landslides and slumps: small ( ), large ( ).

Engineering Structures:

Underground pipes broken ( ).
Railway lines bent ( ).
Damage to: bridges ( ), tunnels ( ), dams ( ). *(Please give details.)*

Waterways:

Please describe any abnormal movements in the sea, rivers, lakes, or small ponds, which may be associated with the earthquake.

. . . .

. . . .

. . . .

. . . .

. . . .

. . . .

. . . .

. . . .

. . . .

. . . .

ISOSEISMAL MAPS OF NEW ZEALAND EARTHQUAKES IN CHRONOLOGICAL ORDER

MAP SYMBOLS

+	- epicentre
x	- epicentre using superseded velocity model
4	- intensity observation
0 (or NF)	- earthquake not felt
? (or F)	- information insufficient to assign intensity
4?	- poorly constrained intensity
4̲	- several observations of the same intensity too closely spaced to be individually marked
IV	- isoseismal line designation
Sarah Ann	- name of vessel reporting earthquake 'felt at sea'

1. ISOSEISMAL MAP OF THE MARLBOROUGH EARTHQUAKE - 1848 OCTOBER 15

DATE:	1848 OCTOBER 15
TIME:	14:10 UT
MAGNITUDE:	7.1 M (I)
EPICENTRE:	41.5°S 173.8°E D (I)
DEPTH:	Shallow

'Historical records show the Marlborough Earthquakes of 1848 to have been centred in the lower Wairau Valley. The principal earthquake, of estimated magnitude 7.1, occurred at 1.40 a.m. local time on Monday October 16, and was followed by a long and well-documented sequence of aftershocks. The shocks produced substantial damage to buildings in Wellington and Nelson, and ground subsidence and faulting in the Wairau. Earlier appraisals suggesting that the earthquakes were associated with movement of the Awatere Fault are based upon faulty historical information.

'The comparatively high intensities near Wellington and their extension northwards, combined with the absence of serious damage in Nelson, might be used to justify an epicentre in Cook Strait, but contemporary evidence and opinion both agree that intensities were higher in Marlborough than in the North Island. There are no reports of serious damage in the Wairarapa, and none of ground fissuring, which would be expected if the epicentre lay to the east or to the north of the position now suggested.' (Eiby 1980)

'The area over which the main shock was felt extends from the neighbourhood of East Cape and northern Taranaki to Banks Peninsula. It was not felt in Auckland or Otago. Damage to buildings occurred in the Wairau, where the intensity reached MM10, and in and about Wellington, where it was extensive, the intensity being not less than MM8. It reached at least MM7 at Nelson, at Wanganui, and in the Wairarapa and Manawatu. Landslides and the liberation of sand and ground water were unusually widespread, but this can be attributed to the period of exceptionally wet and stormy weather that preceded the earthquake.' (Eiby 1973)

The isoseismal map is reproduced from Eiby (1980). An early map of the 1848 earthquake appears in Hayes (1936).

A magnitude of 7.9 has been estimated by Dowrick (1991) using his derived attenuation expressions and Eiby's (1980) map. He notes that it should be treated cautiously because of the need to extrapolate beyond the magnitude range of the data from which the attenuation expressions were derived.

The opinion that the Wairau and not the Awatere Fault moved in 1848 is not universally held. However Eiby (1980) appears to be the only paper that attempts a critical appraisal of all the available historic data. While the epicentre may be open to discussion the intensities on the isoseismal map are unlikely to be altered significantly in the future as there appears to be little probability of finding further contemporary data.

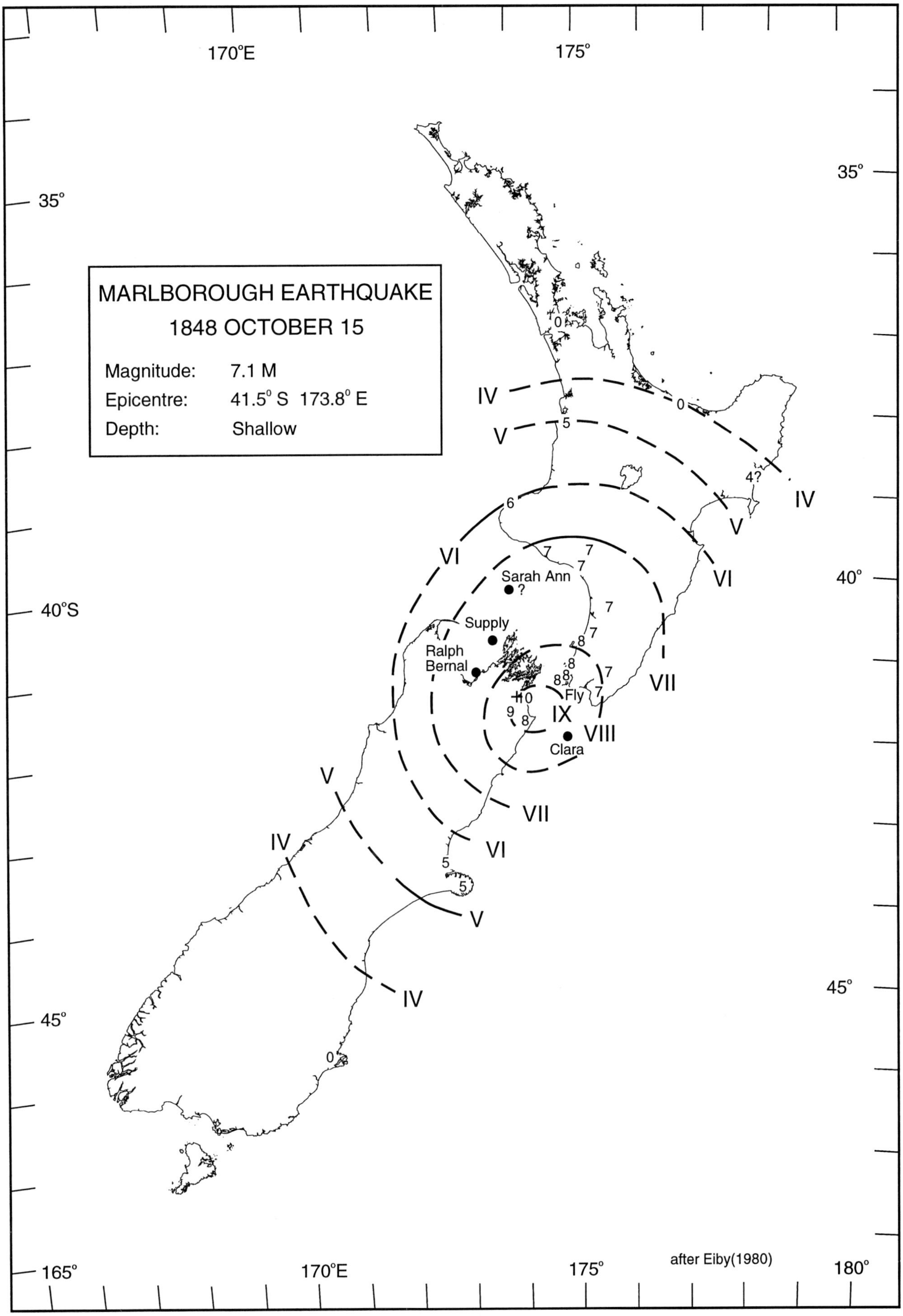
MARLBOROUGH EARTHQUAKE
1848 OCTOBER 15
Magnitude: 7.1 M
Epicentre: 41.5° S 173.8° E
Depth: Shallow
170°E
175°
35°
40°S
40°
45°
165°
180°
Sarah Ann
Supply
Ralph Bernal
Fly
Clara
IV
V
VI
VII
VIII
IX
after Eiby(1980)

2. ISOSEISMAL MAP OF THE WAIRARAPA EARTHQUAKE - 1855 JANUARY 23

DATE:	1855 JANUARY 23
TIME:	09:32 UT
MAGNITUDE:	8.1 - 8.2 M (I)
EPICENTRE:	41.4°S 175.0°E D (I)
DEPTH:	Shallow

The 1855 Wairarapa earthquake is the most severe earthquake known to have occurred in New Zealand since systematic European colonisation began about 1840. The extent of the felt area makes it likely that the magnitude was M8.1-8.2 (Eiby 1989). There was extensive faulting and regional uplift. The felt area of the main shock included almost the whole country, being of destructive intensity in Wellington, MM10, and severely damaging at Wanganui and Kaikoura. There were extensive landslides on both faces of the Rimutaka Ranges and in Wellington and along the Kaikoura coast. Numerous slump cracks occurred in flat areas of Wellington, Hutt Valley, Wairarapa, and in the Manawatu district. Sand craters and liberation of ground water occurred in the same areas. There were many aftershocks, some damaging. Five deaths resulted from the earthquake. There is strong evidence that the earthquake generated a tsunami and it is also possible that small tsunamis accompanied some aftershocks.

A critical appraisal of all the available data has not yet been published. Some aspects are in doubt. Eiby (1989) comments that: "there are no indisputable observations of the faulting. Supposed contemporary accounts do not match observable fault traces. Published estimates of the length of fresh surface breakage are improbably great. The best observations of uplift are at Muka Muka (9ft) [2.7 m] on the western side of Palliser Bay and in Wellington Harbour (4 ft) [1.2 m]. Claimed uplifts at Pauatahanui, Lowry Bay, and to the east of Lake Wairarapa are more reasonably attributed to silting."

Using a forward elastic dislocation modelling technique Darby & Beanland (1992) developed several source models to explain the known deformation data. The moment magnitudes (M_w) of the models lay in the range 8.0 - 8.4.

Research is continuing on this earthquake (R. Grapes (Victoria University of Wellington) and G. Downes (GNS)) and a grant from the Earthquake Commission will allow the publication of all the raw data from contemporary historical papers and manuscripts as well as a summary of the seismological and geological aspects.

The isoseismal map was drawn by Eiby in 1984 as a provisional version during his extensive investigations of this earthquake and is taken from Eiby (1989). A macroseismic epicentre is inferred from Eiby (1989).

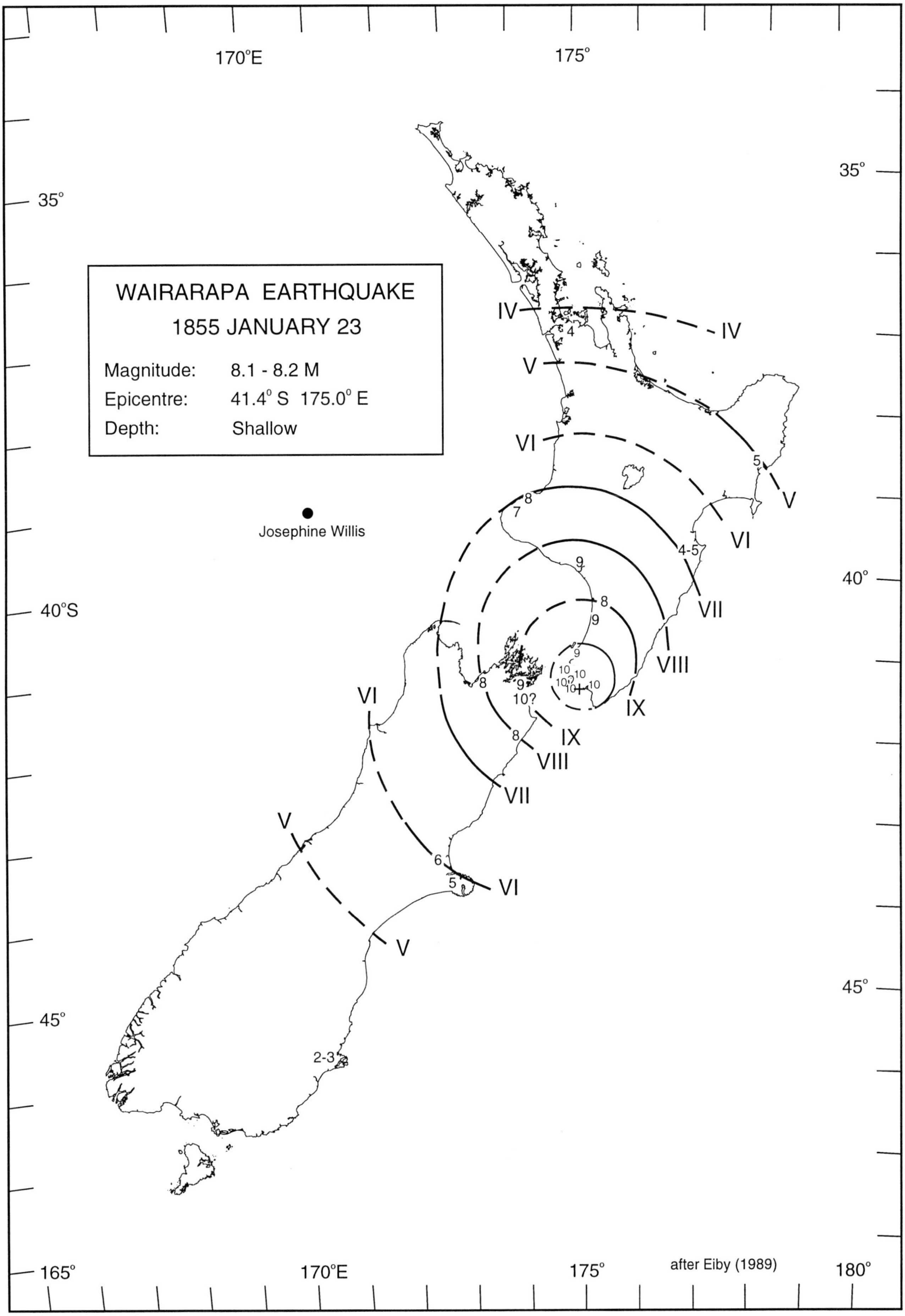

WAIRARAPA EARTHQUAKE
1855 JANUARY 23
Magnitude: 8.1 - 8.2 M
Epicentre: 41.4° S 175.0° E
Depth: Shallow
Josephine Willis
170°E
175°
165°
180°
35°
40°S
40°
45°
IV
V
VI
VII
VIII
IX
2-3
4-5
10?
after Eiby (1989)

eq2b

3. ISOSEISMAL MAP OF THE CAPE FAREWELL EARTHQUAKE - 1868 OCTOBER 18

DATE:	1868 OCTOBER 18
TIME:	12:35 UT
MAGNITUDE:	7.0 - 7.5 M (I)
EPICENTRE:	40.3°S 172.9°E D (I)
DEPTH:	Shallow

'This earthquake, previously assigned locations in the South Taranaki Bight (Smith and Berryman 1986) and Cook Strait (Eiby 1968) because of MM8 intensities reported in Taranaki and Nelson, may be better located to the north of Farewell Spit. From the Collingwood, Pakawau and Farewell Spit area ... there were reports of conspicuous and extensive ground cracking as well as damage to houses, indicating that intensities there were in excess of the MM8 experienced in Taranaki. This evidence suggests an epicentre to the west and south of that determined by Smith and Berryman (1986).

'A possible surface rupture approximately on trend with the Wakamarama fault ... may be indicated in the following description but has not been checked in the field: "Upon once again visiting Cape Farewell and the vicinity I found the effects of the first day's shock ... on a fern and scrub spur rising from the end of the Spit towards Cape Farewell, there is quite a small ravine formed, running nearly north and south, being two or three feet in width. and on the lower side some three or four feet deep. It extends down both sides of the spur, and can be seen at a couple of miles off. Around the Puponga mud-flat several large fissures are to be seen in the roads and in one place the clay soil, and the scrub growing upon it have slipped down, leaving near half an acre of rock bare; and further up the hill the crack continues in a direct course for that on the spur." (from a letter to the Nelson Examiner from J E Fletcher, Pakawau, Oct 27 1868)

'The same correspondent stated that he felt at least one hundred events over the next ten days, at least one of which toppled chimneys previously damaged. At least ten of the aftershocks were also distinctly felt in Hokitika, Nelson, Wellington, Wanganui and New Plymouth, suggesting magnitudes of c.6.' (Anderson et al. 1994)

The extent of the felt area suggests a magnitude of about M7-7.5.

The isoseismal map is taken from Anderson et al. (1994). This map is preliminary, further research continuing. The source material for the felt data was primarily from contemporary newspaper reports, which were quite extensive but not necessarily complete.

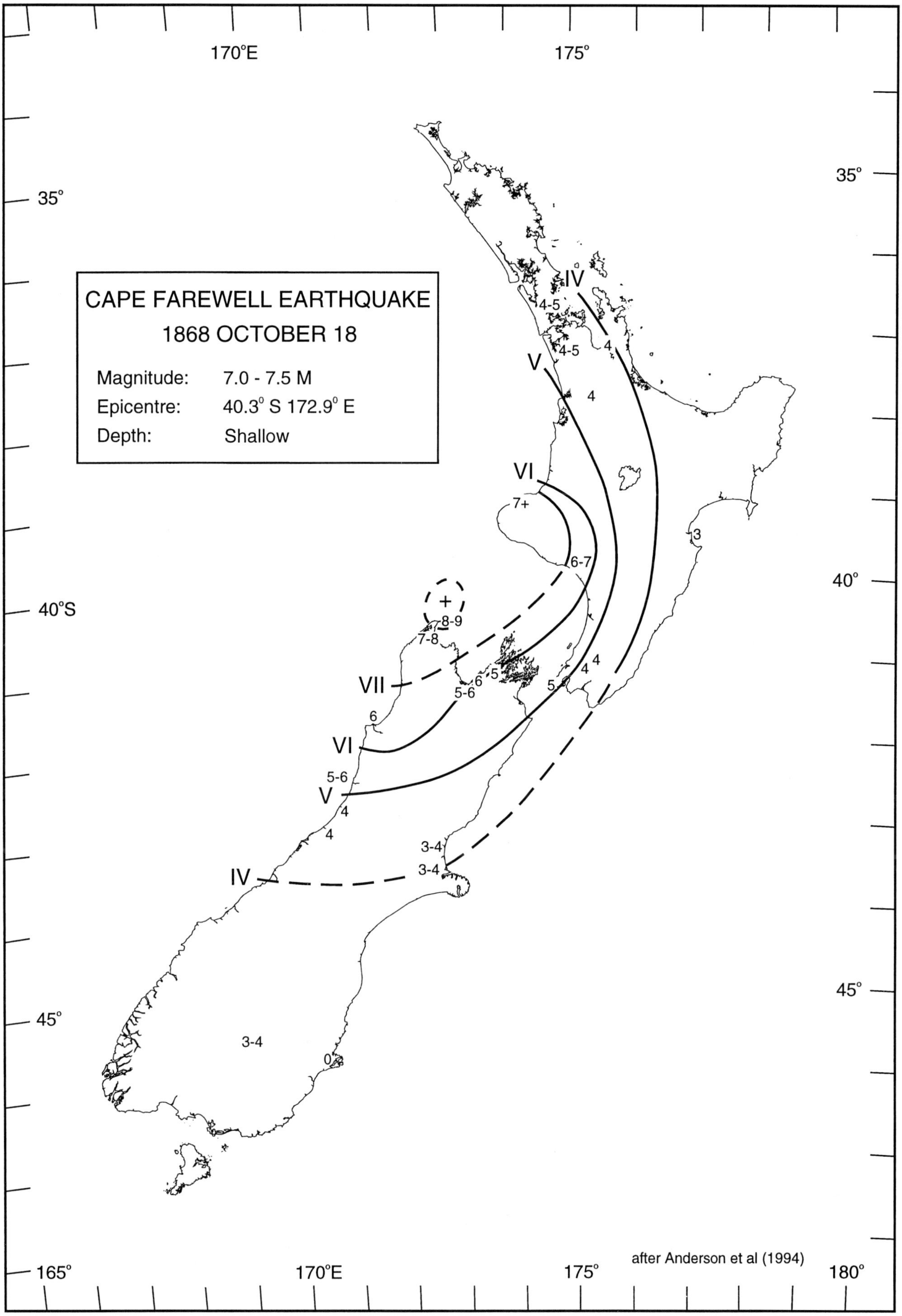

CAPE FAREWELL EARTHQUAKE
1868 OCTOBER 18
Magnitude: 7.0 - 7.5 M
Epicentre: 40.3° S 172.9° E
Depth: Shallow
170°E
175°
35°
40°S
40°
45°
165°
180°
IV
V
VI
VII
4-5
4
7+
6-7
8-9
7-8
6
5-6
5
3
3-4
0
after Anderson et al (1994)

4a. INTENSITY MAP OF THE OAMARU I EARTHQUAKE - 1876 FEBRUARY 25 15:50

DATE:	1876 FEBRUARY 25
TIME:	15:50 UT
MAGNITUDE:	5.75 M (I)
EPICENTRE:	45.2°S 170.9°E C (I)
DEPTH:	Shallow

The earthquakes of February 25 caused significant damage in Oamaru, with reports of 100 chimneys being brought down. Cracking of stone buildings occurred, and a wall of the main building of the hospital required rebuilding. A maximum intensity of MM7 is indicated. Most of the chimneys at Kakanui were also brought down, again indicating MM7. In constructing isoseismal maps for the two events, difficulty was experienced in ascribing descriptions of the shaking to a particular earthquake, as they were closely spaced in time and of similar magnitude.

Most reports from Oamaru suggest the second earthquake was somewhat stronger there. However, the felt area for both events is rather similar, with MM5 being experienced from Timaru to Dunedin. Both earthquakes were followed by a rich aftershock sequence.

The intensity map is taken from Reyners (in prep.).

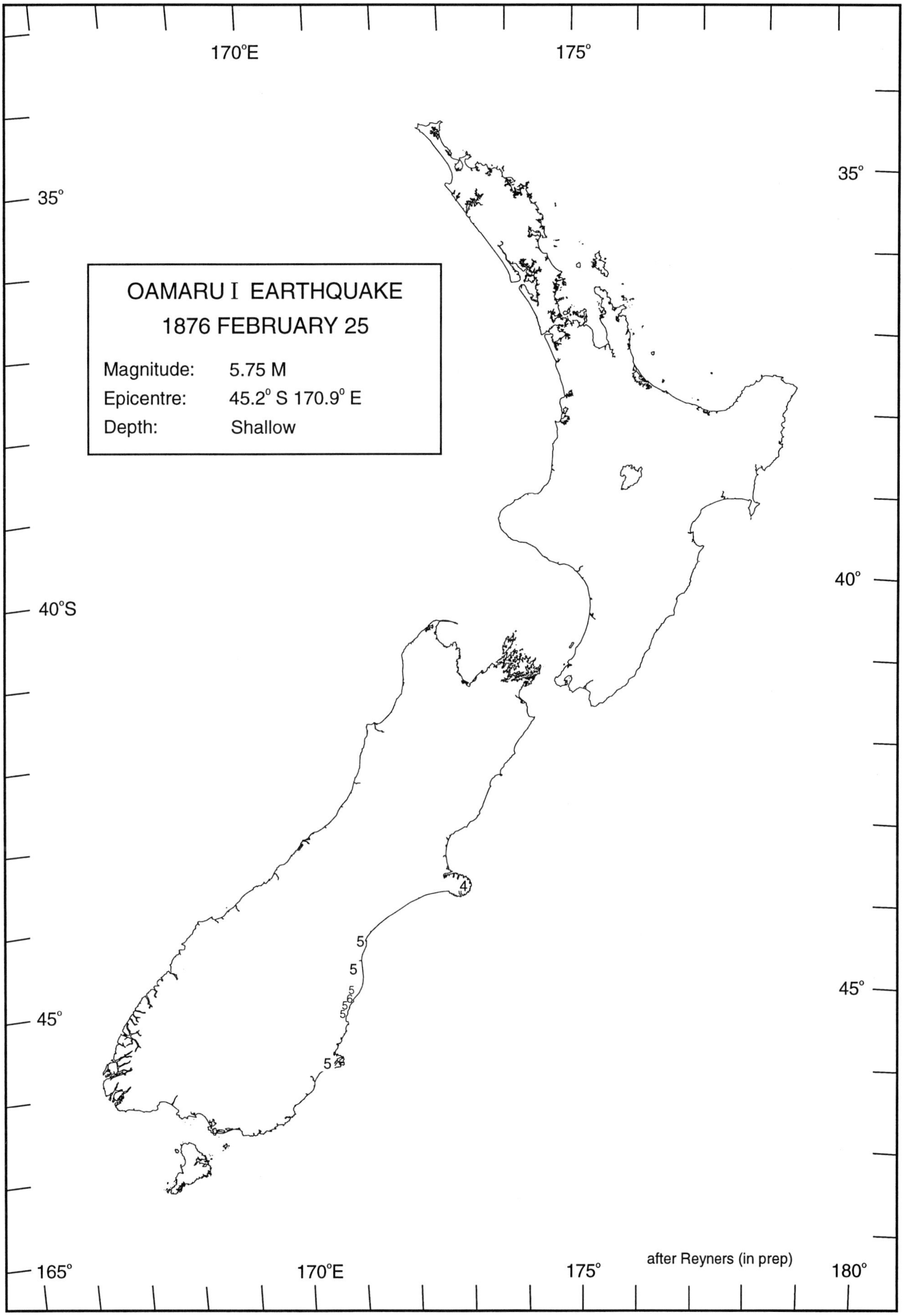

OAMARU I EARTHQUAKE
1876 FEBRUARY 25
Magnitude: 5.75 M
Epicentre: 45.2° S 170.9° E
Depth: Shallow
170°E
175°
35°
40°S
40°
45°
165°
180°
4
5
5
after Reyners (in prep)

4b. INTENSITY MAP OF THE OAMARU II EARTHQUAKE - 1876 FEBRUARY 25 21:15

DATE:	1876 FEBRUARY 25
TIME:	21:15 UT
MAGNITUDE:	5.75 M (I)
EPICENTRE:	45.2°S 170.9°E C (I)
DEPTH:	Shallow

The earthquakes of February 25 caused significant damage in Oamaru, with reports of 100 chimneys being brought down. Cracking of stone buildings occurred, and a wall of the main building of the hospital required rebuilding. A maximum intensity of MM7 is indicated. Most of the chimneys at Kakanui were also brought down, again indicating MM7. In constructing isoseismal maps for the two events, difficulty was experienced in ascribing descriptions of the shaking to a particular earthquake, as they were closely spaced in time and of similar magnitude.

Most reports from Oamaru suggest the second earthquake was somewhat stronger there. However, the felt area for both events is rather similar, with MM5 being experienced from Timaru to Dunedin. Both earthquakes were followed by a rich aftershock sequence.

The intensity map is taken from Reyners (in prep.).

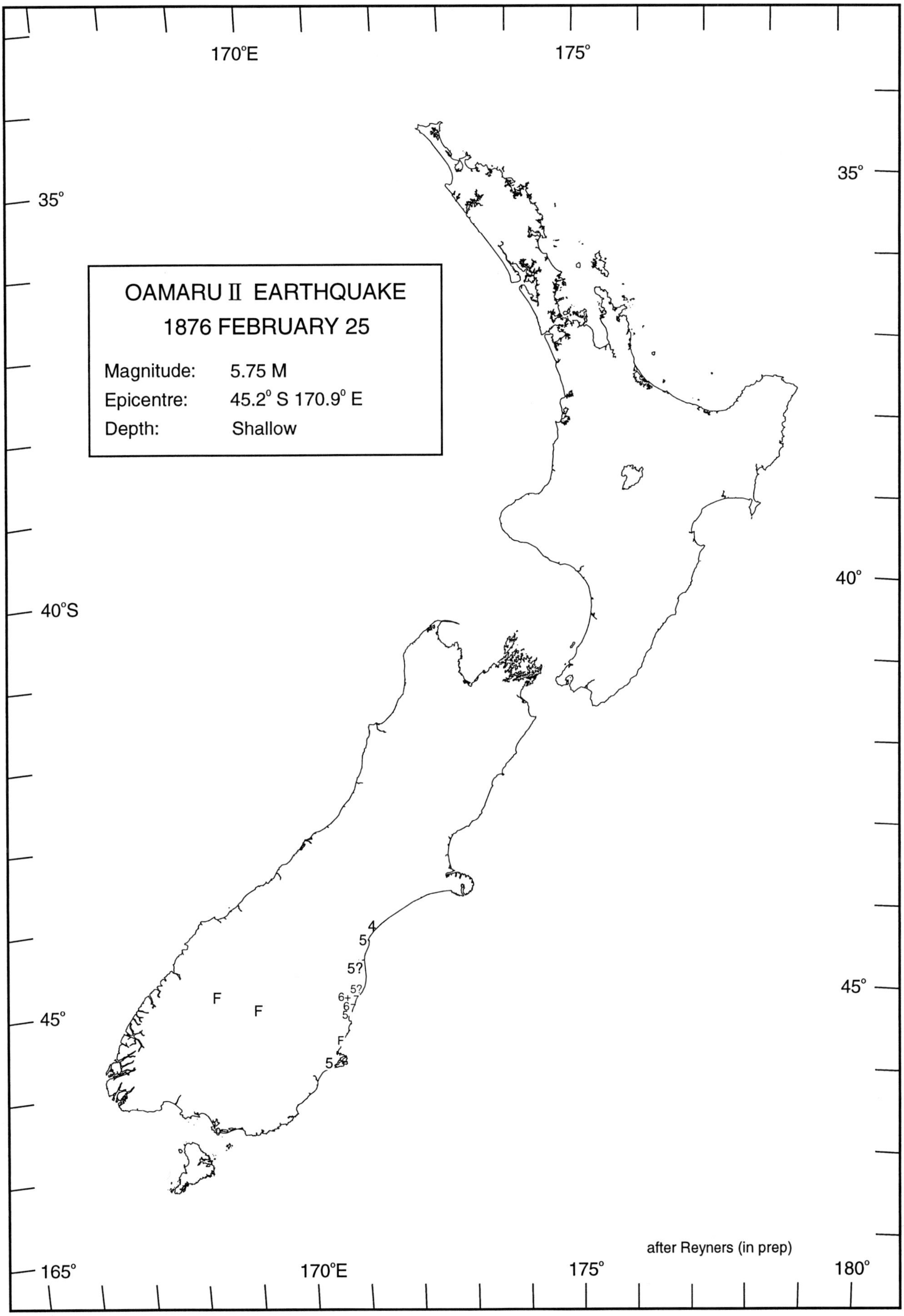

OAMARU II EARTHQUAKE
1876 FEBRUARY 25
Magnitude: 5.75 M
Epicentre: 45.2° S 170.9° E
Depth: Shallow
170°E
175°
35°
40°S
45°
165°
180°
40°
after Reyners (in prep)

4c. INTENSITY MAP OF THE OAMARU III EARTHQUAKE - 1876 APRIL 11

DATE:	1876 APRIL 11
TIME:	00:10 UT
MAGNITUDE:	5.75 M (I)
EPICENTRE:	45.2°S 170.9°E C (I)
DEPTH:	Shallow

The earthquake of 1876 April 11 earthquake is described in contemporary press reports as the heaviest shock ever experienced in Oamaru. Descriptions of the shaking suggest the intensity reached MM7 there. At Kakanui, the event was described as "very severe". At Otepopo (near Herbert), large slips occurred from the cliff facing the river, suggesting that shaking reached MM7. Numerous slips were also reported between Otepopo and Hampden. The earthquake is likely to have occurred at shallow depth, since felt aftershocks continued until at least the end of May.

The intensity map is taken from Reyners (in prep.).

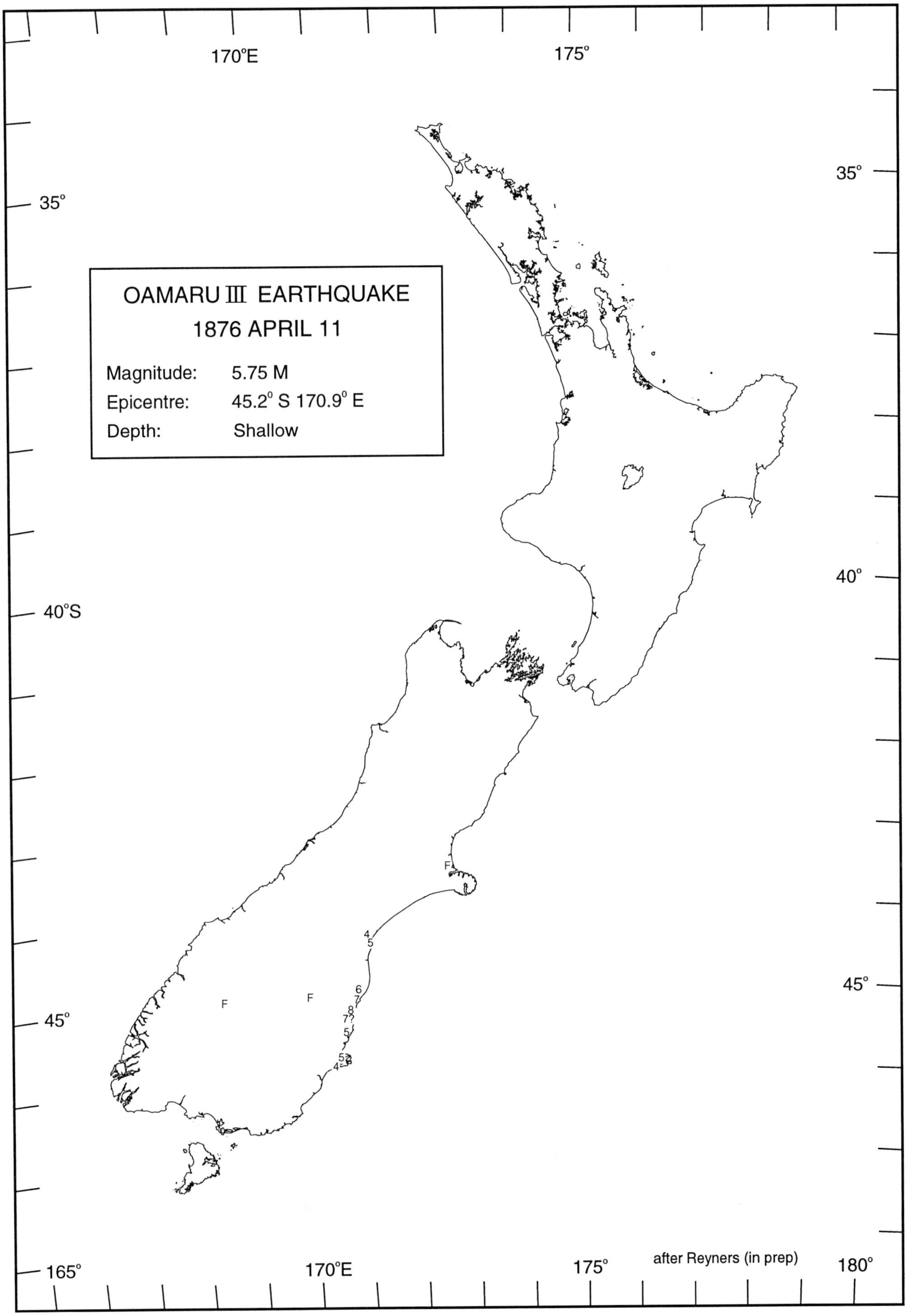
OAMARU III EARTHQUAKE
1876 APRIL 11
Magnitude: 5.75 M
Epicentre: 45.2° S 170.9° E
Depth: Shallow
170°E
175°
35°
40°S
45°
165°
180°
40°
after Reyners (in prep)

5a. ISOSEISMAL MAP OF THE NORTH CANTERBURY EARTHQUAKE - 1888 AUGUST 31

DATE:	1888 AUGUST 31
TIME:	16:45 UT
MAGNITUDE:	7.0-7.3 M (I)
EPICENTRE:	42.6°S 172.4°E C (I)
DEPTH:	Shallow

'On September 1, 1888 [at approximately 0410 NZST], an earthquake of probable magnitude M7-7.3 struck the Amuri District of North Canterbury, 100 km northwest of Christchurch, New Zealand. The earthquake ruptured a segment of the Hope Fault, and damaged buildings over a wide area. The effects of the earthquake indicate Modified Mercalli intensities of MM IX in the epicentral area. High-intensity isoseismals (>MM VII) were strongly elliptical and parallel to faulting, apparently attenuating steeply to the northwest and southeast, respectively. However, in parts of Greymouth and Christchurch, shaking was amplified (to MM VII), presumably by local ground conditions.

'After the earthquake, dextral offsets of between 1.5 and 2.6 m were observed on fencelines crossing the Hope Fault at four localities in the Hope Valley.

'The Hope Fault is an important strike-slip element of the Marlborough Fault Zone.' (Cowan 1991).

Cowan (1991) also notes that empirical formulae using inferred rupture length and maximum observed displacement support the estimates of magnitude M7.0-7.3.

The isoseismal map is from Cowan (1991). A detail map of the epicentral area and further description appears on the following pages.

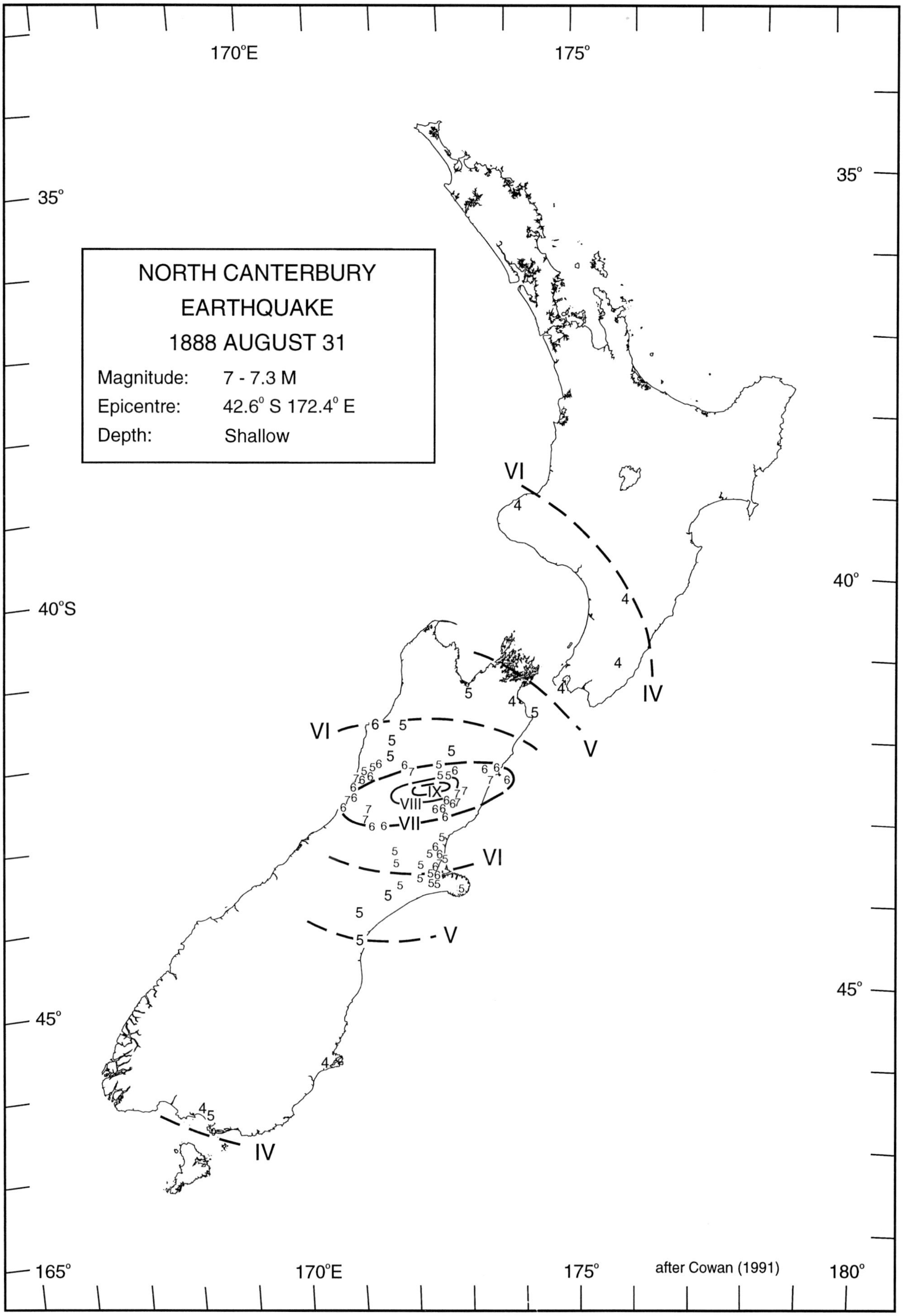
NORTH CANTERBURY
EARTHQUAKE
1888 AUGUST 31
Magnitude: 7 - 7.3 M
Epicentre: 42.6° S 172.4° E
Depth: Shallow
170°E
175°
35°
40°S
40°
45°
165°
180°
IX
VIII
VII
VI
V
IV
after Cowan (1991)

5b. ISOSEISMAL MAP OF THE NORTH CANTERBURY EARTHQUAKE - 1888 AUGUST 31

DATE:	1888 AUGUST 31
TIME:	16:45 UT
MAGNITUDE:	7.0 - 7.3 M (I)
EPICENTRE:	42.6°S 172.4°E C (I)
DEPTH:	Shallow

'In the Hope Valley and on the Hanmer Plain, many cob (rammed earth) and stone buildings were badly damaged or destroyed. Several wooden buildings were distorted or shifted from their foundations...

'Numerous landslides scarred fluvial and lacustrine Quaternary deposits, and large blocks of rock fell from bedrock outcrops. Liquefaction was evident near Glynn Wye, causing the formation or enlargement of large pits and sandblows. Fissuring of the ground was intense along the Hope Fault between the Hope-Boyle Valley junction and Upper Waiau, and numerous fissures were reported on the Hanmer Plain and along the Percival River. At many localities elsewhere in Amuri District there was moderate to severe damage to chimneys and household articles. On the West Coast the strongest shaking was reported from the Otira Gorge, where new springs were observed (three hot and one cold), and a large fissure allegedly formed in Kelly's Creek. In the coastal towns of Hokitika and Greymouth, there were reports of chimney damage and breakage of goods, glass and crockery, but few other localities in Westland reported any damage.' (Cowan 1991)

The isoseismal map is from Cowan (1991). A regional isoseismal map appears on the previous page.

NORTH CANTERBURY
EARTHQUAKE
1888 AUGUST 31

Magnitude: 7 - 7.3 M
Epicentre: 42.6° S 172.4° E
Depth: Shallow

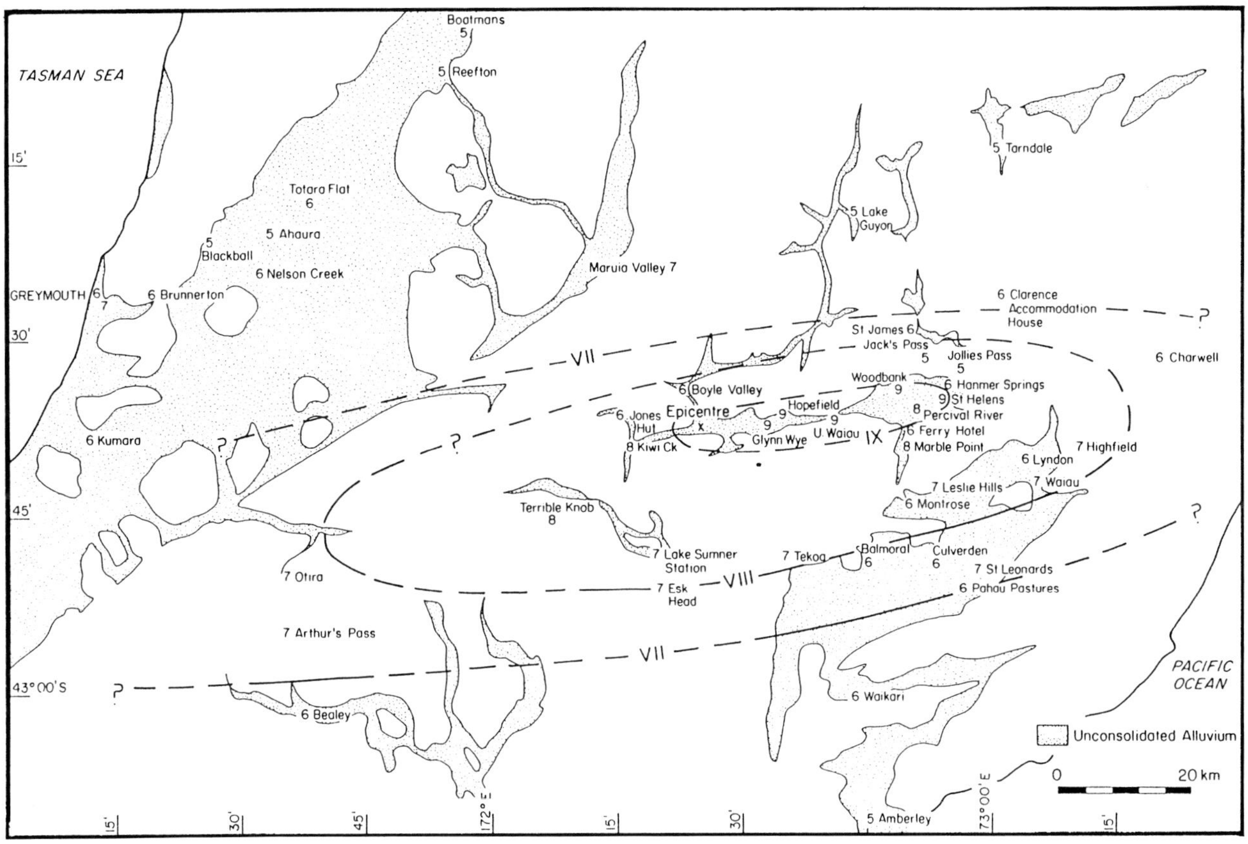

after Cowan (1991)

6. ISOSEISMAL MAP OF THE WAIKATO HEADS EARTHQUAKE - 1891 JUNE 23

DATE:	1891 JUNE 23
TIME:	23:25 UT
MAGNITUDE:	5.5-6.0 (Dowrick, pers. Comm.) (I)
EPICENTRE:	37.4°S 174.6°E D (I)
DEPTH:	Crustal

The earthquake on June 24 1891 at 10:55 a.m. (local time) caused some consternation in Auckland, as the city had experienced few strong earthquakes. However the damage was minimal, being confined to the odd fall of plaster, one chimney in Onehunga and some broken crockery and bottles in Papakura and in several other localities around Auckland. It was felt quite strongly in Helensville. However the maximum intensity seems to have been around the Waikato Heads, where Hogben (1891) reports: "windows broken, bells rung. Children in school thrown off seats."

The isoseismal map is the result of data from Hogben's (1891) paper, several newspaper and lighthouse reports collected and assessed by several Seismological Observatory staff. It is unlikely that more extensive investigation will reveal significantly more detail. A magnitude of about M5.5-6.0 (Dowrick, pers. comm.) seems appropriate. An isoseismal map (Rossi-Forel scale) appears in Hogben (1891), the R-F VII isoseismal differing little from the MM6 presented here.

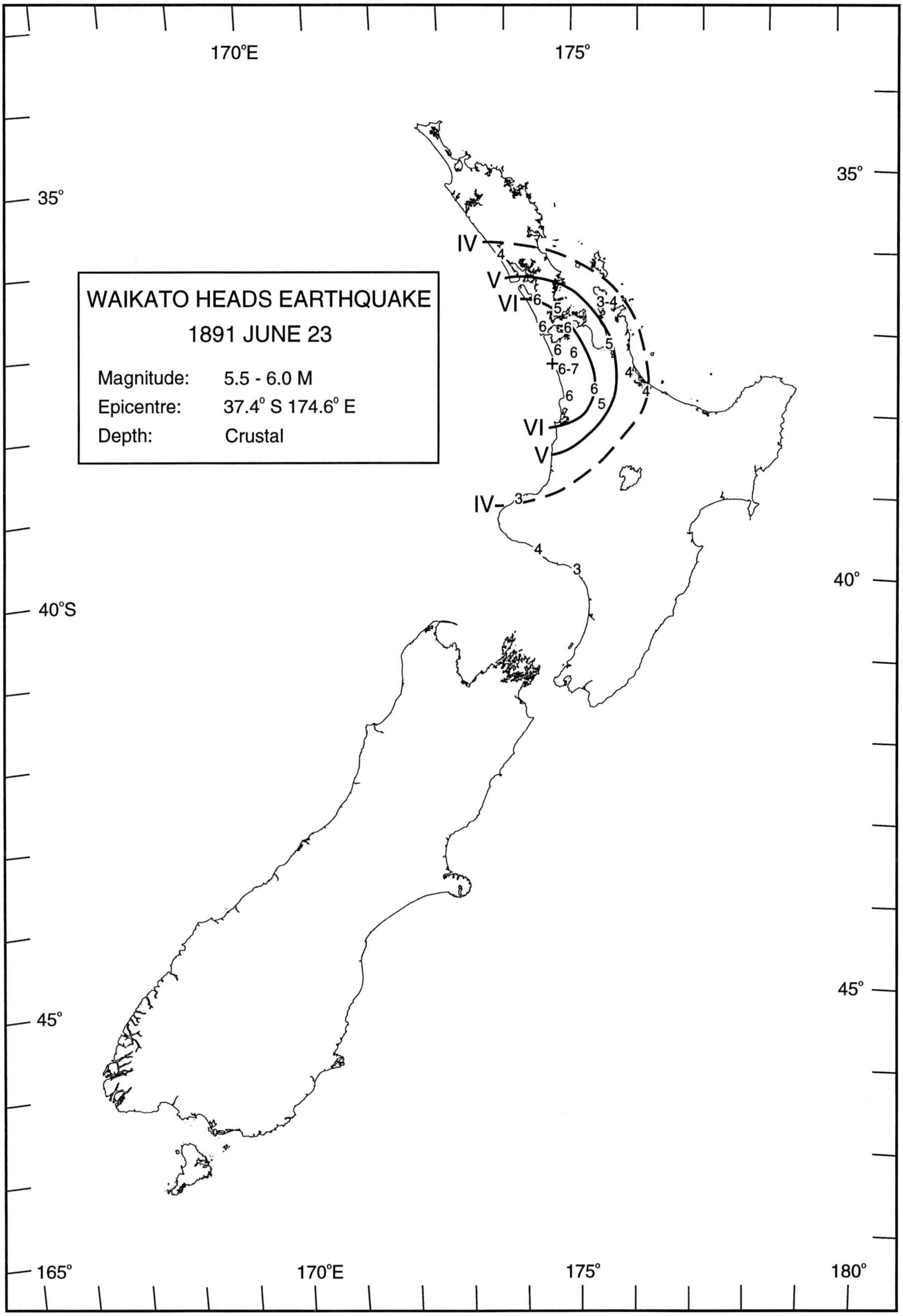

WAIKATO HEADS EARTHQUAKE
1891 JUNE 23
Magnitude: 5.5 - 6.0 M
Epicentre: 37.4° S 174.6° E
Depth: Crustal
170°E
175°
35°
40°S
40°
45°
165°
180°
IV
V
VI
3-4
6-7

7. ISOSEISMAL MAP OF THE NELSON EARTHQUAKE - 1893 FEBRUARY 11

DATE:	1893 FEBRUARY 11
TIME:	20:30 UT
MAGNITUDE:	6.6-6.9 M (I) (Dowrick, pers.comm.)
EPICENTRE:	40.7°S 173.9°E D (I)
DEPTH:	Lower Crustal

'Because of the extensive damage to chimneys and minor structural damage in Nelson, this earthquake was located to the south of Nelson (Hogben, 1893). However evidence of minor chimney damage in Wellington and Wanganui and low intensities on the west coast of the South Island suggest an epicentre to the north of Nelson, possibly in the western Marlborough Sounds. A magnitude of about 7.0 [6.6 - 6.9 (Dowrick pers.comm.)] is indicated by the extent of the localities reporting having felt the earthquake.' (Anderson et al. 1994)

The isoseismal map is taken from Anderson et al. (1994). The felt information has been taken from Hogben (1893) and several newspaper reports, which were quite extensive for Nelson but not elsewhere and the map should be regarded as preliminary and requiring further research.

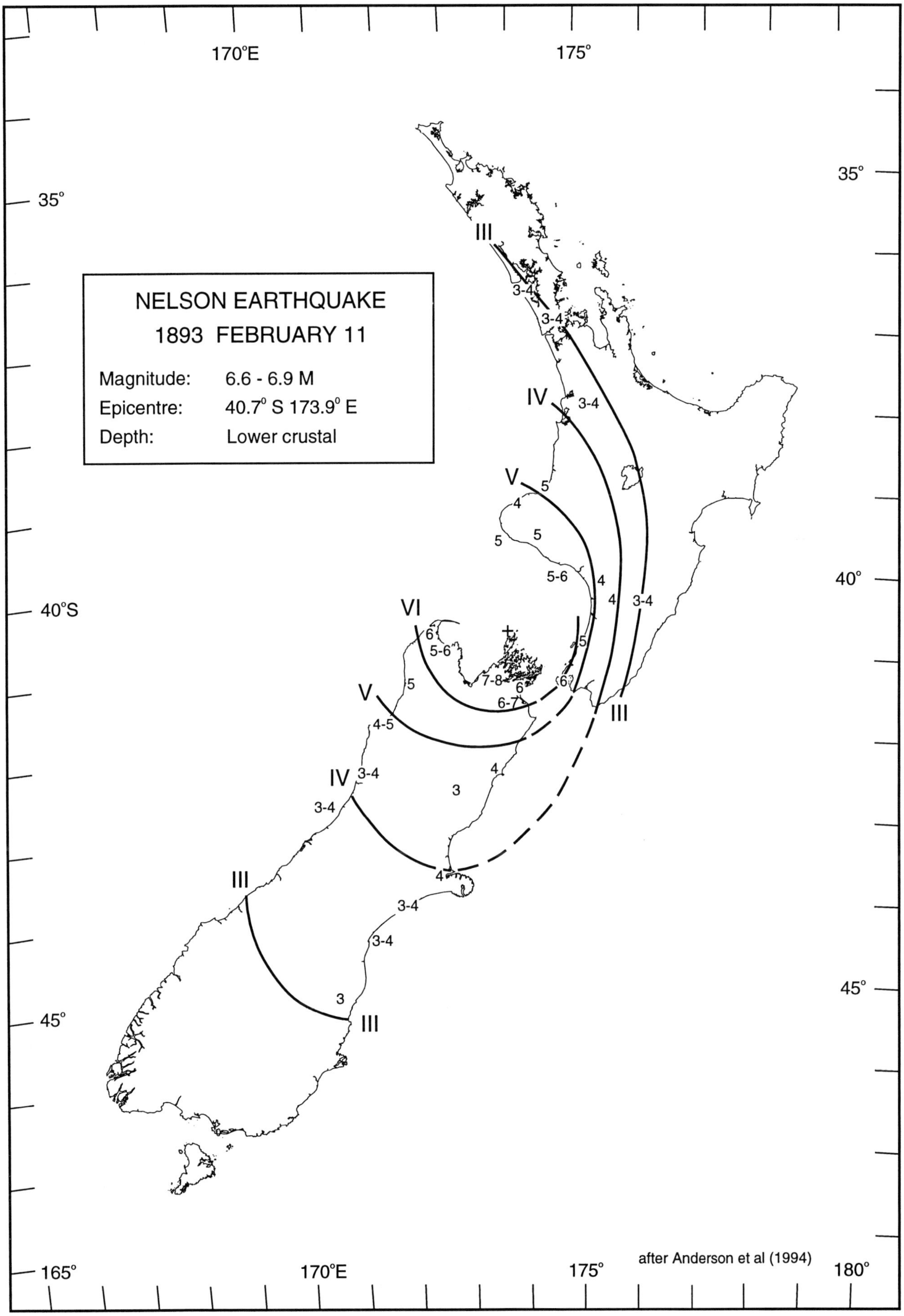

NELSON EARTHQUAKE
1893 FEBRUARY 11
Magnitude: 6.6 - 6.9 M
Epicentre: 40.7° S 173.9° E
Depth: Lower crustal
170°E
175°
35°
40°S
40°
45°
165°
180°
III
IV
V
VI
3-4
4
4-5
5
5-6
6
6-7
7-8
3
after Anderson et al (1994)

8. ISOSEISMAL MAP OF THE CAPE TURNAGAIN EARTHQUAKE - 1904 AUGUST 08

DATE:	1904 AUGUST 08
TIME:	22:50.5 UT
MAGNITUDE:	6.7 M_S (Dowrick & Smith 1990)
EPICENTRE:	40.6°S 176.8°E C (I)
DEPTH:	Lower Crustal

The August 08 1904, M_s 6.7, earthquake caused considerable damage to chimneys from north of Napier to the city of Wellington, where parapets fell from some buildings in the central city area. In the vicinity of Cape Turnagain nearly every chimney was levelled, many water tanks were thrown down and several wooden houses racked and distorted. To the west of the epicentre chimneys were slightly damaged as far as Wanganui. The earthquake was felt from Queenstown to Auckland and on the west coast of the South Island to Hokitika. Sand fountaining was reported from at least ten locations from Napier to Gladstone, a small town south-east of Masterton. A large wave, possibly a tsunami, was reported at Mohaka. There were large landslides reported along the coast north and south of Cape Turnagain and some minor landslides inland. There was possibly a foreshock and the mainshock was followed by few aftershocks, none of damaging or even alarming intensity.

Damage in Wellington varied considerably, from MM5 to MM7 and there is evidence of enhancement of shaking due to the soil structure beneath certain areas of the city.

The isoseismal map is taken from Downes (in prep.), the macroseismic data being the result of extensive searching of newspapers, private diaries and other historical material.

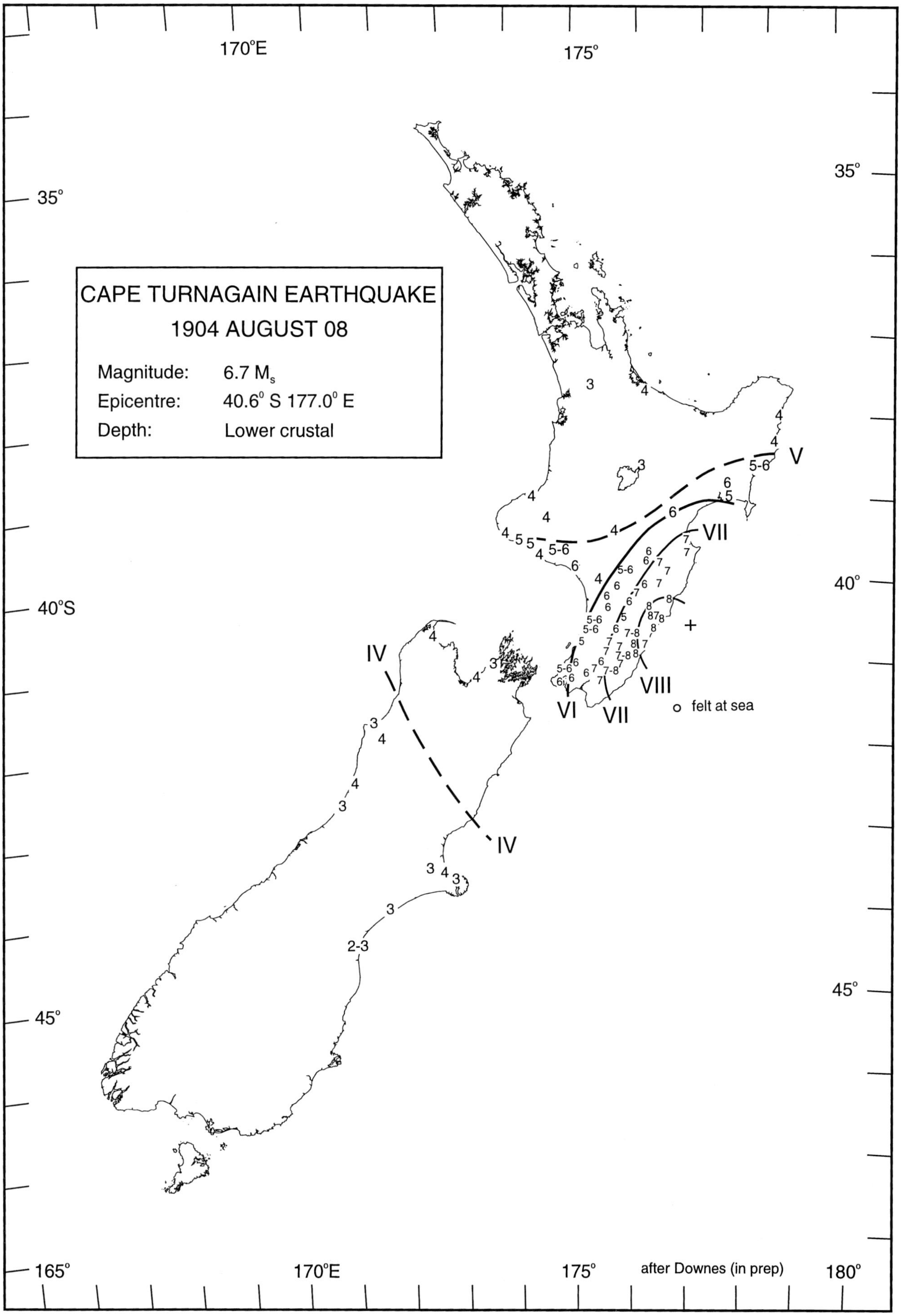
CAPE TURNAGAIN EARTHQUAKE
1904 AUGUST 08
Magnitude: 6.7 M_s
Epicentre: 40.6° S 177.0° E
Depth: Lower crustal
170°E
175°
35°
40°S
40°
45°
165°
180°
IV
V
VI
VII
VIII
felt at sea
after Downes (in prep)

9. ISOSEISMAL MAP OF THE EAST CAPE EARTHQUAKE I - 1914 OCTOBER 06

DATE:	1914 OCTOBER 06
TIME:	19:16 UT
MAGNITUDE:	6.7 M_S (Dowrick & Smith 1990)
EPICENTRE:	37.5°S 178.3°E C (I)
DEPTH:	Shallow

This earthquake was centred near Matakaoa Point, at the northern tip of the Raukumara Peninsula. The Seismological Observatory holds an unusually complete set of first-hand accounts for this earthquake, and the subsequent large shocks felt on the Raukumara Peninsula on 1914 October 28 and 1914 November 22 (Reyners, in prep.). This is largely because the seismologist George Hogben was Secretary for Education at the time, and he instructed all teachers at the Native Schools on the peninsula to furnish reports on the events.

Numerous landslides occurred in the epicentral region, and in one of these near Cape Runaway a shepherd was killed (Morgan 1920). Slips were especially numerous along the course of the Whangaparaoa River (south of Cape Runaway). Numerous fissures opened up along the dry river bed and on the adjacent flats, and in some places earthquake fountains occurred. Earthquake fountains were also reported in the Wharekahika River. Such ground damage indicates intensity reached at least MM9 in the epicentral region.

The distribution of intensity is consistent with the earthquake having occurred at shallow depth. It is also consistent with the surface wave magnitude determined for the event.

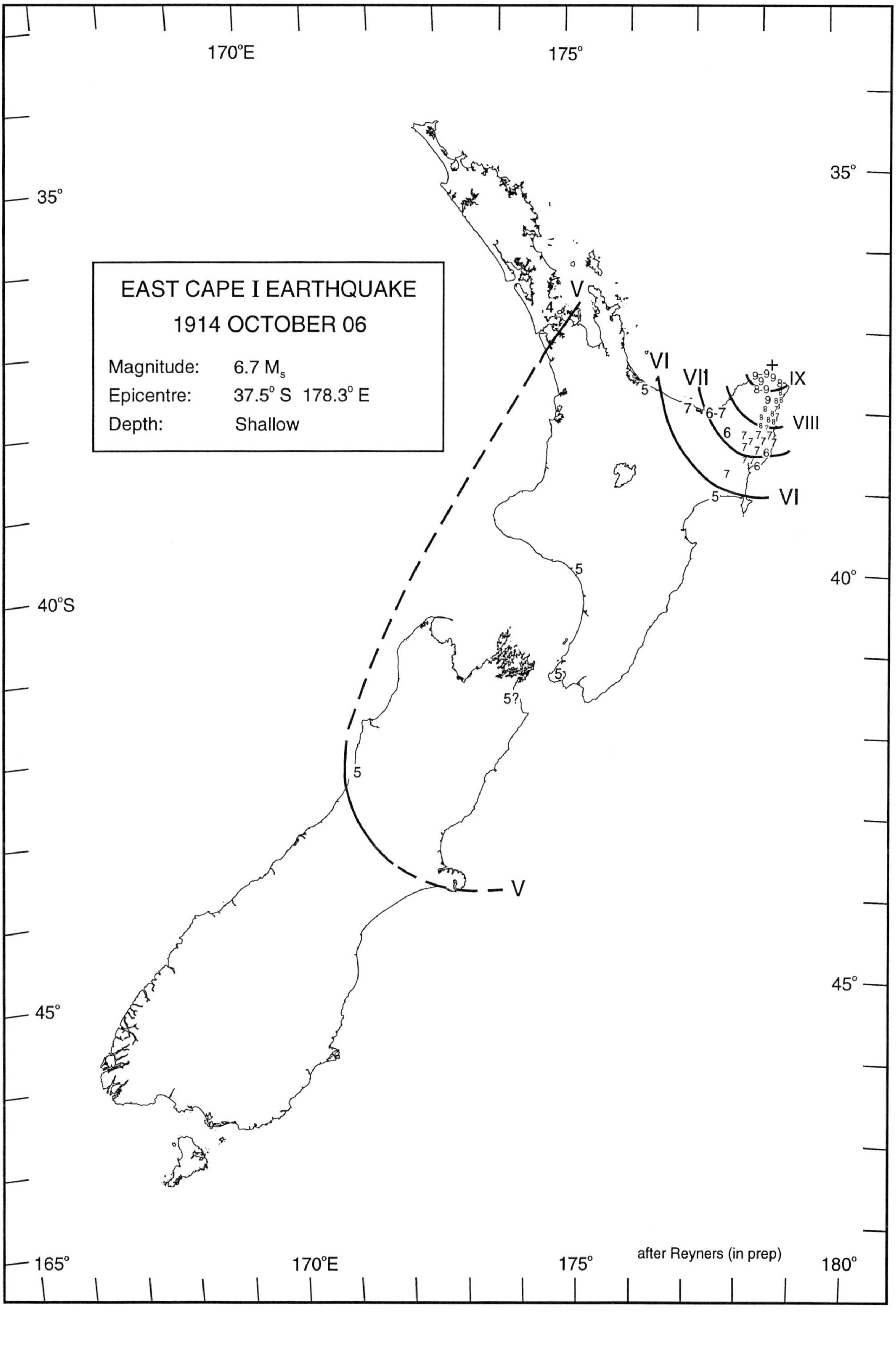

EAST CAPE I EARTHQUAKE
1914 OCTOBER 06
Magnitude: 6.7 Ms
Epicentre: 37.5° S 178.3° E
Depth: Shallow
170°E
175°
35°
40°S
45°
165°
180°
40°
V
VI
VII
VIII
IX
after Reyners (in prep)

10. ISOSEISMAL MAP OF THE EAST CAPE EARTHQUAKE II - 1914 OCTOBER 28

DATE:	1914 OCTOBER 28
TIME:	00:16 UT
MAGNITUDE:	6.5 M_S (Dowrick & Smith 1990)
EPICENTRE:	37.5°S 178.3°E C (I)
DEPTH:	Shallow

This shock appears to have been not quite as severe as that of 1914 October 06 (Morgan 1920). As for that previous shock, there is an unusually complete set of first-hand accounts of felt effects because of reports furnished by teachers at the Native Schools on the Raukumara Peninsula (Reyners, in prep.). There were reports of large slips and the ground "opening up" in the hill country near Te Araroa, and a "huge" landslide was reported at Tawhiti Point (Waipiro Bay).

The region of maximum intensity (MM8) is the northern tip of the Raukumara Peninsula, and it appears that the earthquake occurred close to that of 1914 October 06. The distribution of intensity indicates that the earthquake, like that of October 06, also occurred at shallow depth. This is corroborated by numerous reports of felt aftershocks.

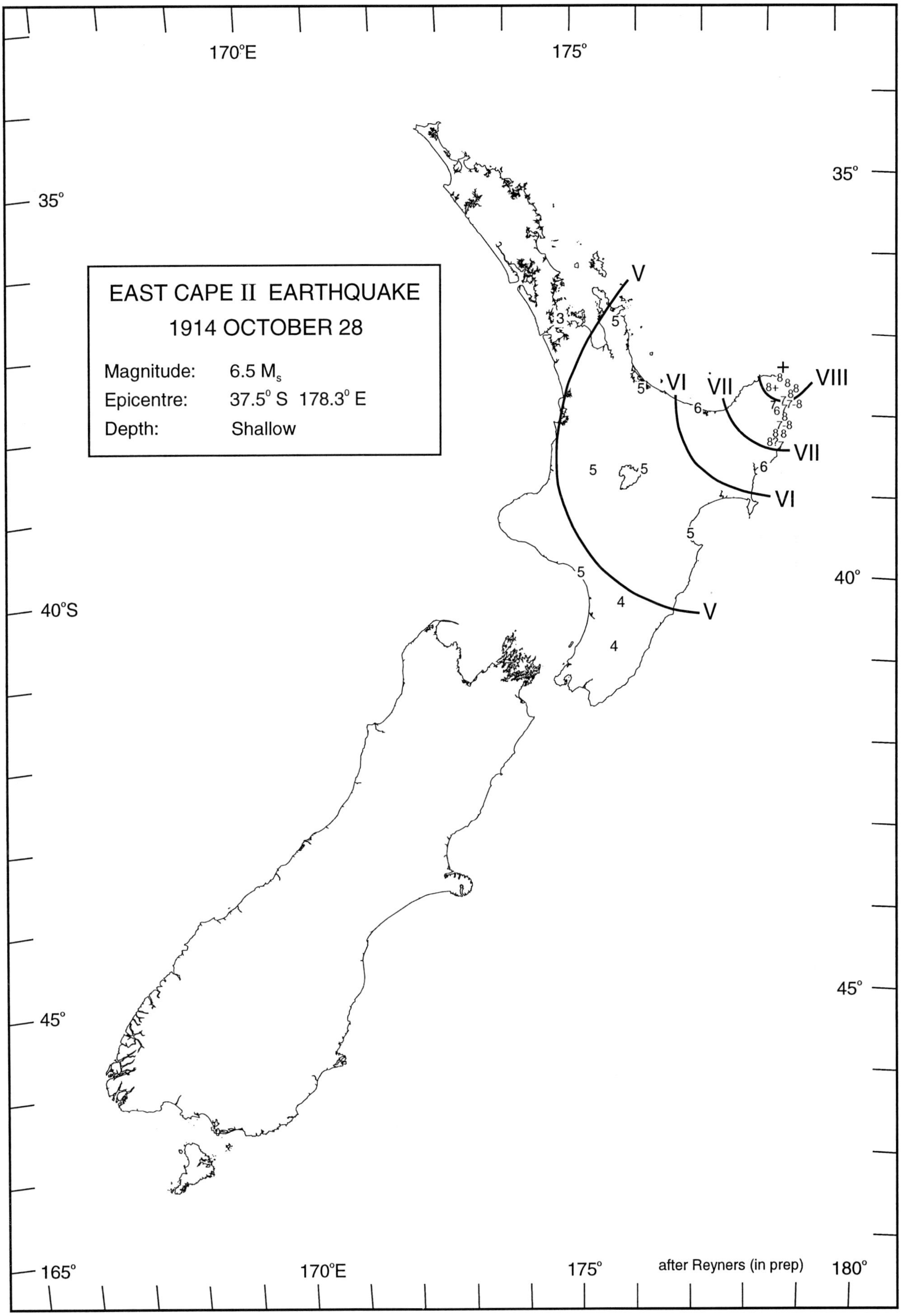
EAST CAPE II EARTHQUAKE
1914 OCTOBER 28
Magnitude: 6.5 M_s
Epicentre: 37.5° S 178.3° E
Depth: Shallow
170°E
175°
35°
40°S
45°
165°
180°
40°
V
VI
VII
VIII
after Reyners (in prep)

11. ISOSEISMAL MAP OF THE BAY OF PLENTY EARTHQUAKE - 1914 NOVEMBER 22

DATE:	1914 NOVEMBER 22
TIME:	08:14 UT
MAGNITUDE:	> 7.2 M (I)
EPICENTRE:	37.5°S 176.5°E D (I)
DEPTH:	~ 300 km

This earthquake was strongly felt at Gisborne, where it was described as "the severest on record" (Morgan 1920). Chimneys, windows and crockery were broken in various parts of the town. Intensities appear to have been greatest near Te Karaka, where cracking of bridge abutments and roads was reported, and some chimneys were "shaken to the foundation" (Reyners, in prep.). The earthquake was felt throughout most of the country, with Auckland and Dunedin reporting slight shocks, and Christchurch a fairly sharp shake. At Wellington the shock was sufficiently violent to shake crockery off shelves, and at Dannevirke chimneys were damaged.

The large area over which this event was felt indicates that it was probably an intermediate-depth or deep shock. Distinct pP and sP phases recorded by the three-component Wiechert seismograph at Apia suggest that the earthquake occurred at about 300 km depth (Reyners, in prep.). Events at this depth in the North Island usually occur in a restricted area under the Bay of Plenty. A modern analogue of the event would be the shock of 1953 September 29 (see isoseismal map #48), which was centred 273 km below the Bay of Plenty and was of magnitude 7.2. Comparison of the intensity distributions of the two events suggests the 1914 event was significantly larger.

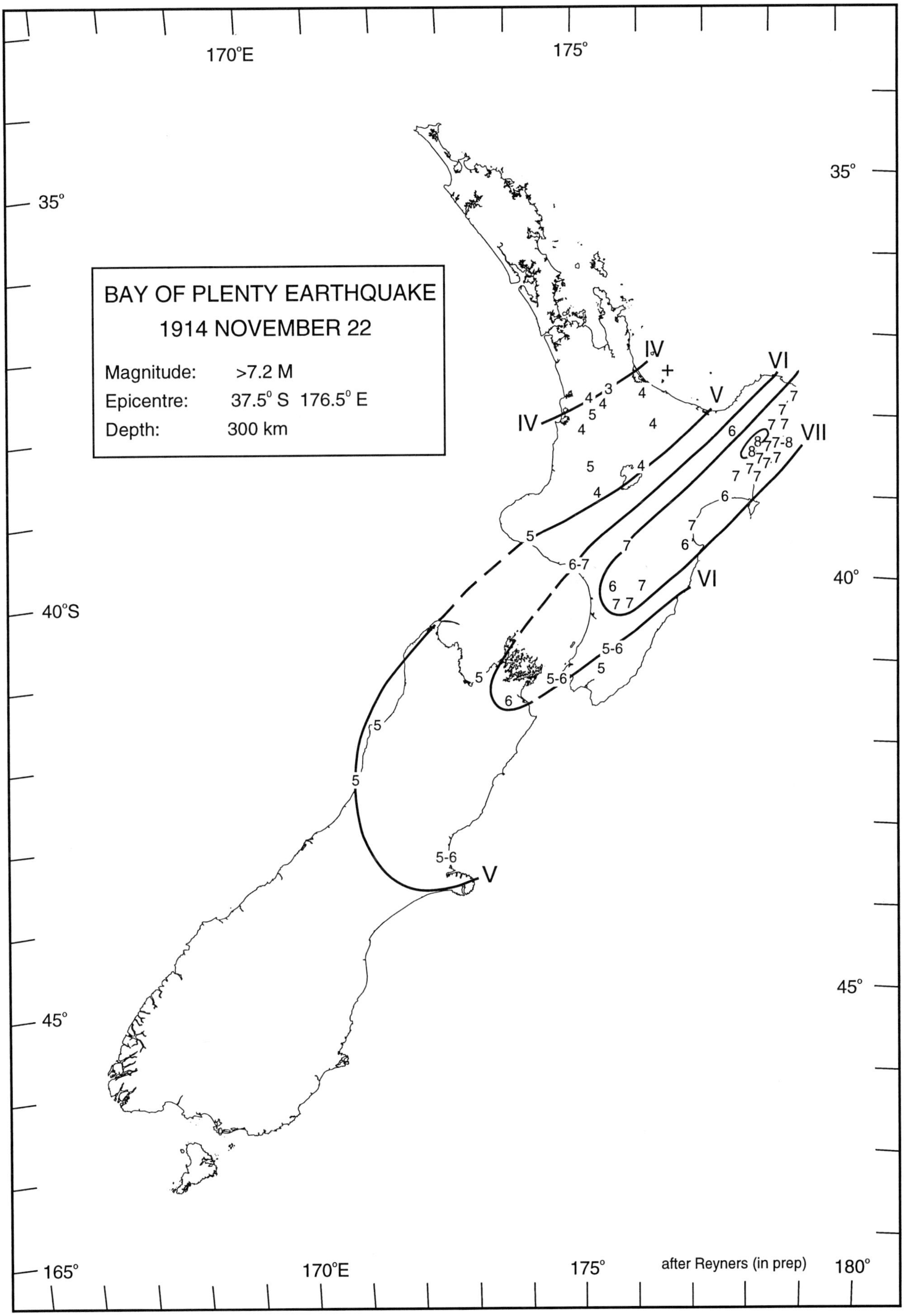

BAY OF PLENTY EARTHQUAKE
1914 NOVEMBER 22
Magnitude: >7.2 M
Epicentre: 37.5° S 176.5° E
Depth: 300 km
170°E
175°
35°
40°S
40°
45°
165°
180°
IV
V
VI
VII
after Reyners (in prep)

12. ISOSEISMAL MAP OF THE HAWKES BAY, 80 KM DEEP EARTHQUAKE - 1921 JUNE 28

DATE:	1921 JUNE 28
TIME:	13:58:54 UT
MAGNITUDE:	7.0 M (Gutenberg & Richter 1949);
	6.4 M_S (Dowrick & Smith 1990)
EPICENTRE:	39.3°S 176.4°E D (Bullen 1937)
DEPTH:	80 km

'An analysis has been made of the Hawke's Bay earthquake of 1921 June 29, using the data of the I.S.S. The most probable position of the epicentre has been found to be the point 39.3°S 176.4°E., which is about 30 miles [50 km] N.W. of Napier, the origin time being 1921 June 28 d. 13 h. 58 m. 54 s. (G.M.C.T.). The epicentre is subject to a standard error of about 0.4. A depth of focus of about 80 km is indicated, and this is confirmed by European observations of the phase P' (or PKP).' (Bullen 1937)

The earthquake, which occurred at about 1.30 a.m. local time, was felt from Auckland, where it woke some residents, to Dunedin. Chimneys were brought down in several locations; Gisborne, Napier, Taihape and Wanganui and one chimney reported damaged in Nelson.

The seismological source material for this event has not been reinterpreted since Bullen's paper in 1937.

Felt summaries held at the Seismological Observatory have been converted from Rossi-Forel scale to Modified Mercalli and the map drawn by Eiby.

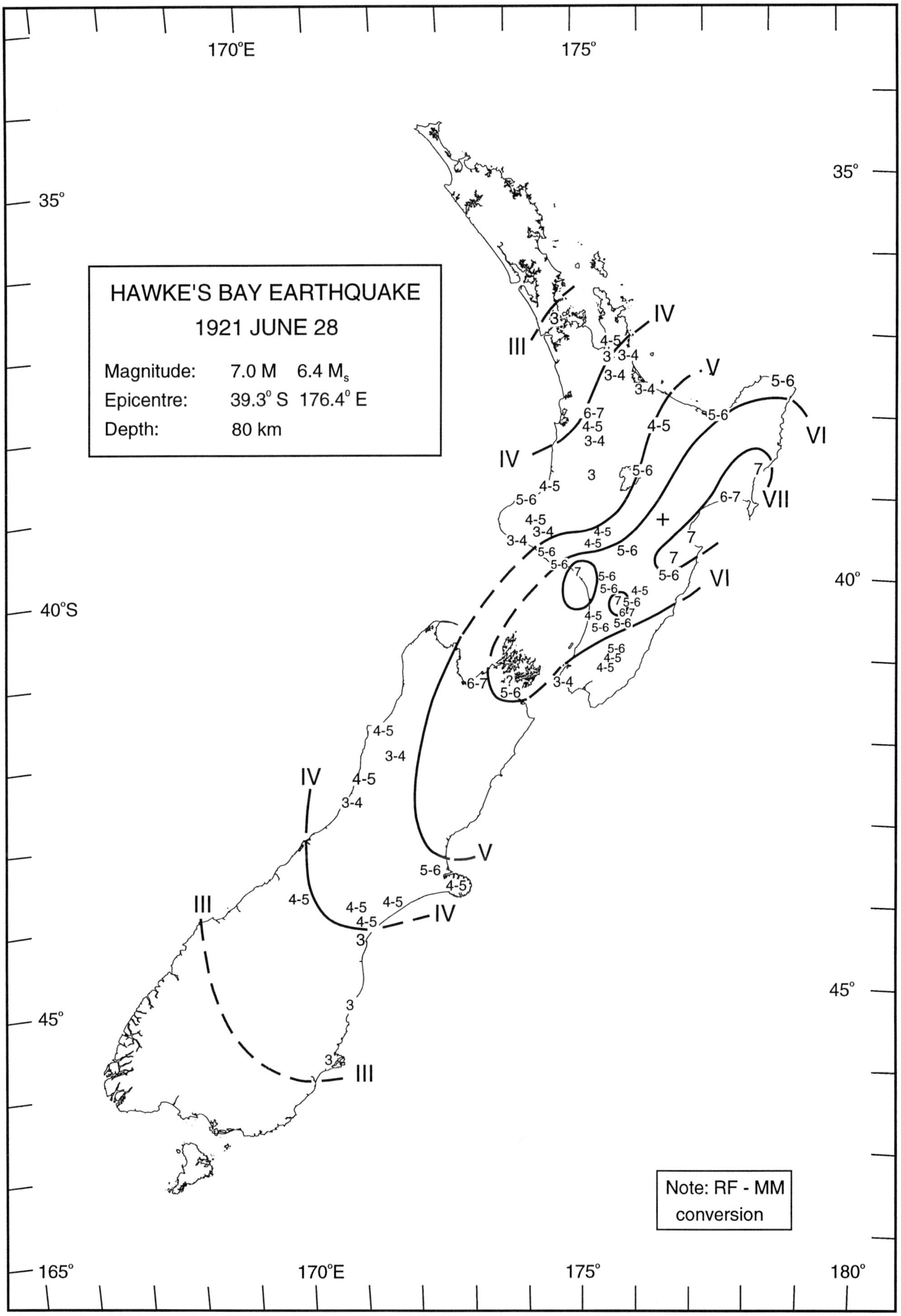

HAWKE'S BAY EARTHQUAKE
1921 JUNE 28
Magnitude: 7.0 M 6.4 Ms
Epicentre: 39.3° S 176.4° E
Depth: 80 km
Note: RF - MM conversion
165°
170°E
175°
180°
35°
40°S
40°
45°
III
IV
V
VI
VII

13a. ISOSEISMAL MAP OF THE MOTUNAU EARTHQUAKE - 1922 DECEMBER 25

DATE:	1922 DECEMBER 25
TIME:	03:33 UT
MAGNITUDE:	6.4 M_S (Dowrick & Smith 1990)
EPICENTRE:	43°S 173°E D (Gutenberg & Richter 1949)
DEPTH:	Crustal

Remembered as the Christmas Day Earthquake, this event occurred just after 3 p.m. when many people were enjoying their Christmas afternoon out of doors. It caused extensive damage in North Canterbury and was felt from Taranaki and Napier in the north to Dunedin in the south and brought down large numbers of chimneys from Cheviot to Rangiora. Isolated chimney damage also occurred on the West Coast at Greymouth. Crockery and household articles fell and were broken in Christchurch. Some liquefaction effects were apparently evident on some North Canterbury beaches and river flats.

Felt summaries held at the Seismological Observatory have been converted from Rossi-Forel scale to Modified Mercalli and the isoseismal map was drawn by Eiby. Newspaper reports and other material are available and this event warrants reinterpretation of the source data.

On the following pages is a detailed map (Rossi-Forel scale) from Skey (1925), which has been included here as it appears to have been based on a different and more extensive set of information from that which was available to the Seismological Observatory in 1922 and therefore to Eiby.

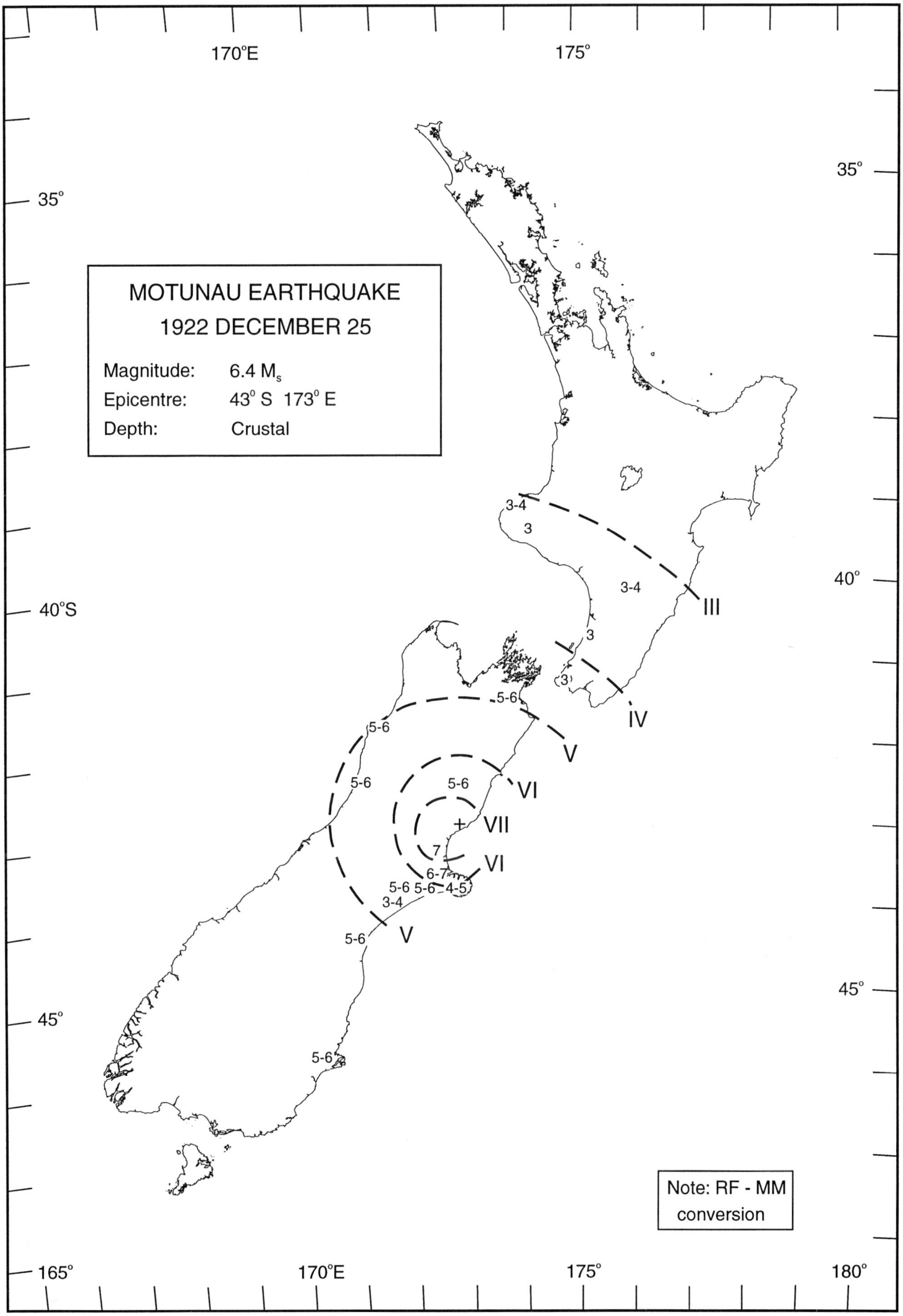
MOTUNAU EARTHQUAKE
1922 DECEMBER 25
Magnitude: 6.4 M_s
Epicentre: 43° S 173° E
Depth: Crustal
Note: RF - MM conversion
170°E
175°
35°
40°S
40°
45°
165°
180°
III
IV
V
VI
VII
3-4
3
5-6
7
6-7
4-5

13b. ISOSEISMAL MAP OF THE MOTUNAU EARTHQUAKE - 1922 DECEMBER 25

DATE:	1922 DECEMBER 25
TIME:	03:33 UT
MAGNITUDE:	6.4 M_S (Dowrick & Smith 1990)
EPICENTRE:	43°S 173°E D (Gutenberg & Richter 1949)
DEPTH:	Crustal

'... an investigation was made of the position of the isoseismals of the above severe earthquake. A postal appeal, chiefly to Postmasters, was made, and a large number of replies received, a circular form being filled as requested.

'... the Rossi-Forel scale values of the effects at the respective stations have been entered upon the accompanying map, and a series of isoseismal lines ranging from 9 to 4 on the scale have been drawn...

'The maximum effect in the epifocal area was of scale 9, and a personal inspection of the effects in the district from Hawarden to Waikari, Scargill, Cheviot, and reports received, confirmed the impression of an elongated epifocal area, surface effects being of appreciable magnitude in the neighbourhood of Cheviot (cracks in made ground, road), and an extension slip in the north bank of the Waikari Creek some two miles below Scargill,.. [causing] an extensive blockage of the river.

'In the epicentral region the chief damage was to the brick chimneys. At Waikari the manse, built of rough stone and mortar, was ruined and rendered uninhabitable, the western wall being thrown out.

'Minute aftershocks were experienced in the Waikari Valley, but no record of these was obtained here [in Christchurch].' (Skey 1925)

The isoseismal map opposite is reproduced from Skey (1925) and the isoseismal lines refer to Rossi-Forel scale. The data appear to have resulted from a survey conducted by Skey and others independent of the Seismological Observatory in Wellington. The map is more detailed in the epicentral area than that on the previous page but it is not possible to ascertain the individual intensities.

MOTUNAU EARTHQUAKE
1922 DECEMBER 25

Magnitude: 6.4 M_s
Epicentre: 43^0 S 173^0 E
Depth: Crustal

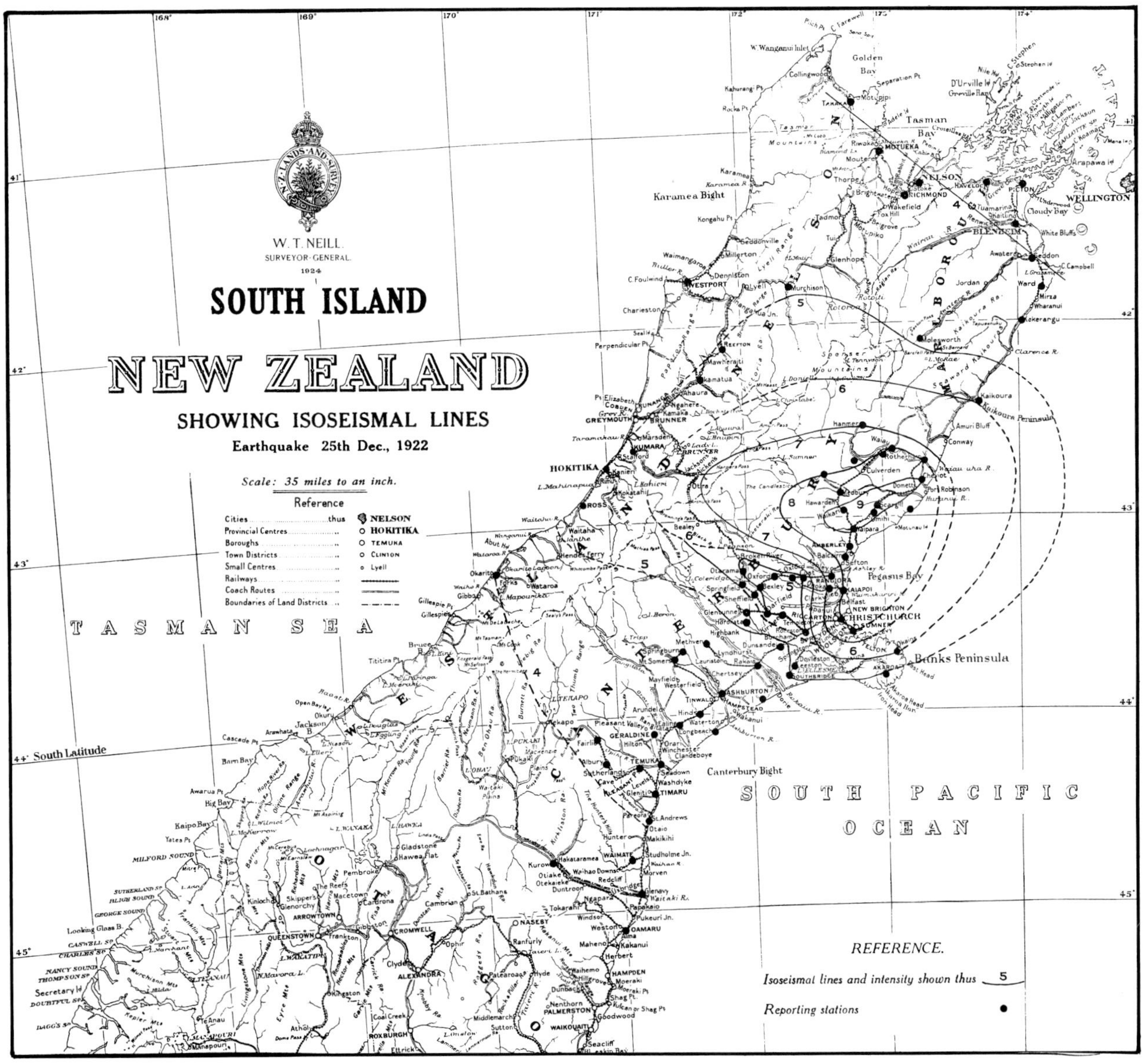

after Skey (1925)

14. ISOSEISMAL MAP OF THE ARTHUR'S PASS EARTHQUAKE - 1929 MARCH 09

DATE:	1929 MARCH 09
TIME:	10:50 UT
MAGNITUDE:	7.1 M_S (Dowrick and Smith 1990)
EPICENTRE:	42.5°S 172.0°E D (ISS)
	42.8°S 171.9°E C (Yang 1992)
DEPTH:	<15 km

At 10:50 p.m. local time on March 9, 1929 a severe earthquake struck the small, isolated communities of Arthur's Pass and Otira, where chimneys toppled and furniture was thrown about. The earthquake was felt throughout the South Island and as far north as Taranaki and Napier in the North Island. Little attention was given to analysing this event as most damage occurred in a sparsely populated area and the effects were completely eclipsed by the Buller earthquake of magnitude M_s 7.8 three months later. However Speight (1933) studied the mountainous area to the north-east of Arthur's Pass some four years after the earthquake and noted the occurrence of numerous and large landslides in a narrow belt about forty kilometres long by four kilometres wide, the occurrence of slides dropping off rapidly outside this belt. Yang (1992) maps these landslides from aerial photographs and confirms this narrow belt of high incidence of what he assumes are earthquake-induced landslides and indicates that they follow the line of the Kakapo Fault, which he proposes as possibly the source of the earthquake. The ISS epicentre is about 30 km north of this.

Felt summaries held at the Seismological Observatory have been converted from Rossi-Forel scale to Modified Mercalli and the isoseismal map has been drawn by Eiby using the ISS epicentre - that is, prior to Yang's (1992) paper. The given intensities are not at variance with an epicentre closer to Arthur's Pass, in particular when the intensity (MM8) at Arthur's Pass, which has apparently been inadvertently omitted by Eiby, is also plotted. Addition of this intensity to the map suggests that some modification of the MM7 and MM8 isoseismals is necessary. However, reinterpretation of the available seismological and macroseismic data is recommended before any attempt to draw a new map and the Kakapo Fault has been adopted as the epicentre of this earthquake until further research suggests otherwise. It is indicated as [+] on Eiby's map.

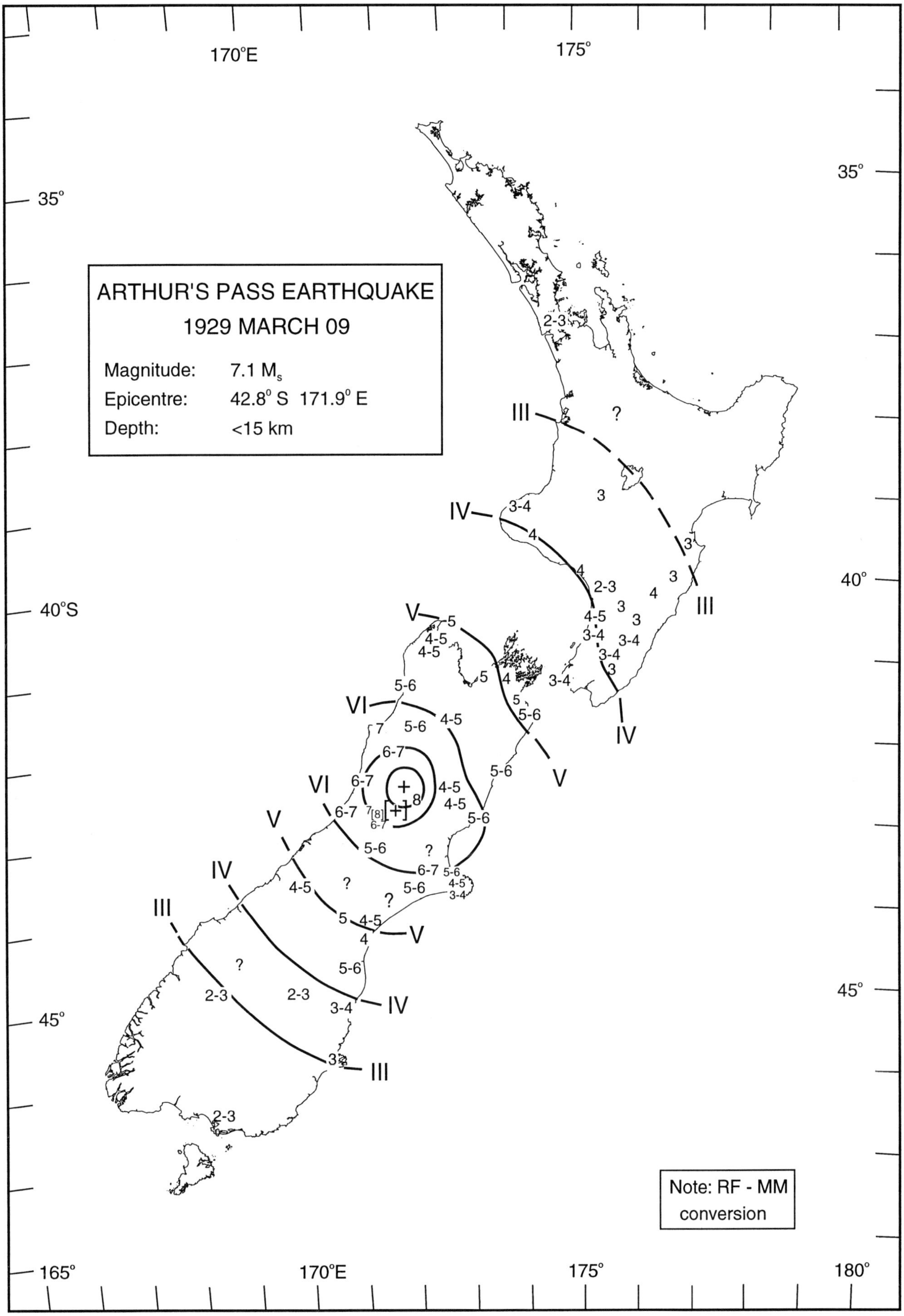

ARTHUR'S PASS EARTHQUAKE
1929 MARCH 09
Magnitude: 7.1 M_s
Epicentre: 42.8° S 171.9° E
Depth: <15 km
Note: RF - MM conversion
165°
170°E
175°
180°
35°
40°S
45°
III
IV
V
VI

15. ISOSEISMAL MAP OF THE BULLER EARTHQUAKE - 1929 JUNE 16
(also referred to as the Murchison earthquake.)

DATE:	1929 JUNE 16
TIME:	22:47:43 UT
MAGNITUDE:	7.8 M_S (Dowrick & Smith 1990)
EPICENTRE:	41.7°S 172.2°E D (I, White Creek Fault)
DEPTH:	20 km

'On June 16, 1929, at 2247 UT, the Murchison earthquake occurred. The epicentre was located at about 41.7°S, 172.2°E and the earthquake had a magnitude of M_s 7.8 (Dowrick & Smith 1990). Shaking was felt over the whole country. The maximum intensity of MM X, accompanied by extensive landslips and faulting, was experienced over a wide area, fortunately sparsely populated. Damaging intensities occurred over much of the Province. Surface faulting of at least 8 km was documented by Fyfe (1929) and Henderson (1937) and more recent work by Berryman (1980) has provided additional fault data. At the Buller River, a 4.5 m high scarp crossed a previously unfaulted 18 ka terrace representing a vertical displacement of 3.0 m, up to the east. A maximum horizontal offset of 2.5 m sinistral was measured by Berryman (1980). Higher terraces are also displaced c. 50 m vertically (260 ka) and c. 35 m (150 ka) at this fault (Suggate 1988a). The strike of the fault near the Buller River is relatively well constrained at about 015° with a near vertical dip (Roder & Suggate 1990). Coseismic uplift reached 4.9 m at a distance of 0.4 km east of the fault and decreased to zero 20 km east of the fault (Suggate 1990).' (Anderson et al. 1994)

'... The earthquake and its many after-shocks, some of which were strong, greatly damaged roads, bridges, buildings, and other structures in the district. Innumerable slips descended from the cliffs and steep slopes; over wide areas fissures opened in the alluvial ground which moved toward the stream channels; and at one point, near White Creek, seven miles from Murchison, a deep-seated rock fracture manifested itself on the surface. In all seventeen persons lost their lives, the majority of them overwhelmed in slips. The area of most intense disturbance extended from Murchison north for forty miles [65 km] to the mouth of the Karamea River. The greater part of this area is quite uninhabited, being densely wooded, mountainous, and difficult of access. Surrounding the intensely shaken part is a much larger region that suffered serious material damage. This extends north to Cape Farewell, south-west to Greymouth, and east to Nelson and the closely settled lowland south and west of that city: the whole region .... covering about 10,000 square miles [26,000 square kilometres]. Tremors were felt at Murchison some hours before the principal shock, and innumerable after-shocks continued for many months .... Epicentres occurred well to the east and west of the Murchison area and as far north as Takaka, eighty miles [130 km] away.' (Henderson 1937)

The isoseismal map has been reproduced from Dowrick (1994). Extensive contemporary newspaper reports, felt reports and personal recollections were studied. The intensities were assessed using the proposed revised Modified Mercalli scale for New Zealand (Study Group 1992). Dowrick (1994) suggests that some further modification of the scale may be advantageous.

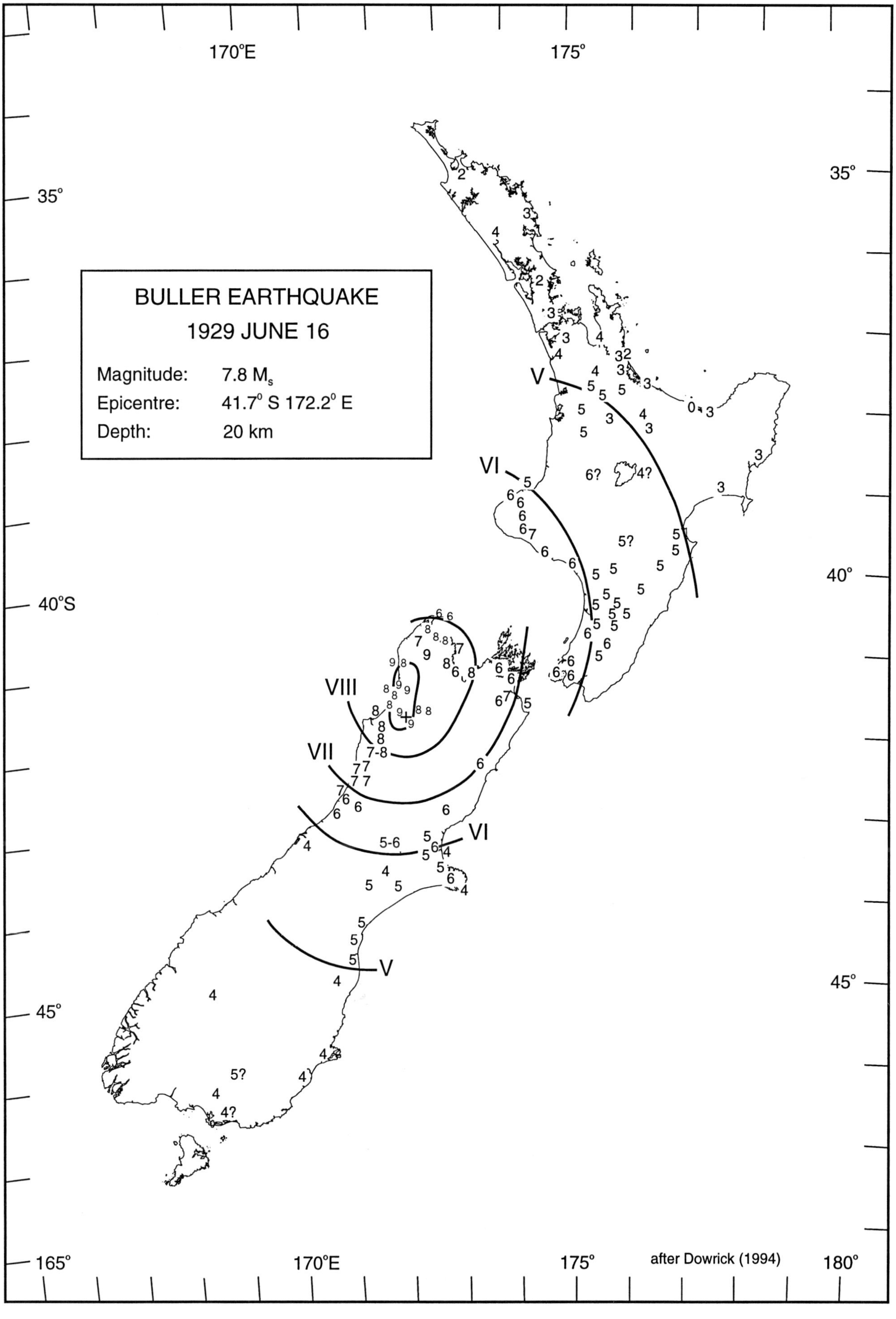
BULLER EARTHQUAKE
1929 JUNE 16
Magnitude: 7.8 M_s
Epicentre: 41.7° S 172.2° E
Depth: 20 km
170°E
175°
35°
40°S
40°
45°
165°
180°
VIII
VII
VI
V
after Dowrick (1994)

16. ISOSEISMAL MAP OF THE HAWKE'S BAY EARTHQUAKE - 1931 FEBRUARY 02

DATE:	1931 FEBRUARY 02
TIME:	22:47 UT
MAGNITUDE:	7.8 M_S (Dowrick & Smith 1990)
EPICENTRE:	39.3°S 177°E D (Bullen 1938)
DEPTH:	30 km

Occurring at 10.47 a.m., New Zealand Daylight Saving Time, on Tuesday, February 3rd, 1931, the Napier earthquake is recorded as having caused the largest loss of life and the most extensive damage ever recorded in New Zealand. The effects of the earthquake were greatest in the towns of Napier and Hastings, but other towns in the Hawke's Bay area also suffered major damage. The official death toll was 256, 161 occurring in Napier, 93 in Hastings and 2 in Wairoa. The earthquake was followed by fires in the business areas of both Napier and Hastings, the fires becoming uncontrollable as the water pressure dwindled because of broken water mains.

The earthquake was felt throughout most of New Zealand except for those areas to the far south and the far north and was followed by many aftershocks, some of large magnitude, and scattered over much of the Hawkes Bay Province. The maximum intensity was MM10, with extensive fissuring, slumping and landslides occurring over the much of Hawkes Bay.

'Surface deformation accompanying the earthquake resulted in a >90 km long, 17 km wide asymmetric dome trending northeast and extending from southwest of Hastings to northeast of the Mohaka River mouth. Maximum uplift of 2.7 m occurred near the mouth of the Aropaoanui River close to the location of major aftershock activity, while maximum subsidence of 1 m was recorded at Hastings, to the southeast of the steeper, southeast-facing side of the dome. Observed surface faulting of about 15 km in length was confined to the southwestern end of the dome where near surface rocks had sufficient strength and suitably oriented pre-existing fractures to accommodate slip. Elsewhere, the elastic nature of the rocks resulted in surface folding above a buried causative fault' (Hull 1990). Of interest to those studying historical earthquakes is the comment by Hull (1990) that, some fifty three years after the earthquake, "only about 3 km of fault trace could be confidently recognised. Present geological techniques for recognition of past earthquakes would fail to determine the extent of surface deformation and the magnitude of the 1931 earthquake."

Preliminary dislocation modelling (Haines & Darby 1987) has indicated that: "The 1931 Napier earthquake possibly involved movement on a major northeast trending submarine fault in Hawke Bay, with a rupture about 80 km long, from 5 km to possibly 35 km deep, dip in the range 59° to 68° northwest, dip-slip in the range 6 m to 8 m, and right-lateral strike-slip in the range 4 m to 8 m."

Felt summaries held at the Seismological Observatory have been converted from Rossi-Forel scale to Modified Mercalli and the isoseismal map has been drawn by Eiby. An early isoseismal map can be found in Adams et al. (1933) but the extent of the information on which their map is based is unknown. It may be more comprehensive than the felt summaries used by Eiby. The contemporary reports of this earthquake are excellent and are being examined by Dowrick (pers. Comm.) in order to prepare a modern MM isoseismal map (Dowrick, in prep.), but there has been no recent reinterpretation of the seismological data.

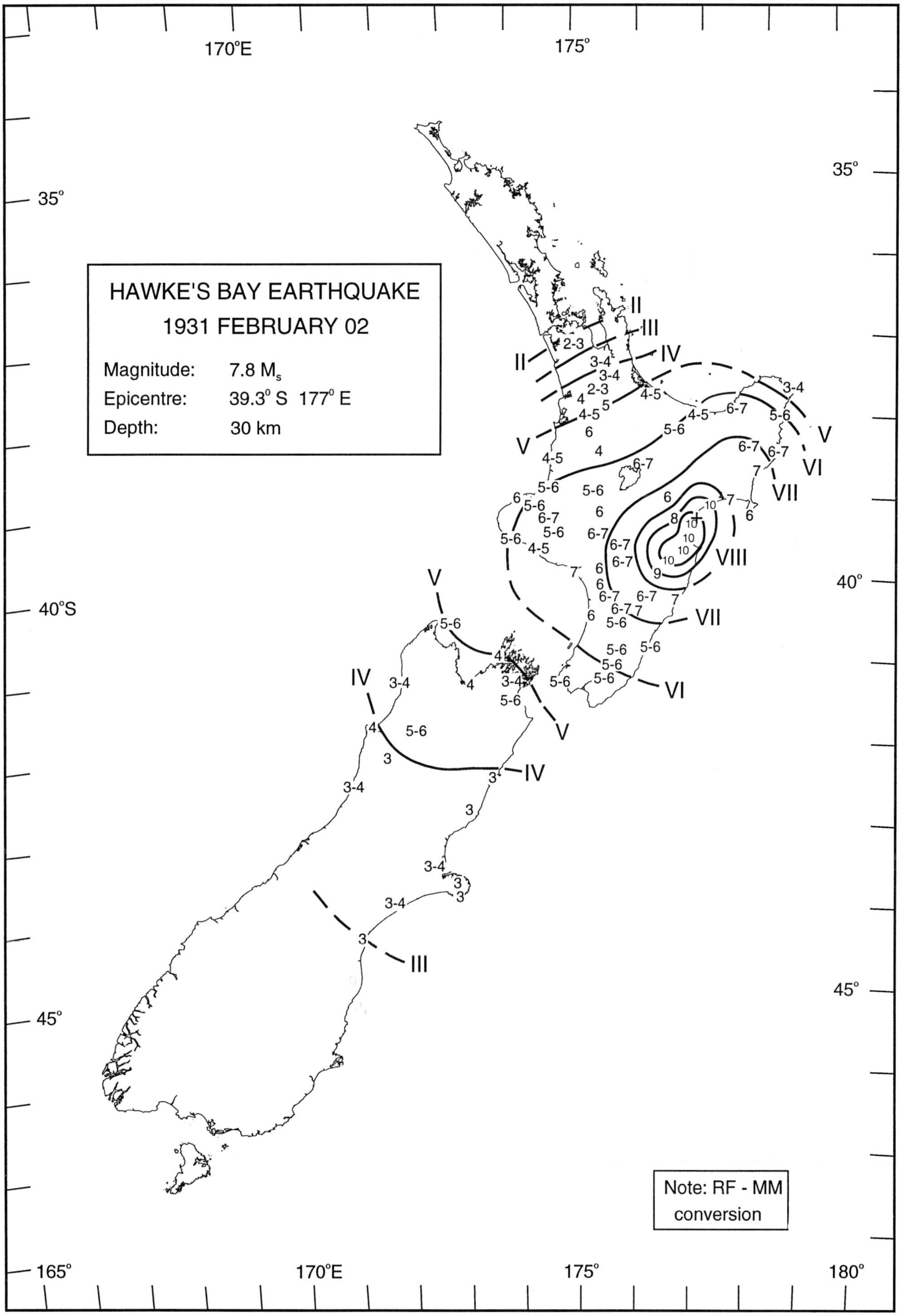
HAWKE'S BAY EARTHQUAKE
1931 FEBRUARY 02
Magnitude: 7.8 Ms
Epicentre: 39.3° S 177° E
Depth: 30 km
Note: RF - MM conversion
165°
170°E
175°
180°
35°
40°S
45°
40°
II
III
IV
V
VI
VII
VIII

17. ISOSEISMAL MAP OF AFTERSHOCK OF THE HAWKE'S BAY EARTHQUAKE - 1931 FEBRUARY 13

DATE:	1931 FEBRUARY 13
TIME:	01:27 UT
MAGNITUDE:	7.3 M_S (Dowrick and Smith 1990)
EPICENTRE:	39.5°S 177.5°E D (I)
DEPTH:	30 km

The earthquake of 1931 February 13 followed ten days after the magnitude M_s 7.8 earthquake which was disastrous in Napier and Hastings. The epicentre has been placed about 50 km from that of the Napier earthquake, to the east and out to sea. The earthquake is little considered in the literature, but was felt widely from Hamilton to south of Christchurch. It was felt with intensity MM7 over much of Hawkes Bay.

'At Napier and Hastings, the movement was much less than that on Feb. 2, but at other localities further inland it was regarded as the more violent of the two.' (Dominion Observatory bulletin E 27.)

Information on the effects of this earthquake and other aftershocks of the Napier earthquake is poor and would benefit from further studies.

Felt summaries held at the Seismological Observatory have been converted from Rossi-Forel scale to Modified Mercalli and the isoseismal map has been drawn by Eiby.

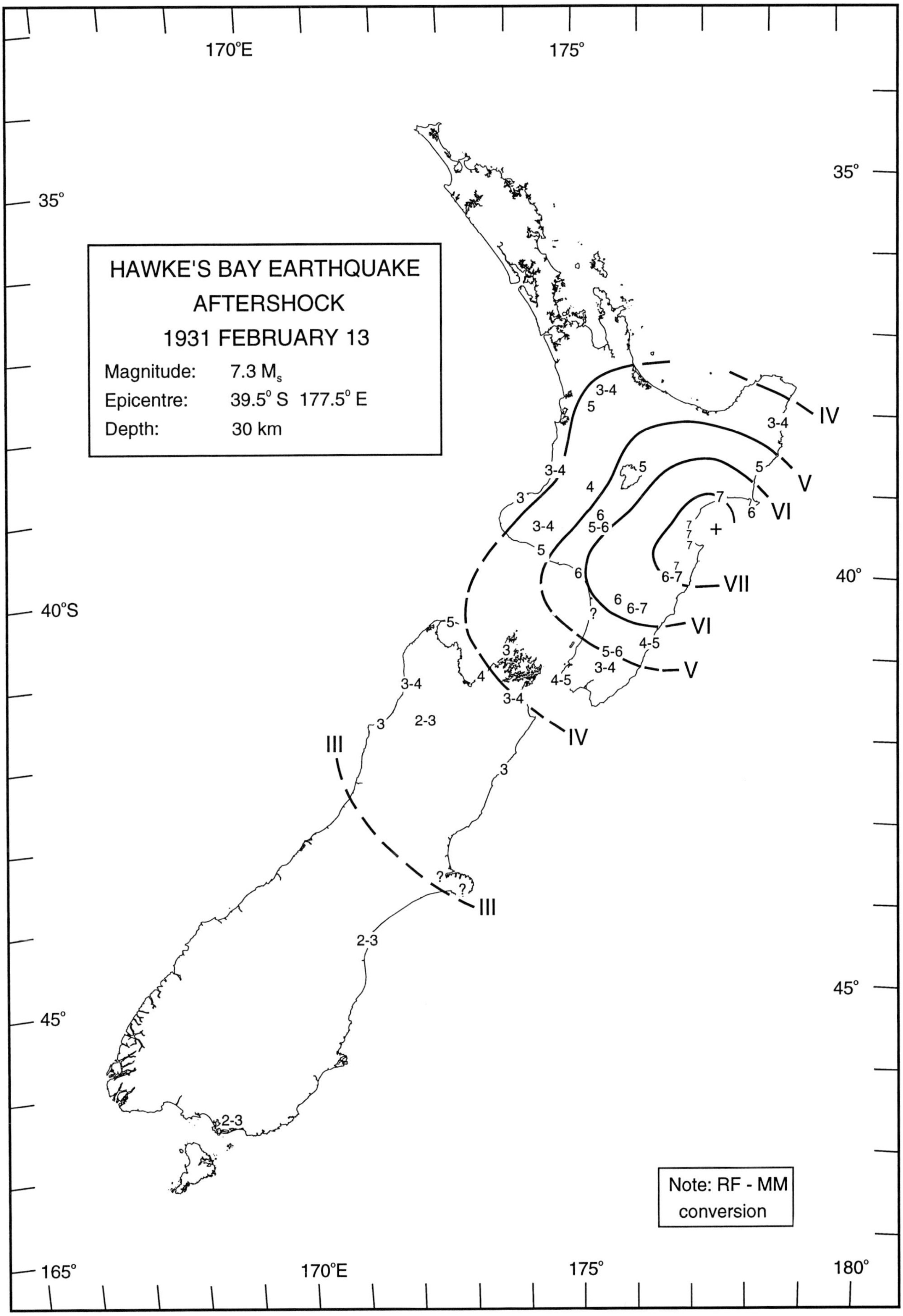
HAWKE'S BAY EARTHQUAKE
AFTERSHOCK
1931 FEBRUARY 13
Magnitude: 7.3 M_s
Epicentre: 39.5° S 177.5° E
Depth: 30 km
Note: RF - MM conversion
170°E
175°
35°
40°S
45°
165°
180°
40°
IV
V
VI
VII
III
3-4
5
4
6
5-6
7
6-7
4-5
3
2-3
?

18. ISOSEISMAL MAP OF THE SOUTHLAND EARTHQUAKE - 1931 SEPTEMBER 15

DATE:	1931 SEPTEMBER 15
TIME:	21:09 UT
MAGNITUDE:	5.4 M_S (Dowrick & Smith 1990)
EPICENTRE:	45°S 168°E D (Gutenberg & Richter 1949)
DEPTH:	<60 km

The earthquake was felt extensively in the South Island, from Hokitika and Christchurch to Stewart Island. The maximum intensity of R-F 5 or MM4-5 was felt in Nightcaps, Tuatapere, Riverton and Awarua.

There has been no recent reinterpretation of the source data.

Felt summaries held at the Seismological Observatory have been converted from Rossi-Forel scale to Modified Mercalli and the isoseismal map has been drawn by Eiby.

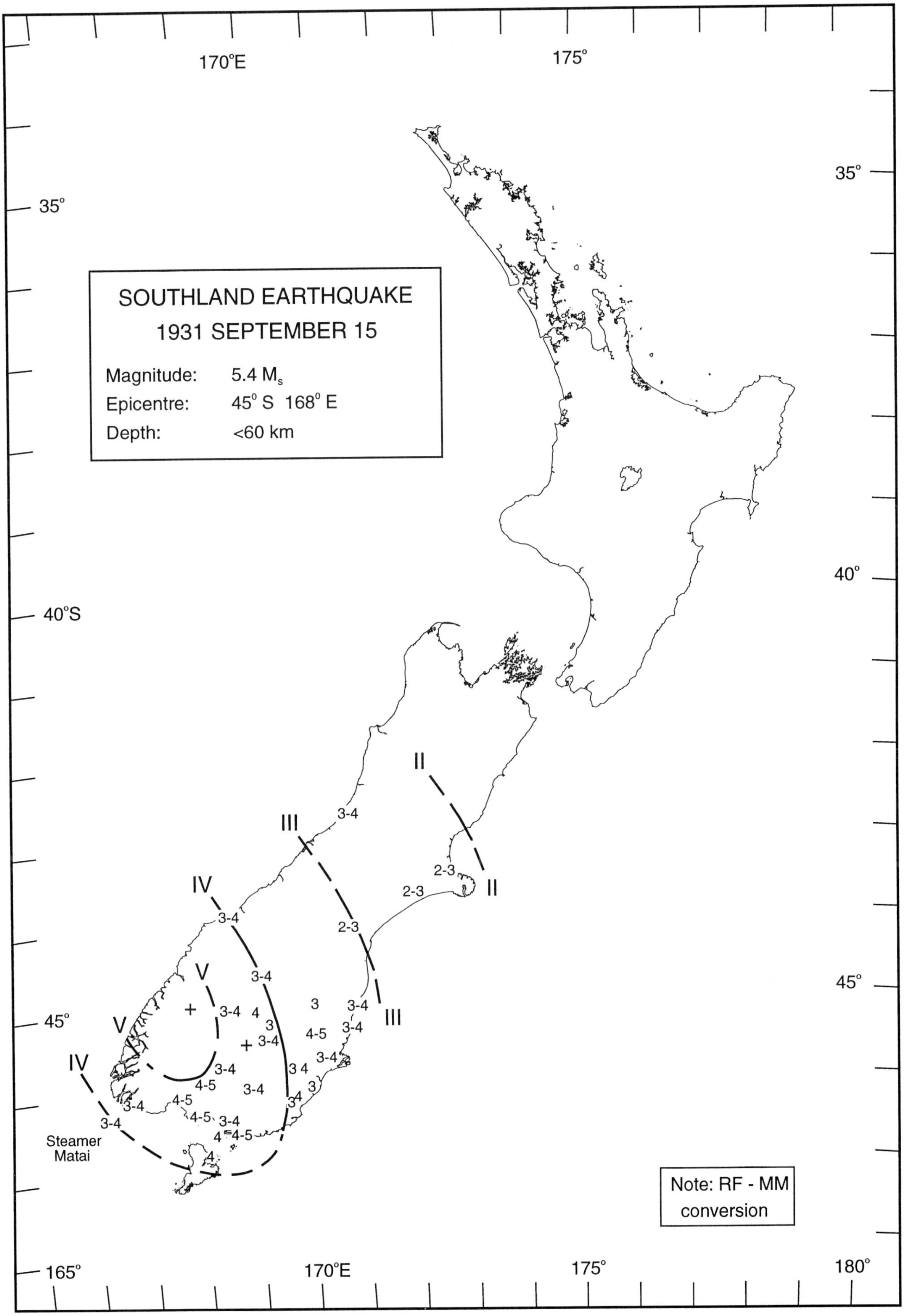

SOUTHLAND EARTHQUAKE
1931 SEPTEMBER 15
Magnitude: 5.4 M_s
Epicentre: 45° S 168° E
Depth: <60 km
170°E
175°
35°
40°S
40°
45°
165°
180°
II
III
IV
V
3-4
2-3
3
4
4-5
Steamer Matai
Note: RF - MM conversion

19. ISOSEISMAL MAP OF THE BAY OF PLENTY EARTHQUAKE - 1931 SEPTEMBER 21

DATE:	1931 SEPTEMBER 21
TIME:	13:34 UT
MAGNITUDE:	6.75 M (Gutenberg & Richter 1949);
	6.1 M_S (Dowrick & Smith 1990)
EPICENTRE:	37.5°S 178°E D (Gutenberg & Richter 1949)
DEPTH:	80 km

The earthquake was felt over most of the North Island south of Auckland and in the South Island north of Greymouth and Christchurch. The maximum reported intensity was MM6-7 at Opotiki, but the earthquake was also felt quite strongly at Hick's Bay, Waipiro Bay and Tolaga Bay on the East Cape peninsula.

As is the case with many of these early moderate magnitude earthquakes, there has been no recent reinterpretation of the source data.

Felt summaries held at the Seismological Observatory have been converted from Rossi-Forel scale to Modified Mercalli and the isoseismal map has been drawn by Eiby.

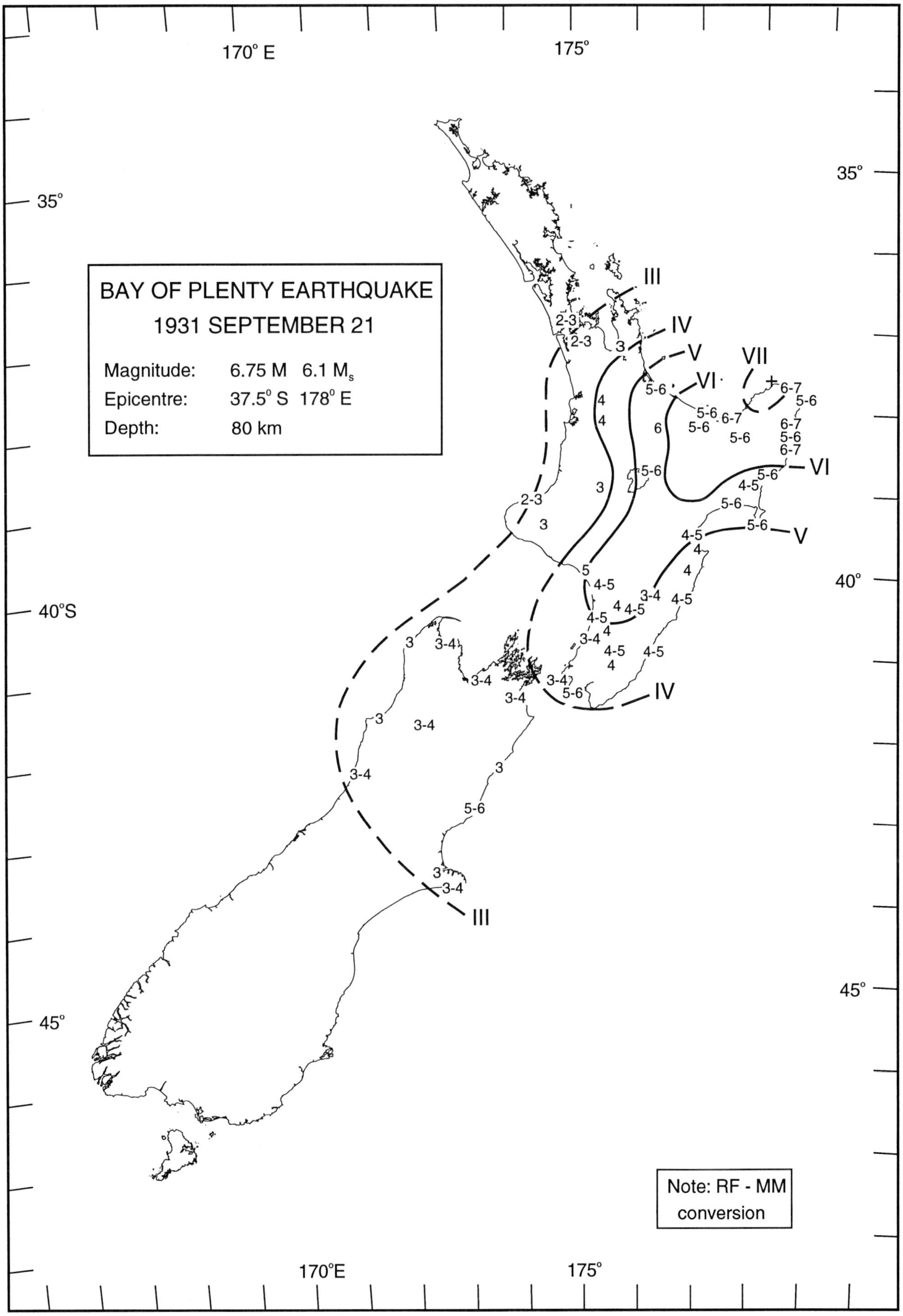

170° E
175°
35°
40°S
45°
40°
BAY OF PLENTY EARTHQUAKE
1931 SEPTEMBER 21
Magnitude: 6.75 M 6.1 M_s
Epicentre: 37.5° S 178° E
Depth: 80 km
III
IV
V
VI
VII
Note: RF - MM conversion
170°E

20. ISOSEISMAL MAP OF THE BAY OF PLENTY EARTHQUAKE - 1932 MARCH 05

DATE:	1932 MARCH 05
TIME:	01:40 UT
MAGNITUDE:	6.0 M_S (Dowrick & Smith 1990)
EPICENTRE:	36.5°S 178°E D (Gutenberg & Richter 1949)
DEPTH:	Lower Crustal

This earthquake was felt in Auckland and Hamilton but the highest intensities were felt in the western Bay of Plenty and Coromandel. The maximum intensity was R-F 6-7, MM6 at Whitianga.

As is the case with many of these early moderate magnitude earthquakes, there has been no recent reinterpretation of the source data, which is quite limited.

Felt summaries held at the Seismological Observatory have been converted from Rossi-Forel scale to Modified Mercalli and the isoseismal map has been drawn by Eiby.

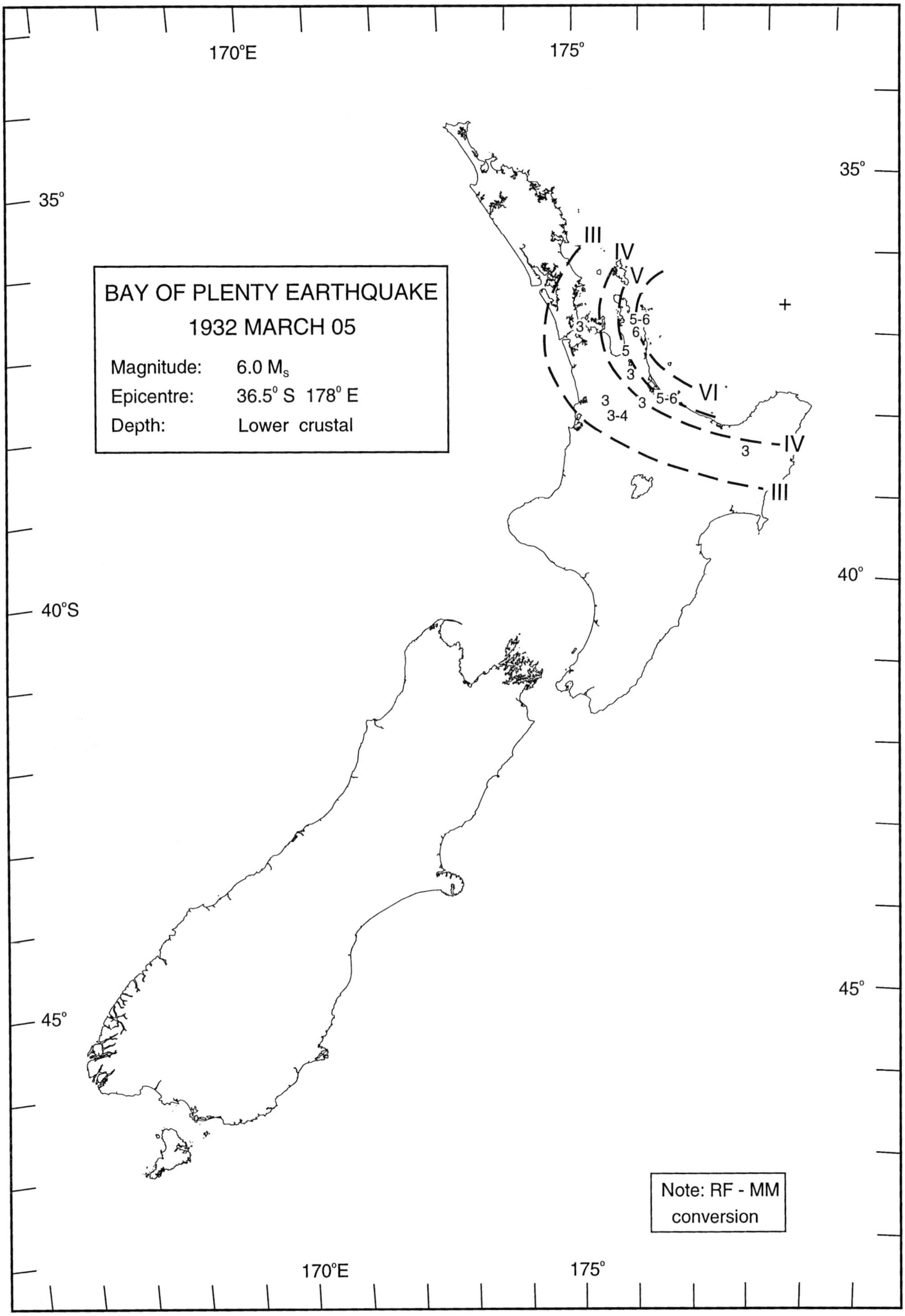

170°E
175°
35°
35°
40°S
40°
45°
45°
BAY OF PLENTY EARTHQUAKE
1932 MARCH 05
Magnitude: 6.0 M$_s$
Epicentre: 36.5° S 178° E
Depth: Lower crustal
III
IV
V
VI
IV
III
3
5-6
6
5
3
3
3-4
3
5-6
3
Note: RF - MM conversion
170°E
175°

21. ISOSEISMAL MAP OF THE HAWKE'S BAY EARTHQUAKE - 1932 MAY 05

DATE:	1932 MAY 05
TIME:	08:23 UT
MAGNITUDE:	5.9 M_S (Dowrick & Smith 1990)
EPICENTRE:	39.5°S 177.5°E D (Gutenberg & Richter 1949)
DEPTH:	Crustal

The earthquake of 1932 May 5 occurred eighteen months after the destructive Napier earthquake. It was one of a number of larger magnitude events which followed the Napier earthquake and had locations in the Hawkes Bay area. This event has been located 50km to the northeast of the epicentre of the Napier earthquake. It was felt over most of the North Island excluding the area north of Auckland. The maximum intensity of MM6-7 was felt in Napier and Taradale.

As is the case with many of these early moderate magnitude earthquakes, there has been no recent reinterpretation of the source data.

Felt summaries held at the Seismological Observatory have been converted from Rossi-Forel scale to Modified Mercalli and the isoseismal map has been drawn by Eiby.

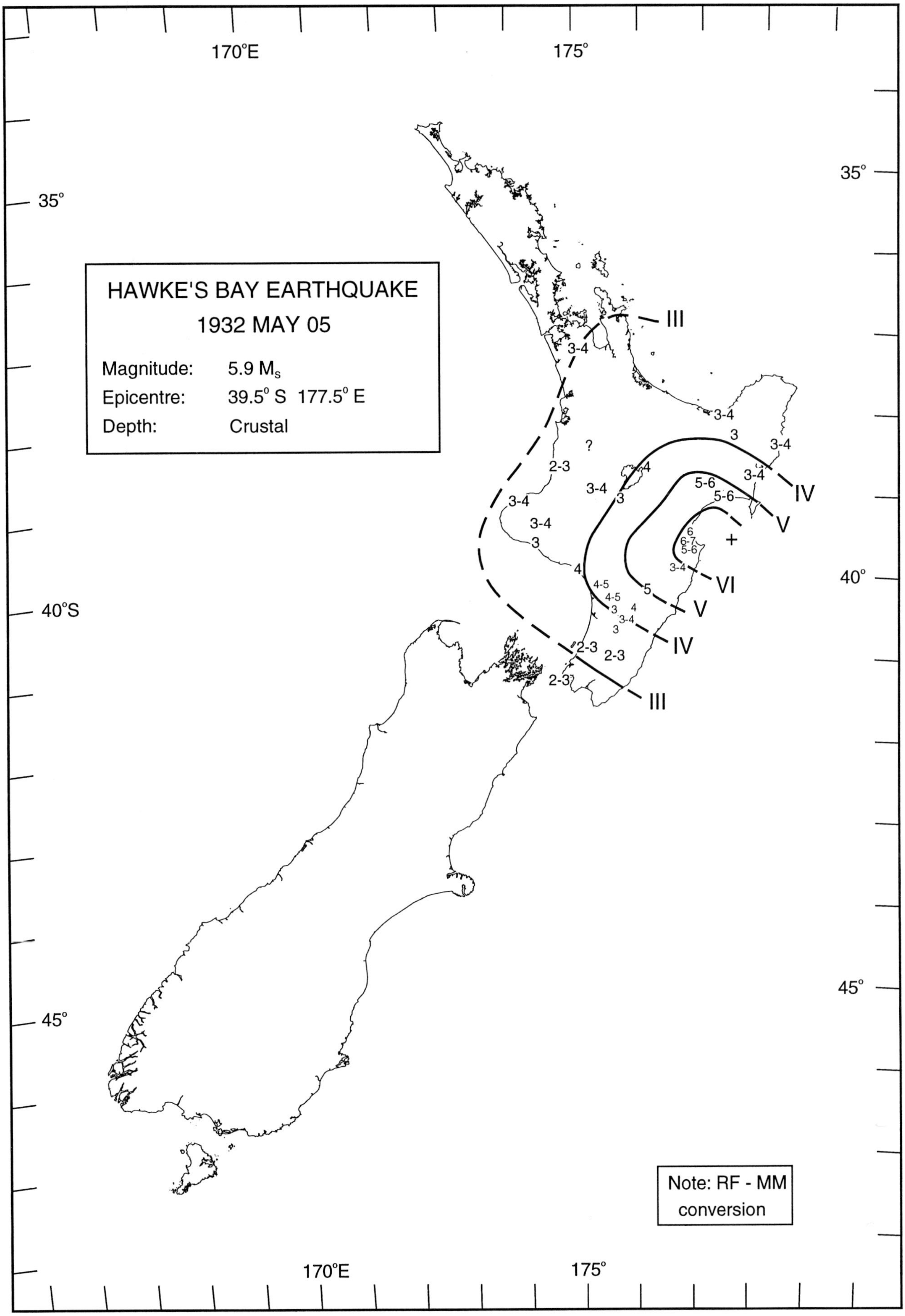

170°E
175°
35°
35°
40°S
40°
45°
45°
HAWKE'S BAY EARTHQUAKE
1932 MAY 05
Magnitude: 5.9 M_s
Epicentre: 39.5° S 177.5° E
Depth: Crustal
III
IV
V
VI
V
IV
III
Note: RF - MM conversion
170°E
175°

22. ISOSEISMAL MAP OF THE TARANAKI BIGHT EARTHQUAKE - 1932 JULY 20

DATE:	1932 JULY 20
TIME:	04:52 UT
MAGNITUDE:	6.3 M (I); 5.1 M_s (Dowrick & Smith 1990)
EPICENTRE:	40°S 174°E D (Gutenberg & Richter 1949)
DEPTH:	60 km

'[The earthquake on 1932 July 20 was] widely felt in New Zealand from Dargaville in the north to Akaroa in the south. R-F7-8 [MM6-7] in Taranaki, and along the coast as far as Wanganui. Very large amplitudes.' (Dominion Observatory bulletin E34.)

Felt summaries held at the Seismological Observatory have been converted from Rossi-Forel scale to Modified Mercalli and the isoseismal map has been drawn by Eiby.

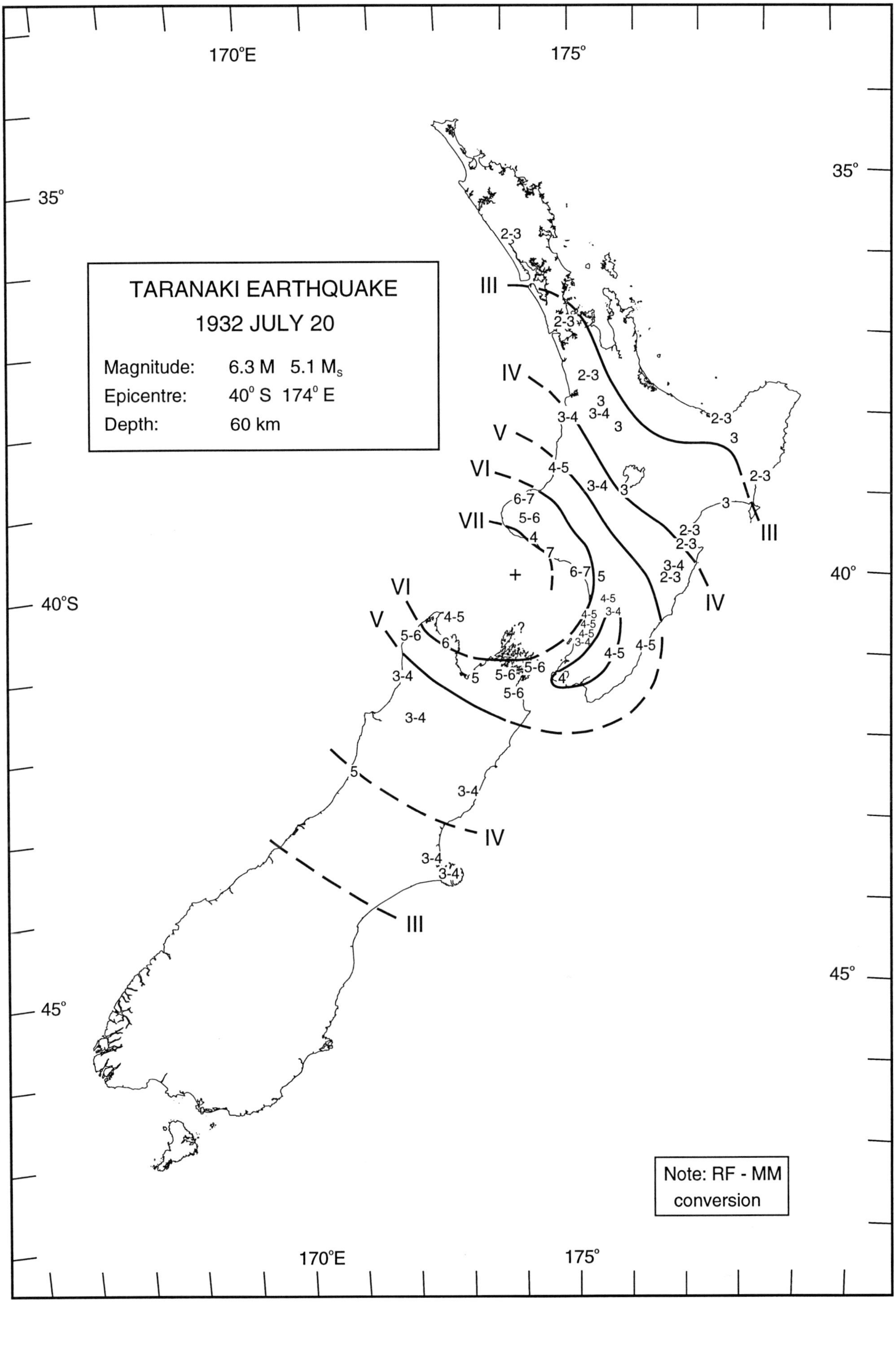
TARANAKI EARTHQUAKE
1932 JULY 20
Magnitude: 6.3 M 5.1 M_s
Epicentre: 40° S 174° E
Depth: 60 km
Note: RF - MM conversion
170°E
175°
35°
40°S
40°
45°
III
IV
V
VI
VII
2-3
3
3-4
4-5
5
5-6
6
6-7
7
4
?

23. ISOSEISMAL MAP OF THE WAIROA EARTHQUAKE - 1932 SEPTEMBER 15

DATE:	1932 SEPTEMBER 15
TIME:	13:55 UT
MAGNITUDE:	6.9 M_S (Dowrick & Smith 1990)
EPICENTRE:	38.9°S 177.6°E D (Bullen 1938)
DEPTH:	30 km

'At 1.25 a.m. on Friday, 16th September, 1932, the earth was shaken so severely at Wairoa and Gisborne as to throw down chimneys, to crack weak buildings, to cause alluvial ground and fillings to open and settle down, to fissure the surface soil on narrow ridges, and to detach steep country in slips.

'... the biggest slip in the district occurred at McCardle's, three miles [5 km] south-west of Wairoa, where the sea-cliff 30 chains [600 m] long slid seaward and the ground came down for a width of 15 chains [300 m], leaving behind it a scar standing 500 ft. [150 m] above sea-level.' (Ongley 1937)

'[In Wairoa and Gisborne] many buildings were badly damaged, and a few collapsed completely in Gisborne. At Wairoa the old river bridge, which was damaged beyond repair by the earthquake of February 3rd, 1931, collapsed, and the new bridge, in course of construction, was damaged. Some subsidence of the ground near the river at Wairoa was reported. Owing probably to the earthquake occurring near the middle of the night, no-one was killed, and only five persons were seriously injured - one at Gisborne and four at Wairoa.

'Reports indicate that portions of the country between Wairoa and Gisborne were badly shattered. The maximum intensity of the earthquake appears therefore to have reached a value between 9 and 10 on the Rossi-Forel scale [MM9-10].' (Hayes 1937)

The earthquake was strongly felt in Hawkes Bay and over most of the North Island, and in the South Island, at Blenheim.

Preliminary dislocation modelling (Haines & Darby 1987) has indicated that: "The 1932 Wairoa earthquake was possibly on a northeastward continuation of this rupture [i.e. that of the 1931 Napier earthquake], from 50 km to 80 km long, from 5 km to possibly 35 km deep, dipping about 68° northwest, with 1 m of either reverse or normal dip-slip, and 1 m right lateral strike-slip."

Felt summaries held at the Seismological Observatory have been converted from Rossi-Forel scale to Modified Mercalli and the isoseismal map has been drawn by Eiby. There has been no recent revision of the seismological data, but a study of the original felt reports is in progress (Dowrick, pers. comm.). Preliminary results indicate an intensity MM8 in Wairoa.

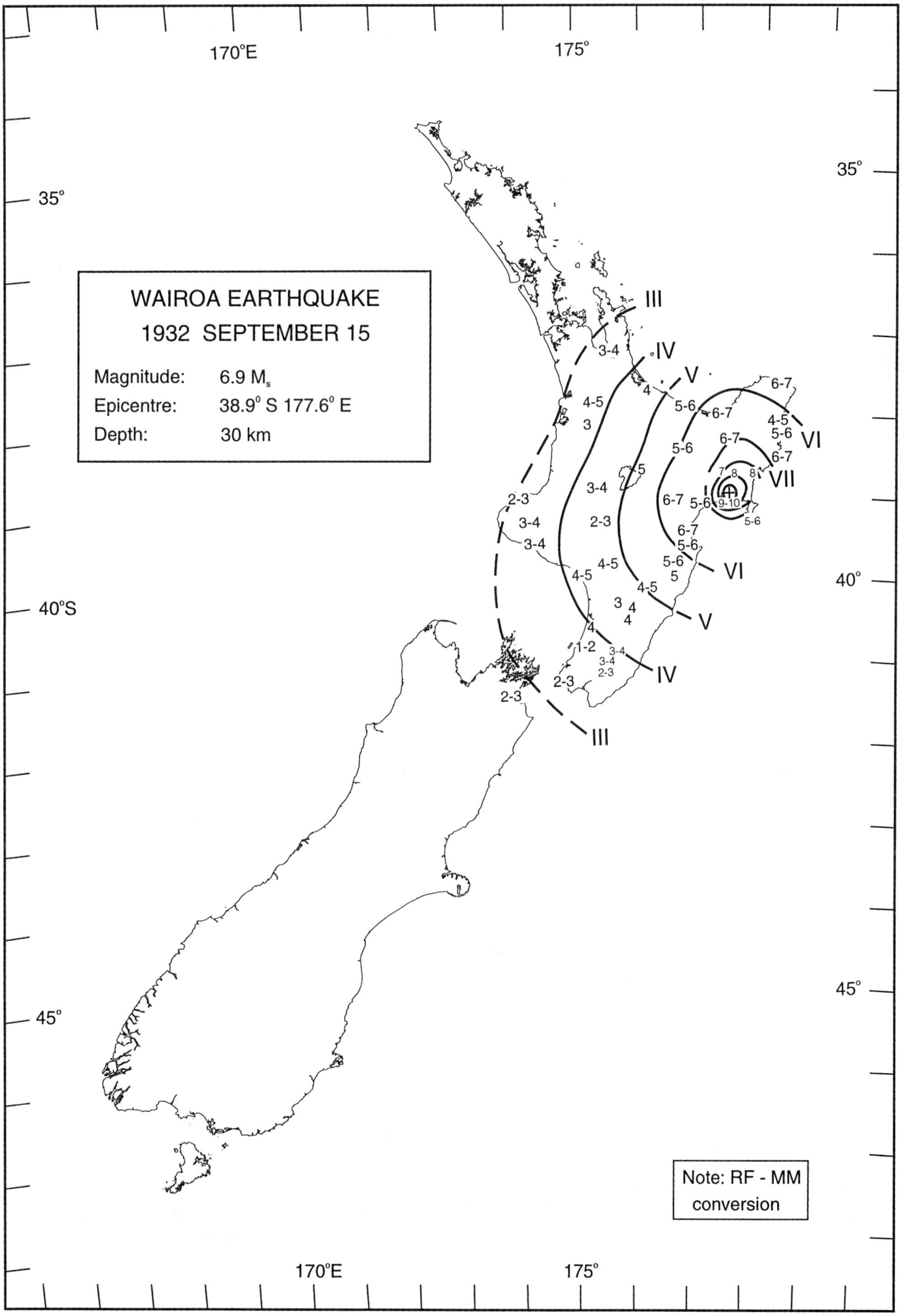

WAIROA EARTHQUAKE
1932 SEPTEMBER 15
Magnitude: 6.9 M_s
Epicentre: 38.9° S 177.6° E
Depth: 30 km
Note: RF - MM conversion
170°E
175°
35°
40°S
40°
45°
III
IV
V
VI
VII

24a. ISOSEISMAL MAP OF THE PAHIATUA EARTHQUAKE - 1934 MARCH 05

DATE:	1934 MARCH 05
TIME:	11:46 UT
MAGNITUDE:	7.6 M_S (Dowrick & Smith 1990)
EPICENTRE:	40.95°S 176.8°E D (Hayes 1935; 1937)
	40.62°S 175.72°E D (Solution B, Bullen 1938)
	40.45°S 175.56°E D (Solution C, Bullen 1938)
DEPTH:	Lower Crustal

'The most important seismic event in the year 1934 was the severe earthquake which occurred on the 5th March at 23h. 16m. N.Z.M.T. This earthquake was felt as far as Auckland in the north and Dunedin in the south.' (Dominion Observatory bulletin S27)

'... The most severely shaken region was in the eastern part of the North Island, between Porangahau and Castlepoint. It appears from the numerous reports that the shock reached intensity R-F 9 [MM9] within a limited area, including parts of the districts of southern Hawke's Bay and northern Wairarapa.

'The isoseismal R-F 8 extended a considerable distance to westward into the Manawatu district, and then to north-westward about as far as Wanganui. This isoseismal marks approximately the limits of the area where the earthquake was of destructive force. This area comprised about 3,400 square miles [8,700 km] of land surface.

'... In the region within the isoseismal R-F 9 [MM9] the damage was considerable, but appears to have chiefly affected the town of Pahiatua, where a number of badly constructed buildings either collapsed or were seriously damaged.

'Between Pahiatua and the east coast of the Island the country is rather sparsely populated, but considerable damage was reported wherever there was settlement. The ground surface was slightly disturbed in places, and the coastal cliffs were scarred by small slips between Cape Turnagain and Castlepoint.

'The earthquake was felt along most of the east coast of the South Island, as far south as Dunedin, but in the mountainous regions farther inland and on the west coast it was felt only in the northern part of the Island.' (Hayes 1937)

The isoseismal map is from Hayes (1937). The isoseismal lines refer to the Rossi-Forel scale. No individual intensities are indicated and the extent of the data on which these lines were based is unknown, but may have exceeded that which was available to Eiby in his interpretation on the following pages. The differences between the two maps and the uncertainty in the epicentre illustrate the need for revision of the source material for this event.

PAHIATUA EARTHQUAKE
1934 MARCH 05

Magnitude:	7.6 M_s
Epicentre :	40.95° S 176.8° E
Depth:	Lower crustal

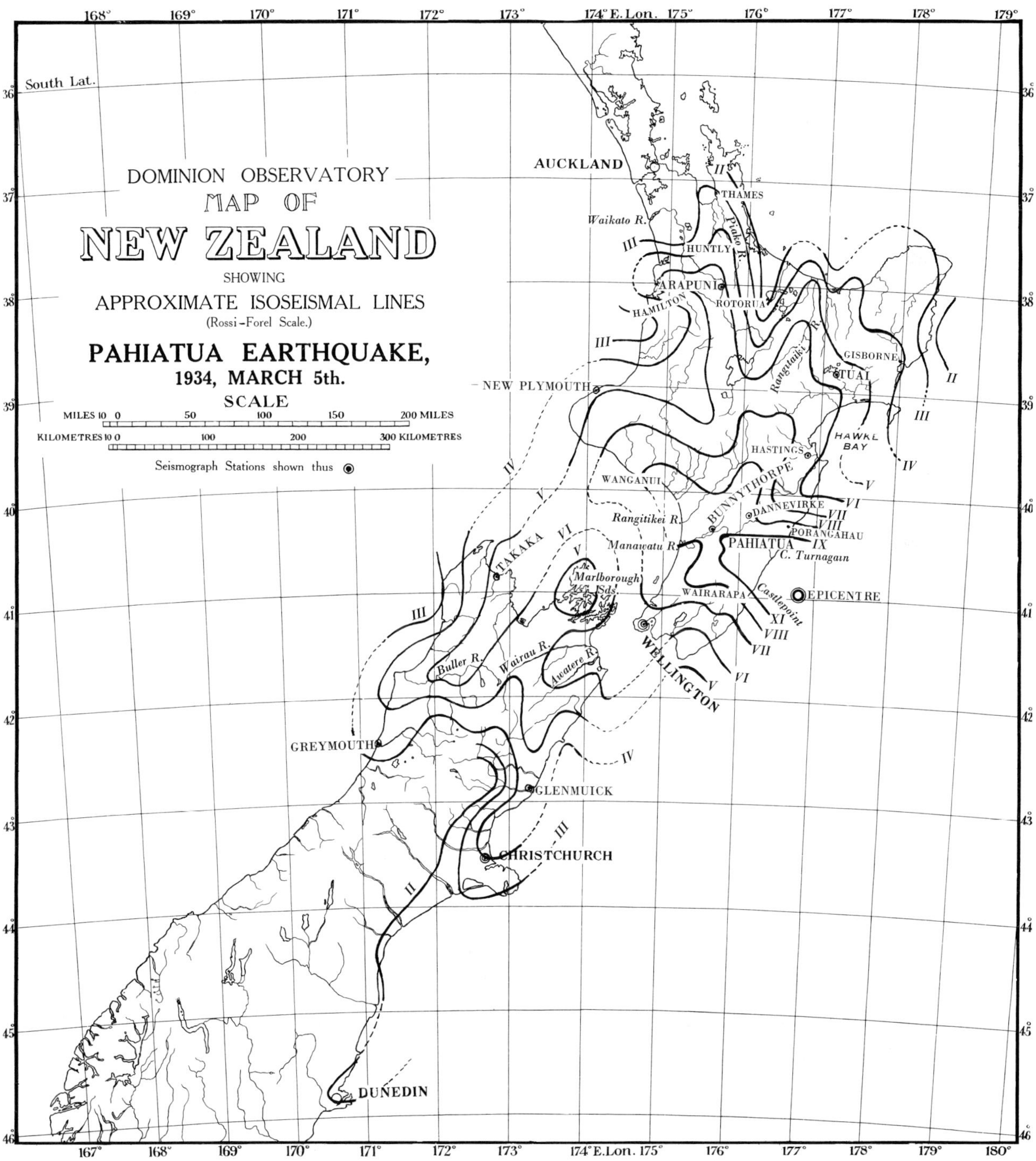

24b. ISOSEISMAL MAP OF THE PAHIATUA EARTHQUAKE - 1934 MARCH 05

DATE:	1934 MARCH 05
TIME:	11:46 UT
MAGNITUDE:	7.6 M_S (Dowrick & Smith 1990)
EPICENTRE:	40.95°S 176.8°E D (Hayes 1935; 1937)
	40.62°S 175.72°E D (Solution B, Bullen 1938)
	40.45°S 175.56°E D (Solution C, Bullen 1938)
DEPTH:	Lower Crustal

'Just after 11.46 p.m. New Zealand summer time (=11.16 p.m. N.Z.M.T.) on 5th March 1934, a violent and prolonged earthquake was experienced in many parts of New Zealand. The most severely shaken region was in the eastern part of the North Island between Porangahau and Castlepoint.' (Hayes 1937)

'The Pahiatua earthquake of 1934, March 5, was among those specially considered in earlier investigations of Hayes (1935) and Bullen (1936) on the problem of crustal structure in the New Zealand region. These two papers both contain solutions for the earthquake based on the readings of the records at New Zealand observatories. On account of the meagreness of the local data, certain assumptions had to be made in both papers concerning the interpretation of this data, so that the results could be regarded as tentative only .... The writer has since examined a preliminary set of P-readings at various stations, distant as well as near, and with the use of the most recent travel-time tables, together with ellipticity tables (1937), has found clear indications that the above epicentre [Hayes (1935); Bullen (1936)] requires a substantial westerly displacement.' (Bullen 1938)

Bullen's solutions B and C result from different weightings assigned to distant stations.

Felt summaries held at the Seismological Observatory have been converted from Rossi-Forel scale to Modified Mercalli and the isoseismal map has been drawn by Eiby. Epicentres B and C are indicated. The felt summaries referred to here comprise assessments of felt information sent in by the Observatory's network of regular felt reporters, including officers of the Post and Telegraph Department, the Marine Department and several private observers, the total number of non-instrumental reporters being about 120 throughout New Zealand. There appear to be no reports from Pahiatua nor from the coast to the east of Pahiatua, areas noted by Hayes (1937) as the most affected. The isoseismal map is clearly limited by this.

The differences between this map and that on the previous page and the uncertainty in the epicentre illustrate the need for revision of the existing source material, both seismological and macroseismic, and for retrieval of further data.

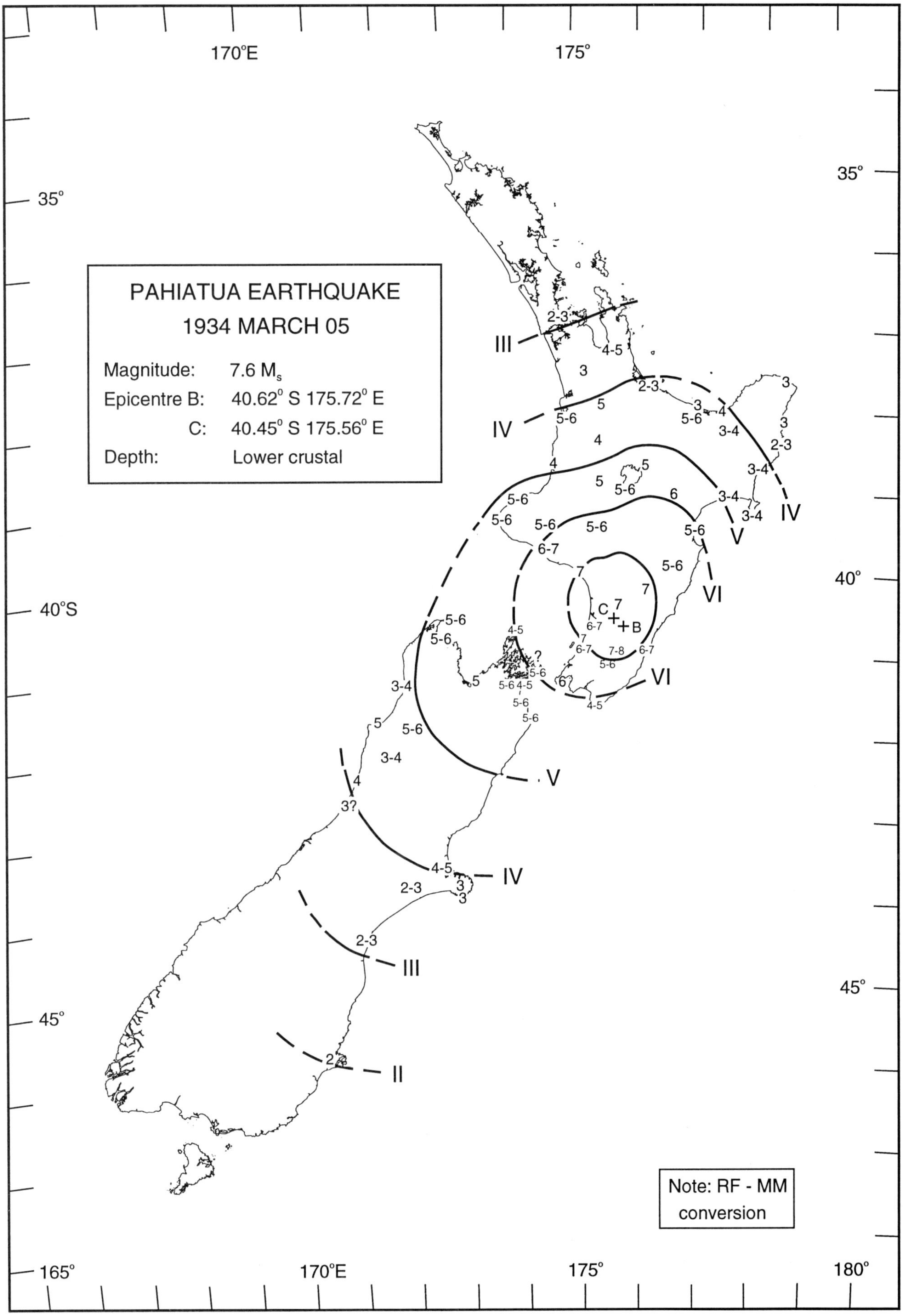

PAHIATUA EARTHQUAKE
1934 MARCH 05
Magnitude: 7.6 M_s
Epicentre B: 40.62° S 175.72° E
C: 40.45° S 175.56° E
Depth: Lower crustal
Note: RF - MM conversion
170°E
175°
165°
180°
35°
40°
40°S
45°
II
III
IV
V
VI

25. ISOSEISMAL MAP OF THE NORTHERN HAWKE'S BAY EARTHQUAKE - 1934 MARCH 15

DATE:	1934 MARCH 15
TIME:	10:46 UT
MAGNITUDE:	6.4 M_S (Dowrick & Smith 1990)
EPICENTRE:	38.9°S 177.2°E D (Gutenberg & Richter 1949)
DEPTH:	≤ 40 km

The earthquake of 1934 March 15 following ten days after the damaging magnitude 7.6 Pahiatua earthquake has attracted little attention. The earthquake was widely felt in Hawke's Bay with a maximum intensity of MM6 at Wairoa and was felt as far as Pukekohe in the north and Westport and Akaroa in the south.

The felt report summaries held at the Seismological Observatory have been converted from Rossi-Forel to Modified Mercalli scale and the isoseismal map was drawn by Eiby.

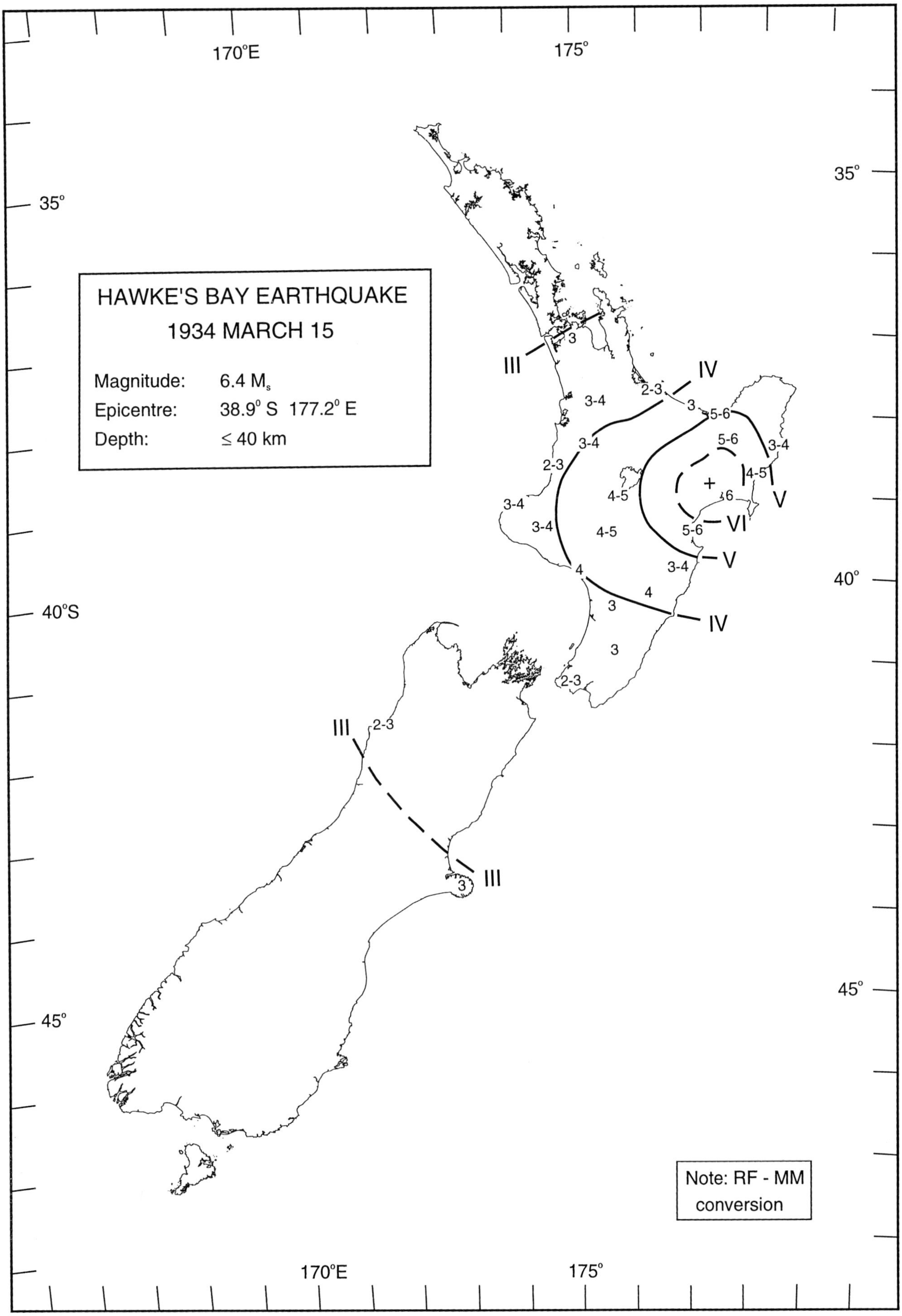
170°E
175°
35°
35°
40°S
40°
45°
45°
HAWKE'S BAY EARTHQUAKE
1934 MARCH 15
Magnitude: 6.4 M_s
Epicentre: 38.9° S 177.2° E
Depth: ≤ 40 km
III
IV
V
VI
V
IV
III
III
3
2-3
3-4
3
5-6
5-6
3-4
4-5
3-4
2-3
3-4
4-5
6
3-4
4-5
5-6
4
3-4
4
3
3
2-3
2-3
3
Note: RF - MM conversion
170°E
175°

26. ISOSEISMAL MAP OF THE SOUTHERN HAWKES BAY EARTHQUAKE - 1938 DECEMBER 15

DATE:	1938 DECEMBER 15
TIME:	09.12 UT
MAGNITUDE:	5.6 M_S (Dowrick & Smith 1990)
EPICENTRE:	40.3°S 176.4°E D (I)
DEPTH:	< 45 km

'The earthquake of 1938 December 15 was felt over the whole of the North Island with the exception of the Auckland Peninsula, with a maximum of MM7 in southern Hawke's Bay at Porangahau and Waipawa. It was also reported felt at isolated places in the South Island as far south as Greymouth and Banks Peninsula and at the Chatham Islands.' (Dominion Observatory bulletin S56)

This event appears to have been followed by another, probably of lesser magnitude, at a similar location on 1938 December 30.

The felt report summaries held at the Seismological Observatory have been converted from Rossi-Forel to Modified Mercalli scale and the isoseismal map was drawn by Eiby.

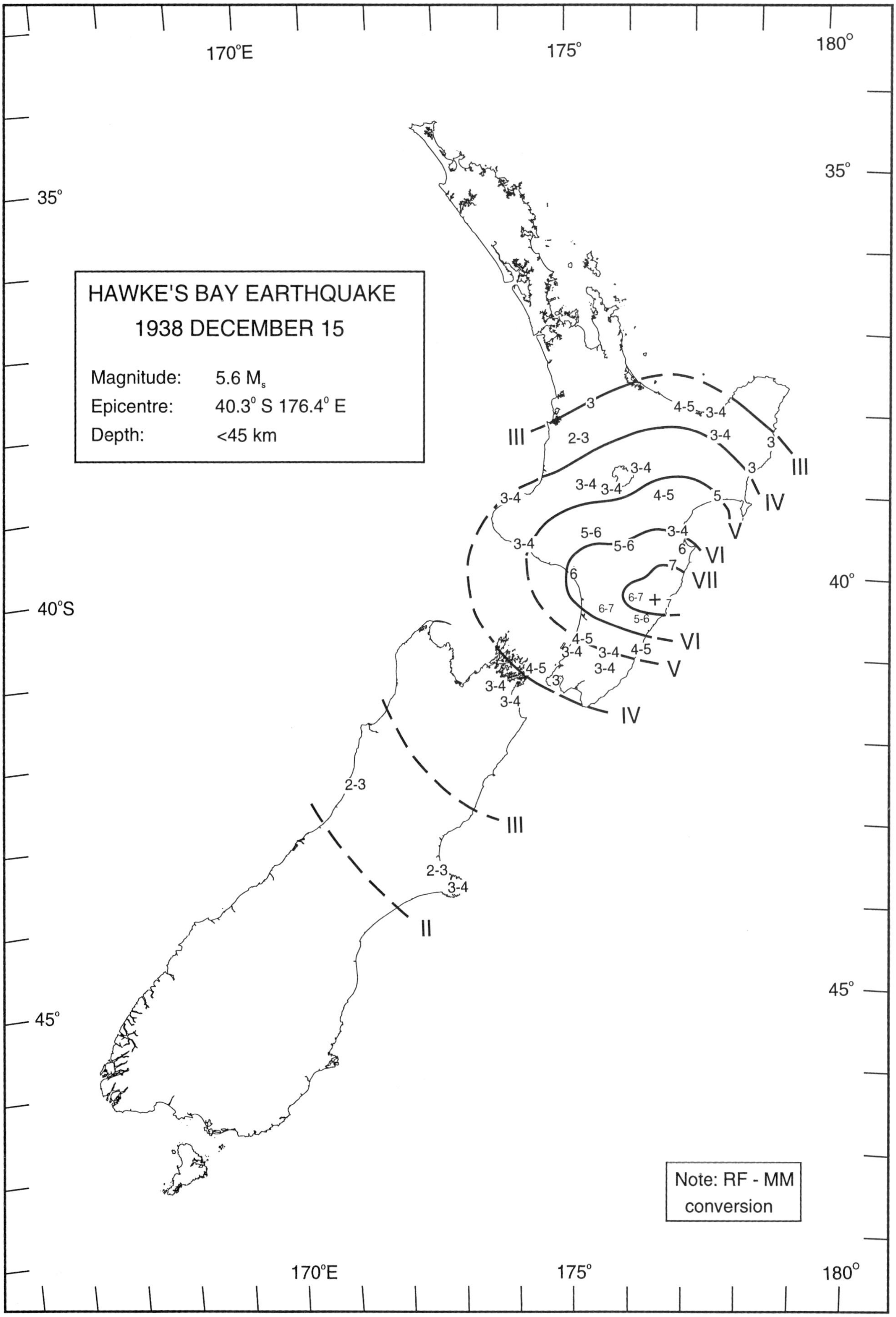

HAWKE'S BAY EARTHQUAKE
1938 DECEMBER 15
Magnitude: 5.6 Ms
Epicentre: 40.3° S 176.4° E
Depth: <45 km
170°E
175°
180°
35°
40°S
40°
45°
III
IV
V
VI
VII
II
Note: RF - MM conversion

27. ISOSEISMAL MAP OF THE CHARLES SOUND EARTHQUAKE - 1938 DECEMBER 16

DATE:	1938 DECEMBER 16
TIME:	17:21 UT
MAGNITUDE:	7.0 M_S (Dowrick & Smith 1990)
EPICENTRE:	45°S 167°E D (Gutenberg & Richter 1949)
DEPTH:	60 km

'After a long period of comparative quiet, the south-west portion of the South Island was shaken by a powerful disturbance on December 17 [NZ local time], the maximum intensity reported being R-F 6-7 [MM6] at Queenstown. The origin of this shock was deeper than normal, and an intensity R-F 6 [MM5] was experienced in most parts of western Otago and Southland. It was followed by numerous aftershocks during the latter part of December, and some activity continued well into the following year. One hundred and seventy-five shocks were recorded on the Jaggar seismograph at Monowai up to the end of December, 1938.' (Dominion Observatory bulletin S56)

The felt report summaries held at the Seismological Observatory have been converted from Rossi-Forel to Modified Mercalli and the isoseismal map was drawn by Eiby.

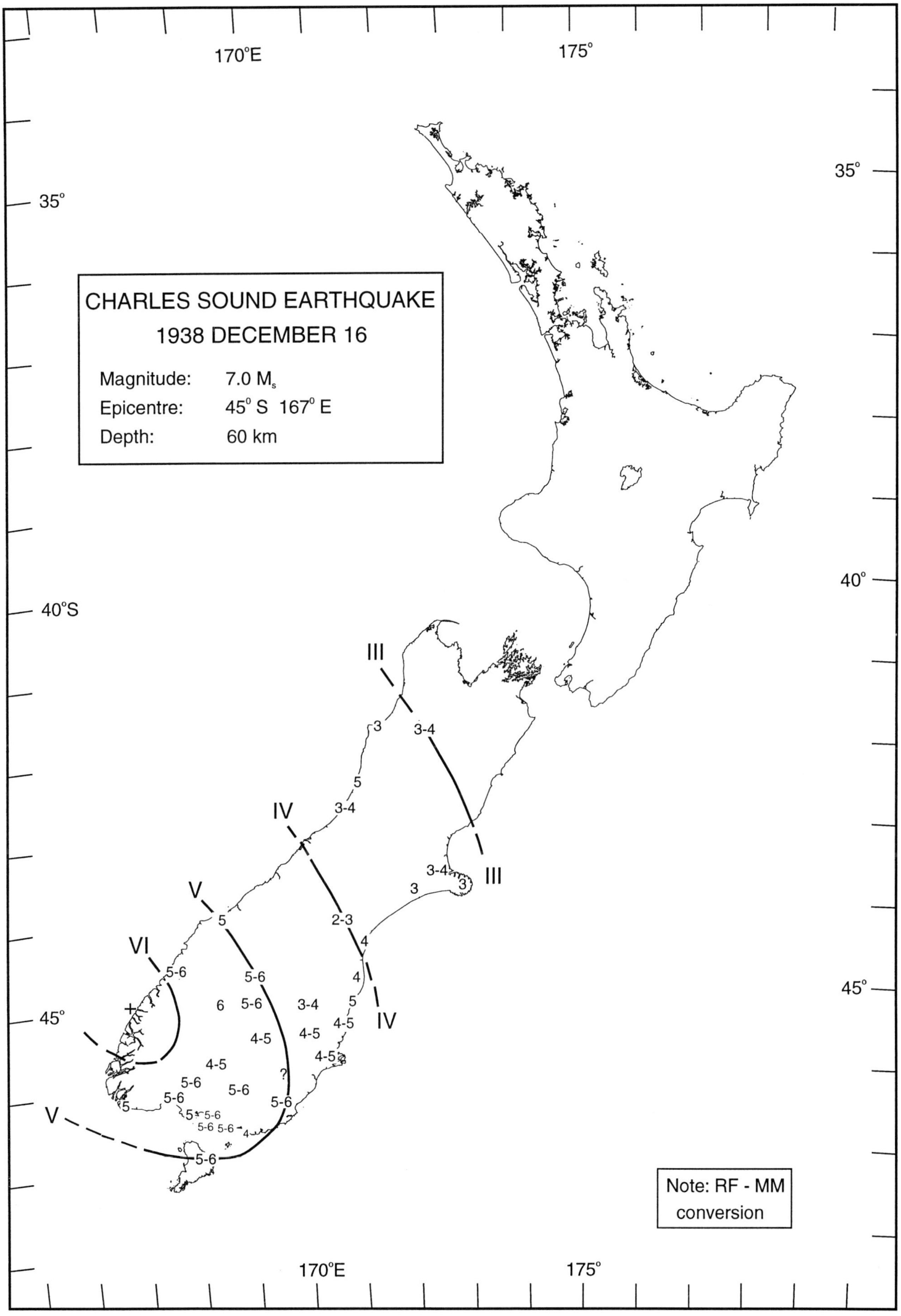

CHARLES SOUND EARTHQUAKE
1938 DECEMBER 16
Magnitude: 7.0 M_s
Epicentre: 45° S 167° E
Depth: 60 km
170°E
175°
35°
40°S
40°
45°
III
IV
V
VI
3-4
3
5
2-3
4
4-5
5-6
6
?
Note: RF - MM conversion

28. ISOSEISMAL MAP OF THE WAIRARAPA I EARTHQUAKE - 1942 JUNE 24

DATE:	1942 JUNE 24
TIME:	11:16:26 UT
MAGNITUDE:	7.2 M_S (Dowrick & Smith 1990); 7.0 M_S(Abe 1981, 1983)
EPICENTRE:	40.9°S 175.9°E B GRAPHICAL
DEPTH:	15 km (Webb 1989)

'Following a rather strong fore-shock at 8:14 p.m. on 24th June, the major shock occurred without further warning at 11h 16.5m p.m. New Zealand Daylight Saving Time [12 hours in advance of G.M.T.]. Press reports on the morning of the 25th June indicated considerable damage in Masterton and other Wairarapa towns. Later investigations showed that old, or poorly constructed buildings had suffered more or less severely; in striking contrast with the apparent absence of damage to modern buildings. No fatal casualties have been reported, but a few persons received minor injuries from falling debris. The macroseismic data, together with the seismograph records at Wellington, indicated that the epicentre could not be far from Masterton; where the intensity was evidently about R-F 8 [MM7], and possibly 8+ [MM8+] at some points. Intensity 8 [MM7] was reached over a large part of the Wairarapa, and also at some points on the West Coast from Manawatu to Wellington. In Wellington City and surroundings the intensity varied between 7 and 8 [MM6-7], according to the type of ground formation. Some minor structural damage occurred. Reconnaissance of the Wairarapa district a few days after the shock revealed surface evidence of fault movement at a point 8 miles [13 km] east of Masterton, with a narrow belt of severe surface damage running from about 10 miles [16 km] north-north-east of this point to about 30 miles [48 km] south of it. At points in this belt, the shock must have exceeded R-F 9 [MM8], judging by the severe shattering of the ground surface. The total extent of the felt area was comparable with other major New Zealand earthquakes. It extended from about Auckland in the north to Dunedin and Queenstown in the south; the latter place being 470 miles [750 km] from the epicentre.' (Hayes 1942)

The damage to both private dwellings and public buildings within Wellington City was so widespread that the Wellington City Council undertook to organise the repair work to household chimneys under The Emergency Precautions Scheme. It was estimated that some 20,000 chimneys had been damaged. In the following months, the Wellington City Council Engineer's Department carried out a building by building survey in the City area noting damage and reporting on areas of the structure that needed to be strengthened or were considered dangerous. The results of this survey are found in Aked (1945) and Luke (1943). Northey (1974) discusses the correlation of the distribution of the claims with the underlying soil structure.

Webb (1989) discusses the modelling of long-period data from Pasadena, California, and La Paz, Bolivia, for this earthquake and that on August 1.

The felt report summaries held at the Seismological Observatory have been converted from Rossi-Forel to Modified Mercalli and a new map was drawn by Eiby. Much could be gained by reinterpreting the original felt reports and detailing the effects of the earthquake from these, newspaper reports and other contemporary data. The extent of surface rupture observed by Ongley (1943) is not as large as might be expected from an earthquake of this magnitude.

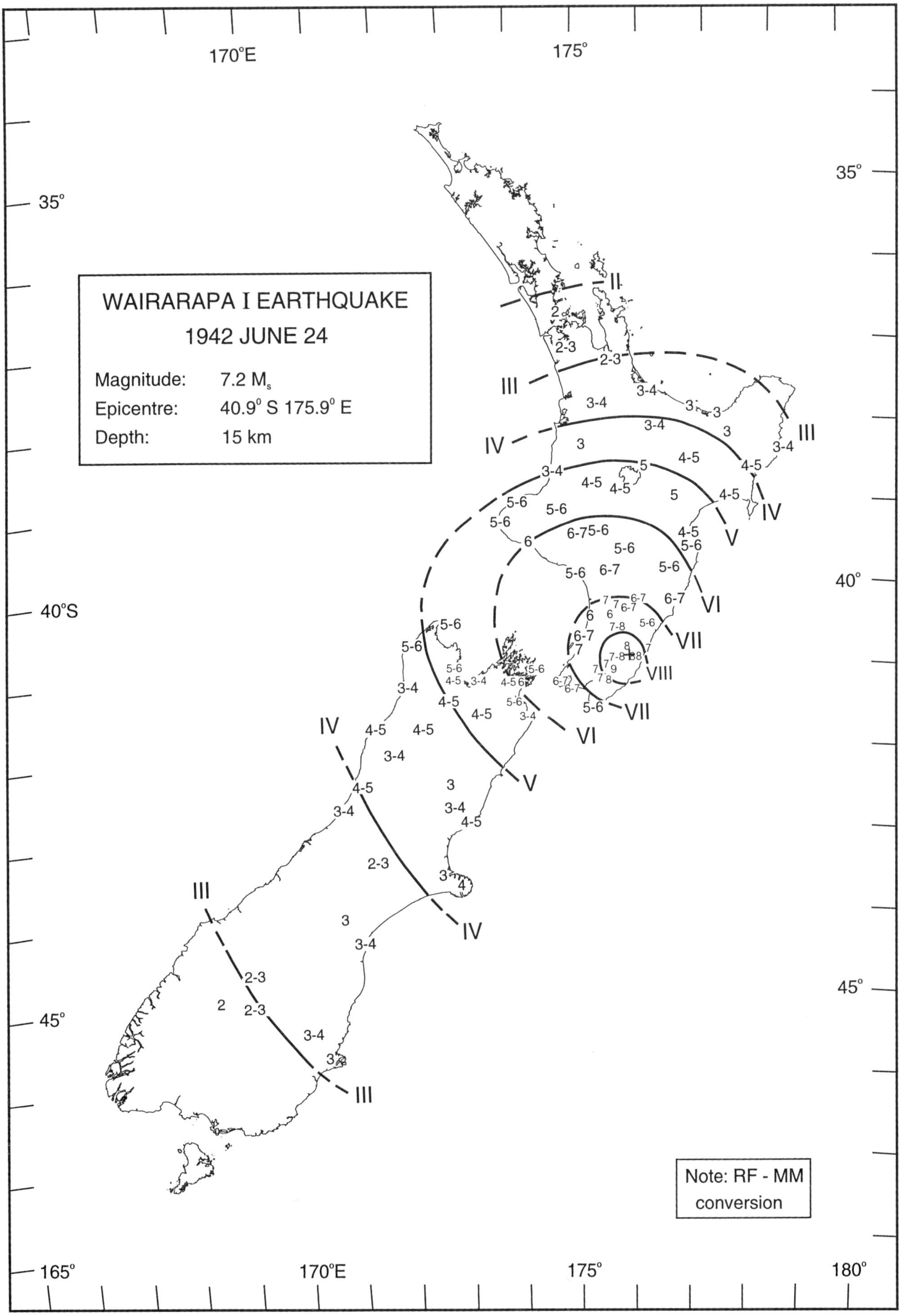

WAIRARAPA I EARTHQUAKE
1942 JUNE 24
Magnitude: 7.2 M_s
Epicentre: 40.9° S 175.9° E
Depth: 15 km
Note: RF - MM conversion
165°
170°E
175°
180°
35°
40°S
40°
45°
II
III
IV
V
VI
VII
VIII

29. ISOSEISMAL MAP OF THE WAIRARAPA II EARTHQUAKE - 1942 AUGUST 01

DATE:	1942 AUGUST 01
TIME:	12:34:05 UT
MAGNITUDE:	7.0 M_S (Dowrick & Smith 1990);
	7.3 (Richter 1958); 6.9 (Abe 1981, 1983)
EPICENTRE:	41.0°S 175.8°E A GRAPHICAL
DEPTH:	43 km (Webb 1989)

The shock on August 1 UT (August 2 at 00:34 NZT) was nearly as severe as the disastrous June 24 earthquake, just five weeks earlier. Its epicentre was close to that of the previous event but at a greater depth. The felt effects of the two earthquakes differed.

'No fresh surface faulting was found on 2nd August, and the maximum intensity in the epicentral area was less concentrated. The maximum Rossi-Forel intensity was probably between 8 and 9 [MM8]. Some further damage was caused, particularly to chimneys repaired after the previous shock, in which the mortar had not had time to set. In Wellington, many chimneys were damaged or dislodged by this shock; and following it some further damage to buildings was revealed. There is no doubt that a large part of the damage revealed after the shock on 2nd August must have originated in the initial major shock of 24th June.' (Hayes 1942)

Neef (1976), however, suggests that minor faulting may have occurred near Alfredton. Berryman (1977) reports that movement on a small fault in the Aorangi Mountains north of Cape Palliser was witnessed during a large earthquake in 1942 (date unknown). Association with one of the events in 1942 needs further investigation.

Webb (1989) discusses the modelling of long-period data from Pasadena, California, and La Paz, Bolivia, for this earthquake and that on June 24 and states that: 'The August 1 event shows clear depth phases pP and sS on the Pasadena records, and it can be modelled as a simple source with a 0.5, 3.0, 0.5 second trapezoidal time function at 43 km depth in Robinson's (1986) velocity model. The mechanism shows predominantly normal faulting with one nodal plane dipping 45°S to the NW with a strike of 40°E. The source is thus located within the high-velocity part of the subducted Pacific Plate.' (Webb 1989)

A further shock of magnitude 6.0 occurred on December 2, with its own set of aftershocks. Its epicentre was located about 30 km to the south of the June 24 epicentre and 20 km to the south of the August 1 event. A small number of chimneys were brought down to the east of Masterton and several in Wellington. This earthquake has received little attention in the literature.

The felt report summaries held at the Seismological Observatory have been converted from Rossi-Forel to Modified Mercalli and a new map was drawn by Eiby. However much could be gained by reinterpreting the original felt reports on the MM scale and detailing the effects of the earthquake from these, newspaper reports and other contemporary data.

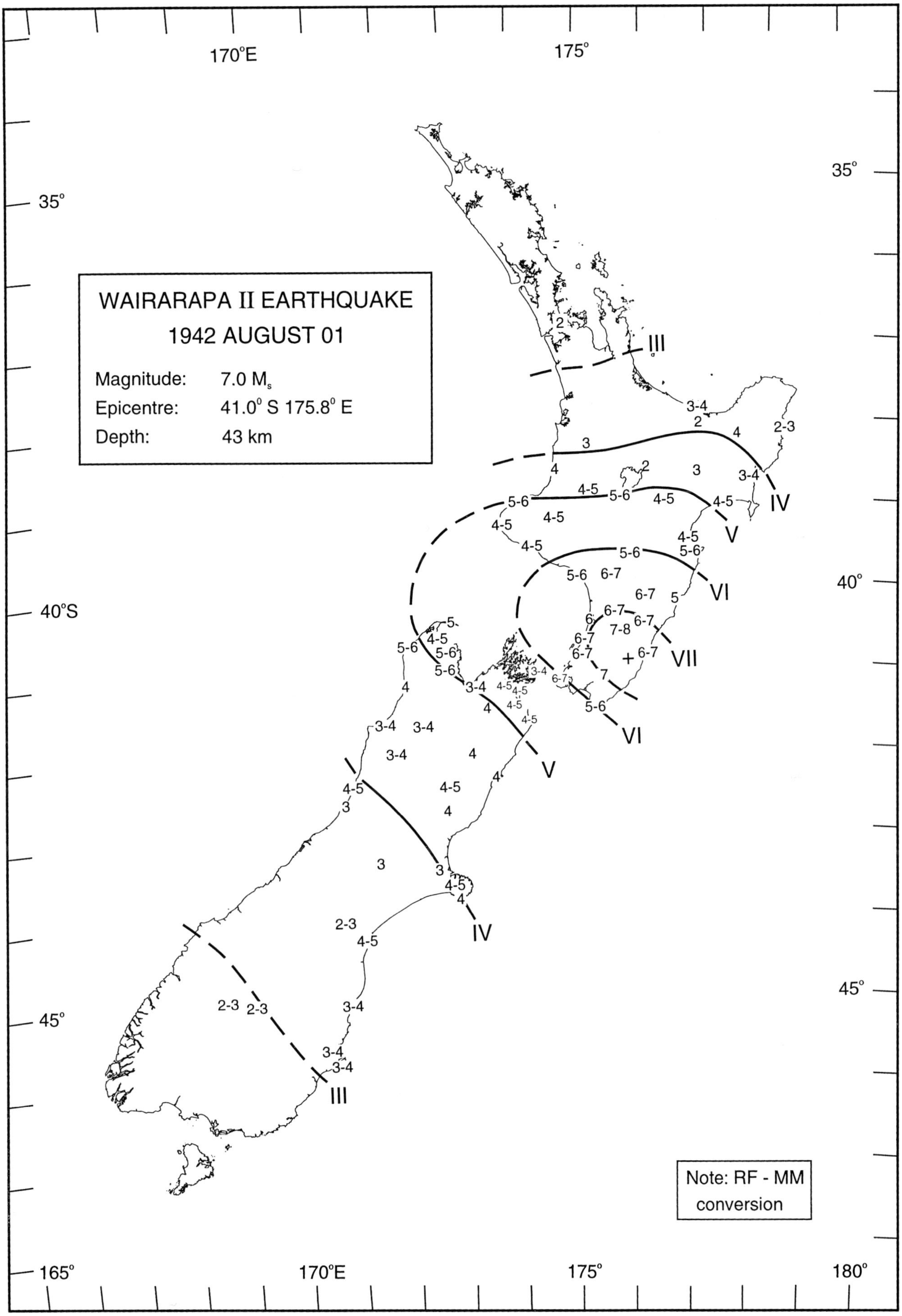
WAIRARAPA II EARTHQUAKE
1942 AUGUST 01
Magnitude: 7.0 M_s
Epicentre: 41.0° S 175.8° E
Depth: 43 km
Note: RF - MM conversion
170°E
175°
35°
40°S
40°
45°
165°
180°
III
IV
V
VI
VII

30. INTENSITY MAP OF THE PUYSEGUR BANK EARTHQUAKE - 1945 SEPTEMBER 01

DATE:	1945 SEPTEMBER 01
TIME:	22:44:08 UT
MAGNITUDE:	6.5M_L; 7.0 M_S (Dowrick & Smith 1990)
EPICENTRE:	47.47°S 166.12°E (1)
DEPTH:	12 km (restricted)

'The largest shallow earthquake in 1945 had a magnitude of 6.5. It occurred on September 1, and was felt throughout Otago and Southland, with reports extending northwards to Timaru and Jackson's Bay. The epicentre lay about 100km south-west of Stewart Island, but the lack of detail in the reports suggests that the intensity was not great. An aftershock of magnitude 5.7 has been placed some 50 km to the north-west. The International Seismological Summary allots both shocks a common epicentre 25 km farther in the same direction. At this period, the only seismograph in the southern part of the South Island was a Jaggar shock-recorder without absolute timing at Monowai, but the Australian stations at Brisbane and Riverview both have negative residuals, and do not favour the ISS solution.' (Eiby 1983))

The map detailing felt intensities, resulting from the conversion of felt summaries from Rossi-Forel to MM, is from the GNS files. There is clearly insufficient data to suggest isoseismals, but the earthquake is included because of its magnitude and location close to the New Zealand coast.

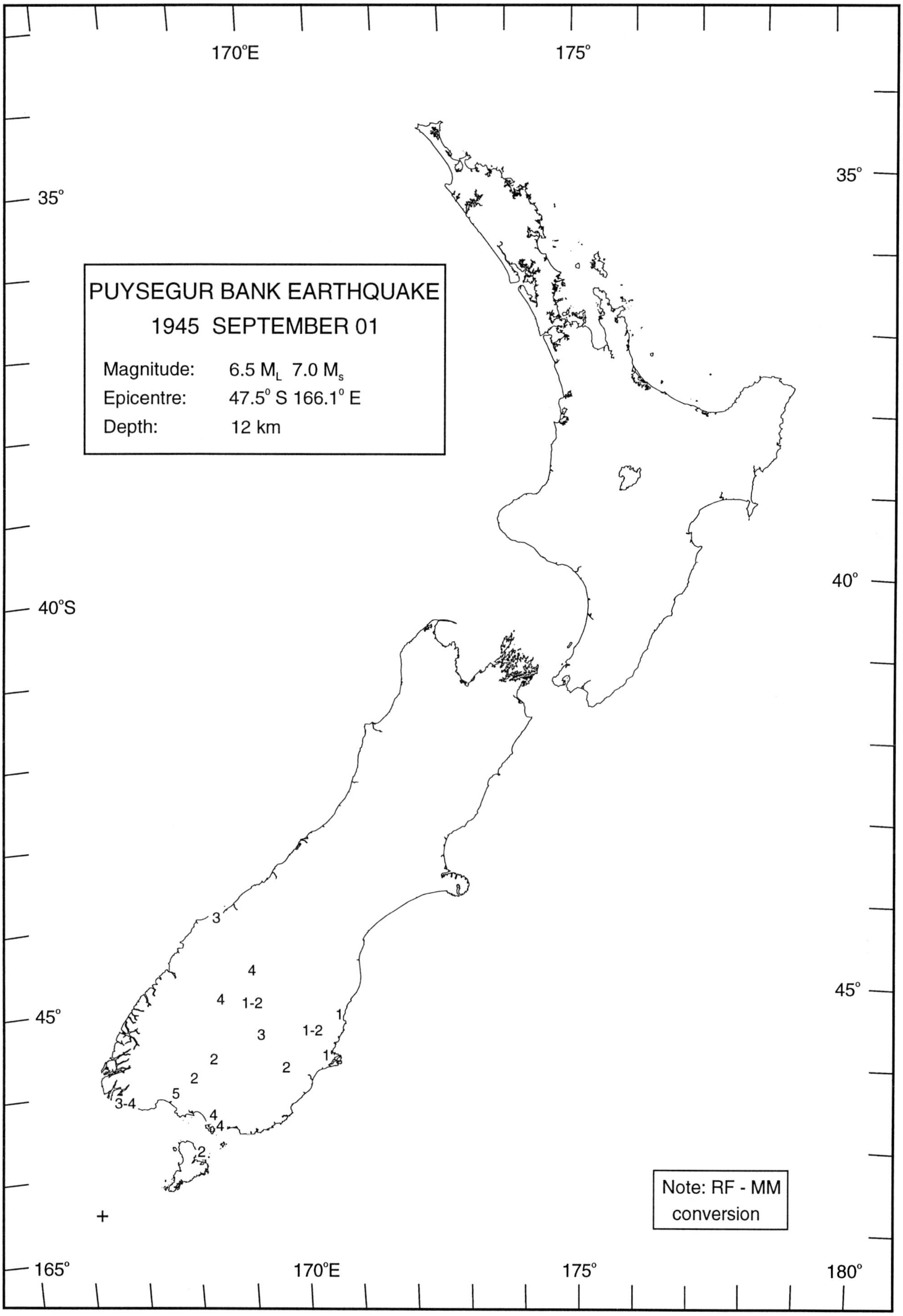

170°E
175°
35°
35°
40°
40°S
45°
45°
165°
170°E
175°
180°
PUYSEGUR BANK EARTHQUAKE
1945 SEPTEMBER 01
Magnitude: 6.5 M_L 7.0 M_s
Epicentre: 47.5° S 166.1° E
Depth: 12 km
3
4
4
1-2
1
3
1-2
1
2
2
2
5
3-4
4
4
2
Note: RF - MM conversion

31. ISOSEISMAL MAP OF THE LAKE COLERIDGE EARTHQUAKE - 1946 JUNE 26

DATE:	1946 JUNE 26
TIME:	12:34:39 UT
MAGNITUDE:	6.2 M_L; 6.4 M_s (Dowrick pers. comm.)
EPICENTRE:	43.18°S 171.68°E (1)
DEPTH:	12 km (restricted)

'The magnitude (M_L) 6.2 earthquake of 1946 June 26d 12h 34m 39s U.T., with an epicentre at 43.18°S 171.68°E, near Lake Coleridge, is the largest known event in the Central Seismic Region of New Zealand. It was felt over the greater part of the South Island, and caused minor structural damage to homesteads in the Upper Rakaia basin, and at the Lake Coleridge hydro-electric power station. There were also numerous landslides and changes to watercourses. It was preceded by two foreshocks and followed by numerous aftershocks, the largest of which had a magnitude of 5.8. These persisted until the end of 1949.

'... shows Modified Mercalli isoseismals for the principal shock ... based on questionnaires returned by the Observatory's regular reporters together with information contained in the newspapers of South Canterbury and Westland, an unpublished two-page report to the Acting-Director of the Dominion Observatory by Mr. H. F. Baird of the Magnetic Survey, Christchurch, the log book of the Lake Coleridge power station, and the records of the Regional Engineer of the N. Z. Electricity Department, Christchurch.' (Eiby 1990)

The isoseismal map has been reproduced from Eiby (1990) in which there is a full description of the effects in the epicentral area.

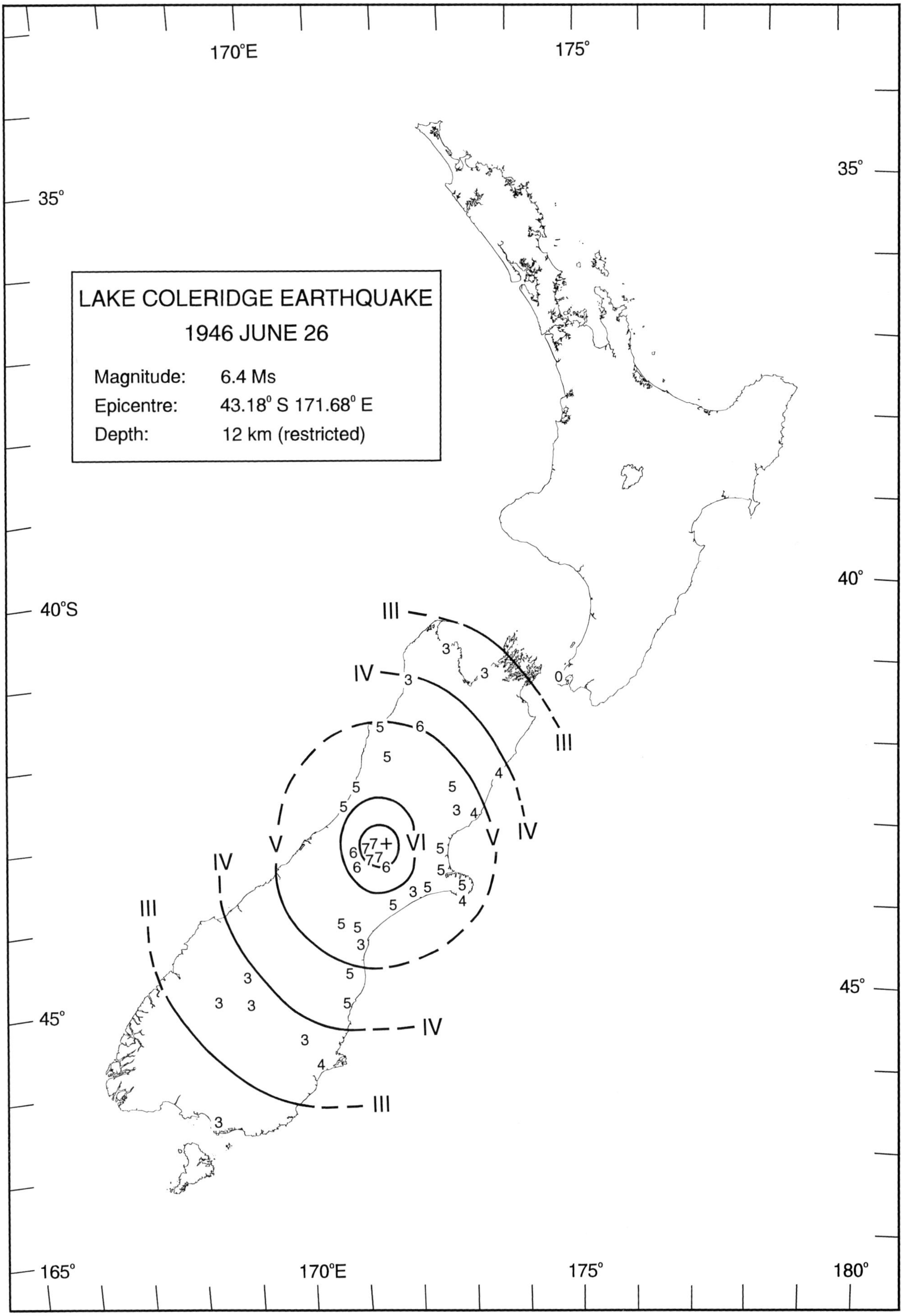
LAKE COLERIDGE EARTHQUAKE
1946 JUNE 26
Magnitude: 6.4 Ms
Epicentre: 43.18° S 171.68° E
Depth: 12 km (restricted)
170°E
175°
165°
180°
35°
40°S
40°
45°
III
IV
V
VI

32. ISOSEISMAL MAP OF THE WAIAU EARTHQUAKE - 1948 MAY 22

DATE:	1948 MAY 22
TIME:	19:21:27 UT
MAGNITUDE:	6.4 M_L*
EPICENTRE:	42.50°S 173.00°E (1)
DEPTH:	12 km (restricted)

'The most serious of these [shallow shocks recorded in the period 1948-1950] were the events in the Hanmer-Waiau district in May 1948. The largest shock, on May 23 [local time], had a magnitude of 6.4, and was accompanied by a foreshock and numerous aftershocks, several of which had magnitudes closer to that of the main shock than is usual. Minor activity continued for some months. An account of these events, including an isoseismal map and an associated study of crustal structure, has been published by Eiby (1953). The shock was felt over an area that included the northern half of the South Island and extended across Cook Strait to Wellington; but damage was confined to the settlements of Hanmer and Waiau and to the countryside between. Chimneys fell, and other minor structural damage occurred, indicating a maximum intensity of about MM VIII.

'The foreshock, of magnitude 5.9, occurred only 24 minutes before the main event; and within the next two hours there was an aftershock of magnitude 6.2, two of magnitude 5.7, and one of magnitude 5.8. No comparable grouping of large shallow earthquakes in New Zealand can be recalled. The remainder of the sequence has the normal pattern of aftershocks for a large shallow earthquake.' (Eiby 1982)

The isoseismal map is from Eiby (1953).

* Location and local magnitude have been revised since Hayes (1949). Refer to introduction.

WAIAU EARTHQUAKE 1948 MAY 22	
Magnitude:	6.4 M_L
Epicentre:	42.50° S 173.00° E
Depth:	12 km (restricted)

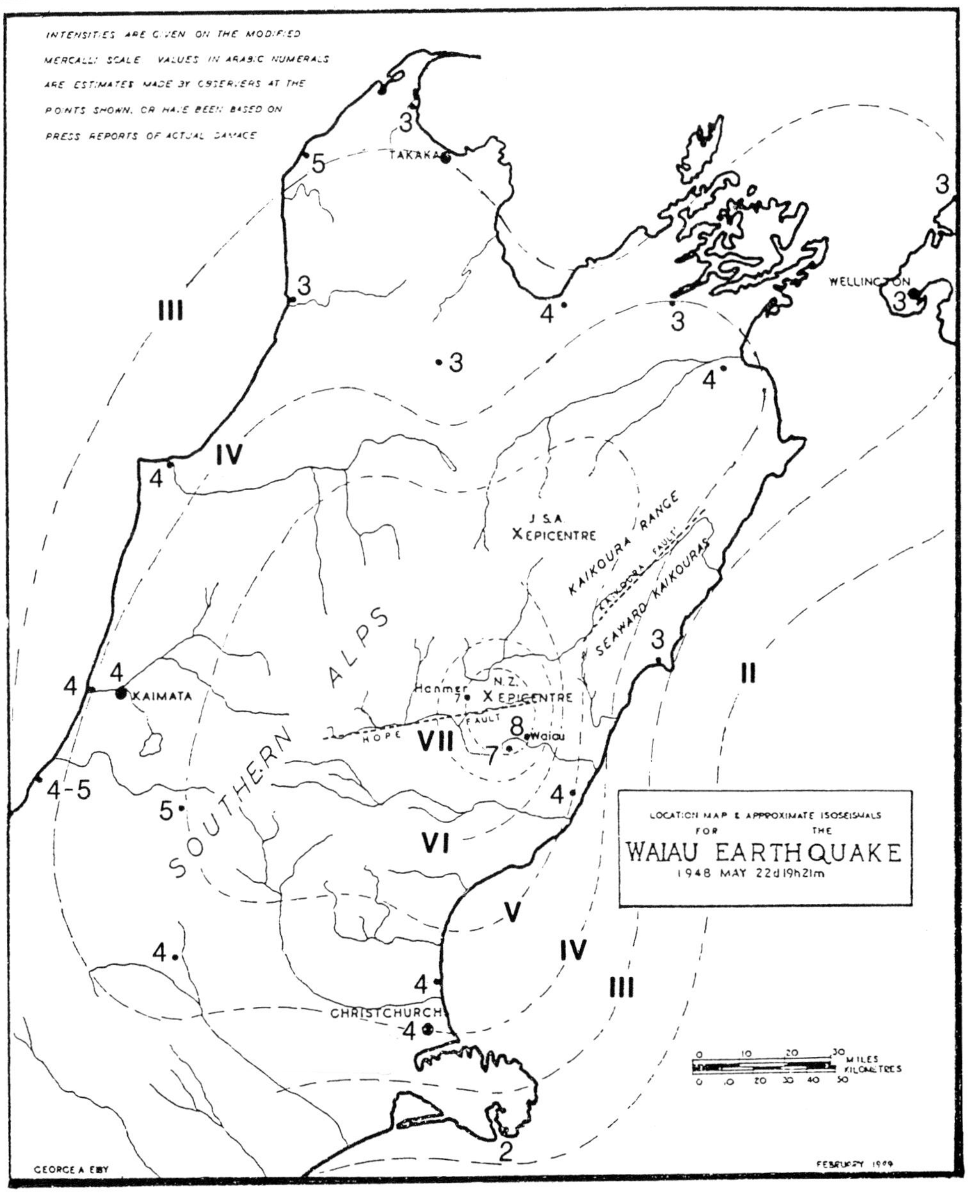

after Eiby (1953)

33. ISOSEISMAL MAP OF THE COOK STRAIT I EARTHQUAKE - 1950 JANUARY 07

DATE:	1950 JANUARY 07
TIME:	14:21:03 UT
MAGNITUDE:	5.6 M_L *
EPICENTRE:	41.14°S 174.56°E (1)
DEPTH:	33 km (restricted)

'During January and most of February 1950, a series of earthquakes occurred in the Cook Strait region, four of which were sufficiently strong to cause some alarm in Wellington. Two shocks on 13 January resulted in some minor damage in the City. The activity commenced with a rather strong shock on 8 January, and continued intermittently for a period of about six weeks. In addition to eighteen perceptible shocks, more than three hundred minor shocks and tremors were recorded at the Seismological Observatory.

'The largest shocks on 8 January, 13 January and 4 February were each followed by a series of aftershocks ...' (Hayes 1952) [N. B. local time quoted throughout]

The felt reports held at the Seismological Observatory have been reinterpreted by Eiby and a new map was drawn.

* Location and local magnitude have been revised since the publication of the Seismological Observatory bulletin S 96 and Hayes (1952). Refer to introduction.

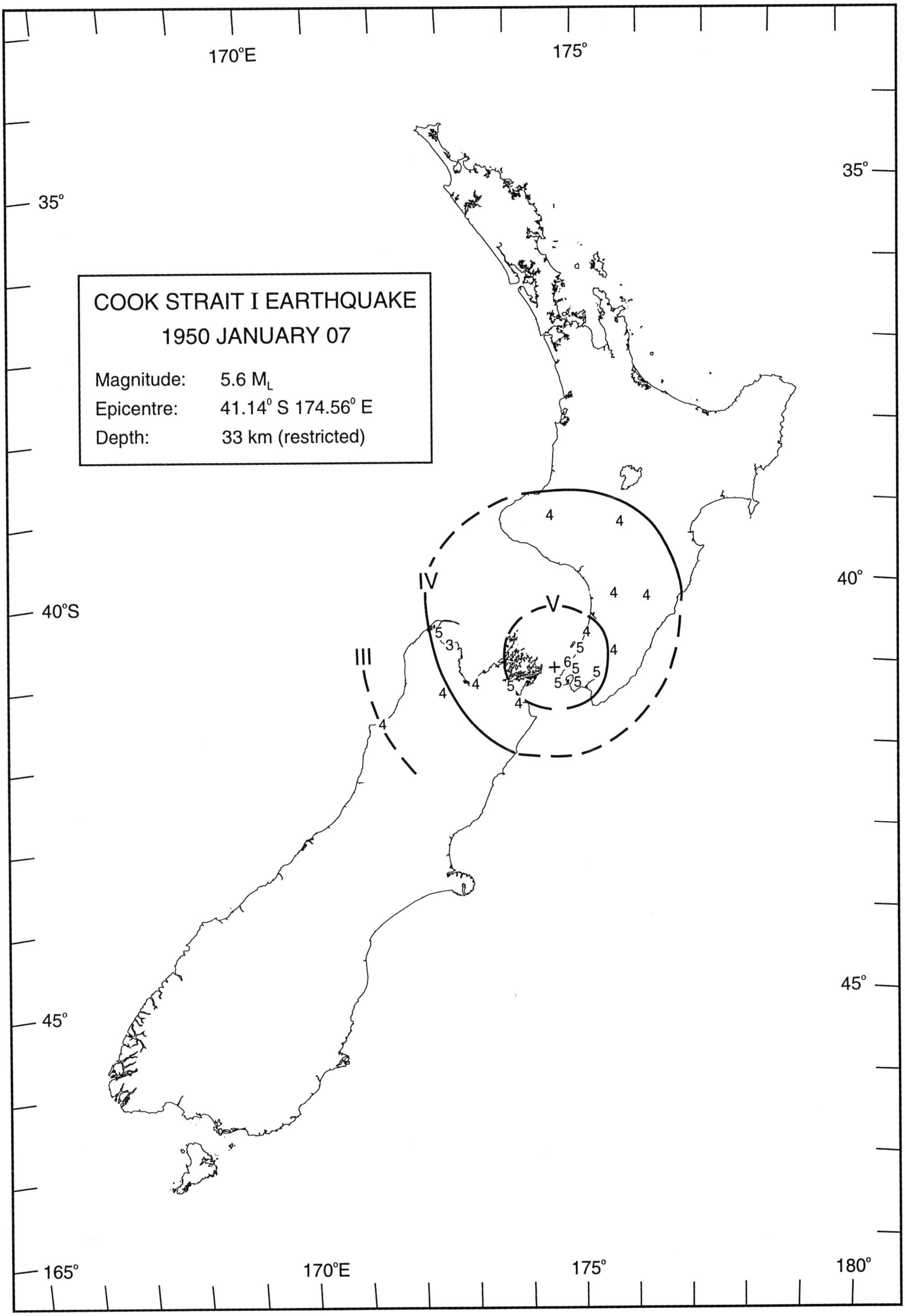

COOK STRAIT I EARTHQUAKE
1950 JANUARY 07
Magnitude: 5.6 M_L
Epicentre: 41.14° S 174.56° E
Depth: 33 km (restricted)
170°E
175°
35°
40°S
40°
45°
165°
180°
III
IV
V

34. ISOSEISMAL MAP OF THE COOK STRAIT III EARTHQUAKE - 1950 JANUARY 12 20:49

DATE:	1950 JANUARY 12
TIME:	20:48:20 UT
MAGNITUDE:	5.7 M_L*
EPICENTRE:	41.06°S 174.36°E (1)
DEPTH:	33 km (restricted)

DATE:	1950 JANUARY 12
TIME:	20:49:50 UT
MAGNITUDE:	5.6 M_L*
EPICENTRE:	41.15°S 174.61°E (1)
DEPTH:	33 km (restricted)

'During January and most of February 1950, a series of earthquakes occurred in the Cook Strait region, four of which were sufficiently strong to cause some alarm in Wellington. Two shocks on 13 January resulted in some minor damage in the City. The activity commenced with a rather strong shock on 8 January, and continued intermittently for a period of about six weeks. In addition to eighteen perceptible shocks, more than three hundred minor shocks and tremors were recorded at the Seismological Observatory.

'The largest shocks on 8 January, 13 January and 4 February were each followed by a series of aftershocks ...' (Hayes 1952) [N. B. local time quoted throughout this passage]

'The latter two shocks [i.e. the two on January 12 occurring 1.5 minutes apart] resulted in some minor damage [in Wellington], but it was insufficient to justify assigning an intensity above MM V.' (Eiby 1982)

The felt reports held at the Seismological Observatory have been reinterpreted by Eiby and new maps have been drawn. There is little difference in the isoseismal maps for these two earthquakes. The isoseismal map for the second earthquake at 20:49 UT is given.

* Location and local magnitude have been revised since the publication of the Seismological Observatory bulletin S 96 and Hayes (1952). Refer to introduction.

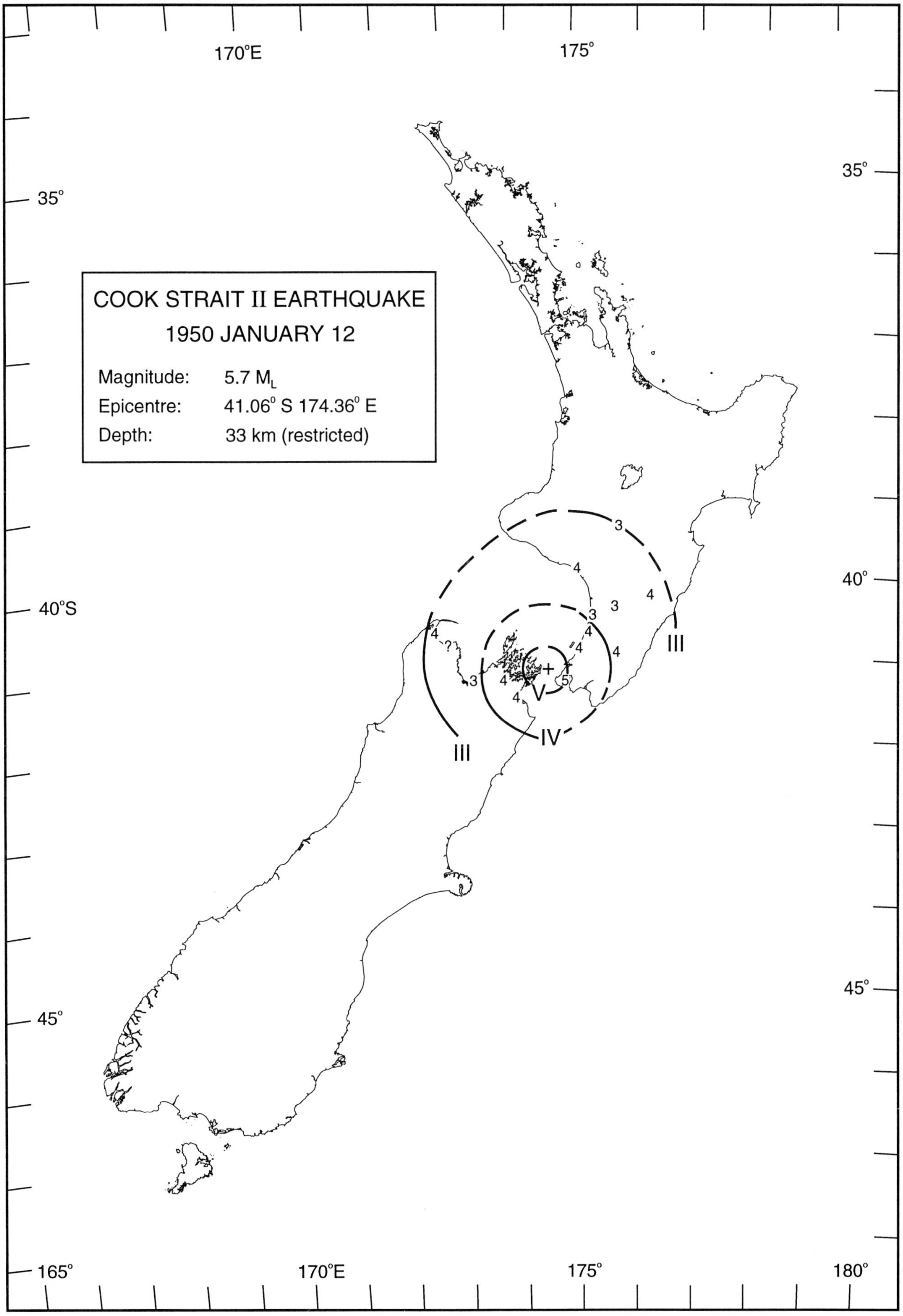

170°E
175°
35°
35°
40°S
40°
45°
45°
165°
170°E
175°
180°
COOK STRAIT II EARTHQUAKE
1950 JANUARY 12
Magnitude: 5.7 M_L
Epicentre: 41.06° S 174.36° E
Depth: 33 km (restricted)
3
4
4
3
3
4
4
4
4
?
3
4
4
5
III
III
IV
V

35. ISOSEISMAL MAP OF THE HUIARAU RANGE EARTHQUAKE - 1950 FEBRUARY 28

DATE:	1950 FEBRUARY 28
TIME:	18:58:44 UT
MAGNITUDE:	5.7 M_L*
EPICENTRE:	38.88°S 176.70°E (1)
DEPTH:	30 km

'On February 28 a shock of magnitude 5.7 centred midway between Lakes Taupo and Waikaremoana was felt over the greater part of the North Island, and is reported to have caused some damage at Te Whaiti. Hayes (1952) gives an isoseismal map.' (Eiby 1982)

The felt reports held at the Seismological Observatory have been reinterpreted by Eiby and a new map was drawn. However the isoseismal lines are poorly constrained.

---

* Location and local magnitude have been revised since the publication of the Seismological Observatory bulletin S 96 and Hayes(1952). Refer to introduction.

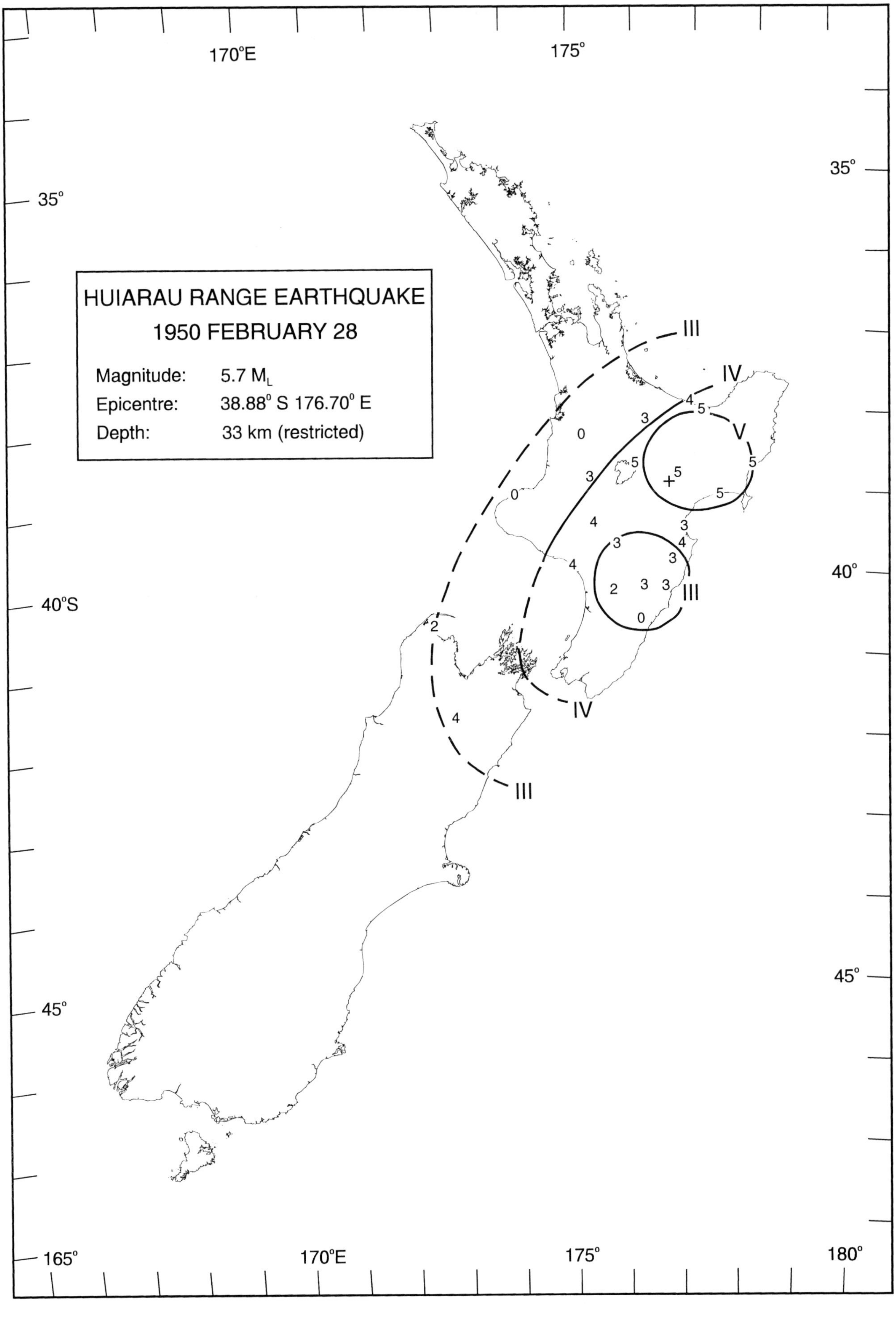
HUIARAU RANGE EARTHQUAKE
1950 FEBRUARY 28
Magnitude: 5.7 M_L
Epicentre: 38.88° S 176.70° E
Depth: 33 km (restricted)
170°E
175°
35°
40°S
40°
45°
165°
180°
III
IV
V
0
2
3
4
5

36. ISOSEISMAL MAP OF THE STEPHENS ISLAND EARTHQUAKE - 1950 MARCH 13

DATE:	1950 MARCH 13
TIME:	09:38:28 UT
MAGNITUDE:	5.7 M_L*
EPICENTRE:	40.55°S 174.04°E (1)
DEPTH:	33 km (restricted)

During the first few months of 1950 Wellington and coastal settlements to the north experienced a number of moderate and small earthquakes mostly centred in Cook Strait. The earthquake of 1950 March 13 was another which was felt in these places. It was located north of Stephens Island and was felt widely with an intensity not exceeding MM4 from Taranaki and Hawke's Bay to Christchurch with a few reports from the west coast of the South Island.

The felt reports held at the Seismological Observatory have been reinterpreted by Eiby and a new map was drawn.

---

* Location and local magnitude have been revised since the publication of the Seismological Observatory bulletin S 96 and Hayes(1952). Refer to introduction.

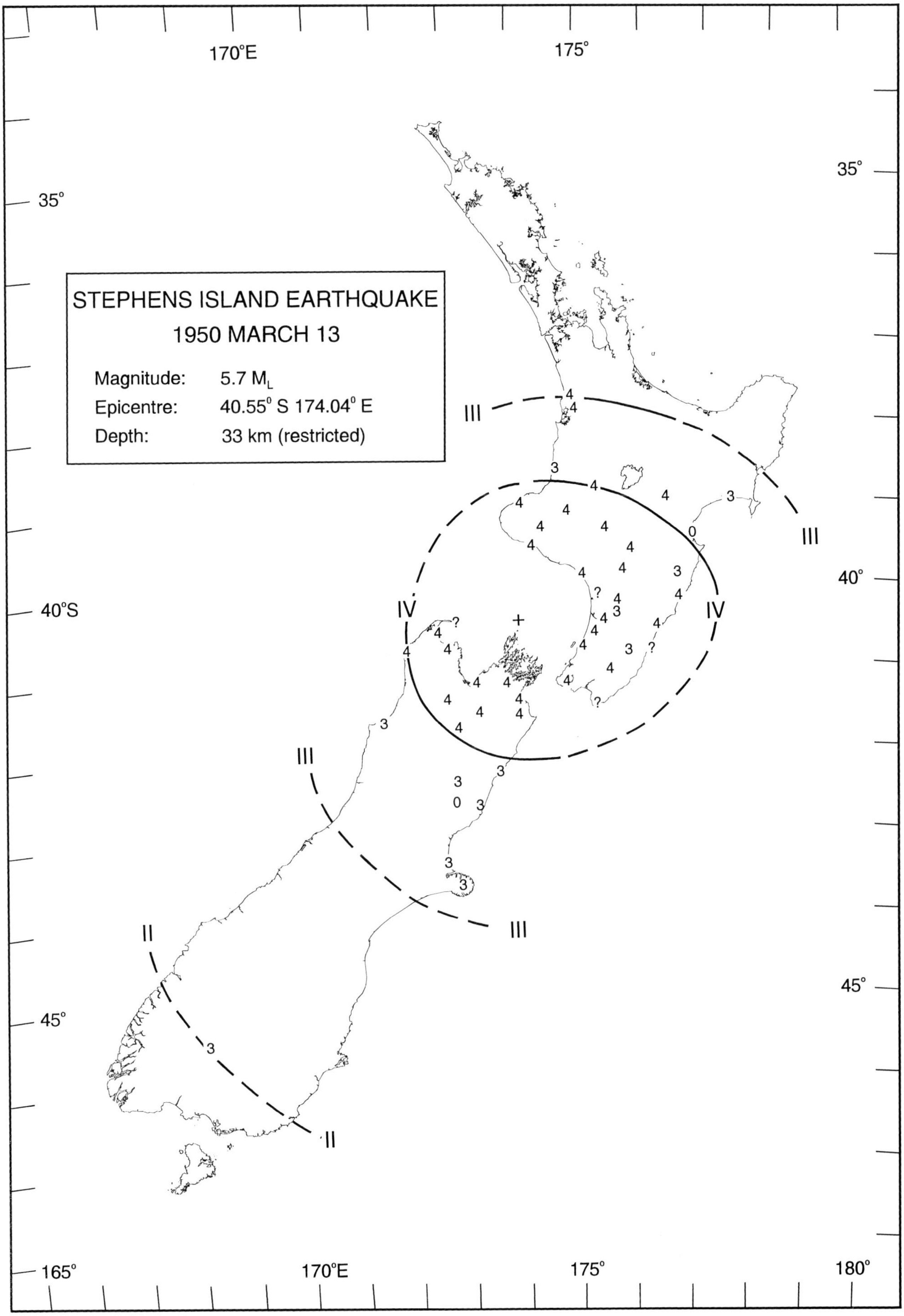

STEPHENS ISLAND EARTHQUAKE
1950 MARCH 13
Magnitude: 5.7 M_L
Epicentre: 40.55° S 174.04° E
Depth: 33 km (restricted)
170°E
175°
165°
180°
35°
40°S
40°
45°
III
IV
II

37. ISOSEISMAL MAP OF THE NORTH ISLAND EARTHQUAKE - 1950 JUNE 17

DATE:	1950 JUNE 17
TIME:	15:56:33 UT
MAGNITUDE:	5.9 M_L*
EPICENTRE:	38.92°S 175.25°E (1)
DEPTH:	185 km

'A shock on 1950 June 17 was felt in eastern parts of both islands from the Bay of Plenty to Banks Peninsula, but not at places in the west. Its focal depth was 185 km, and its epicentre close to Taumarunui, where the felt intensity was MM IV [reassessed to be MM3]. This displacement of the felt region and its extension along the axis of the country is usual in deep focus shocks ...' (Eiby 1982))

The felt reports held at the Seismological Observatory have been reinterpreted by Eiby and a new map was drawn.

* Location and local magnitude have been revised since the publication of the Seismological Observatory bulletin S 96. Refer to introduction.

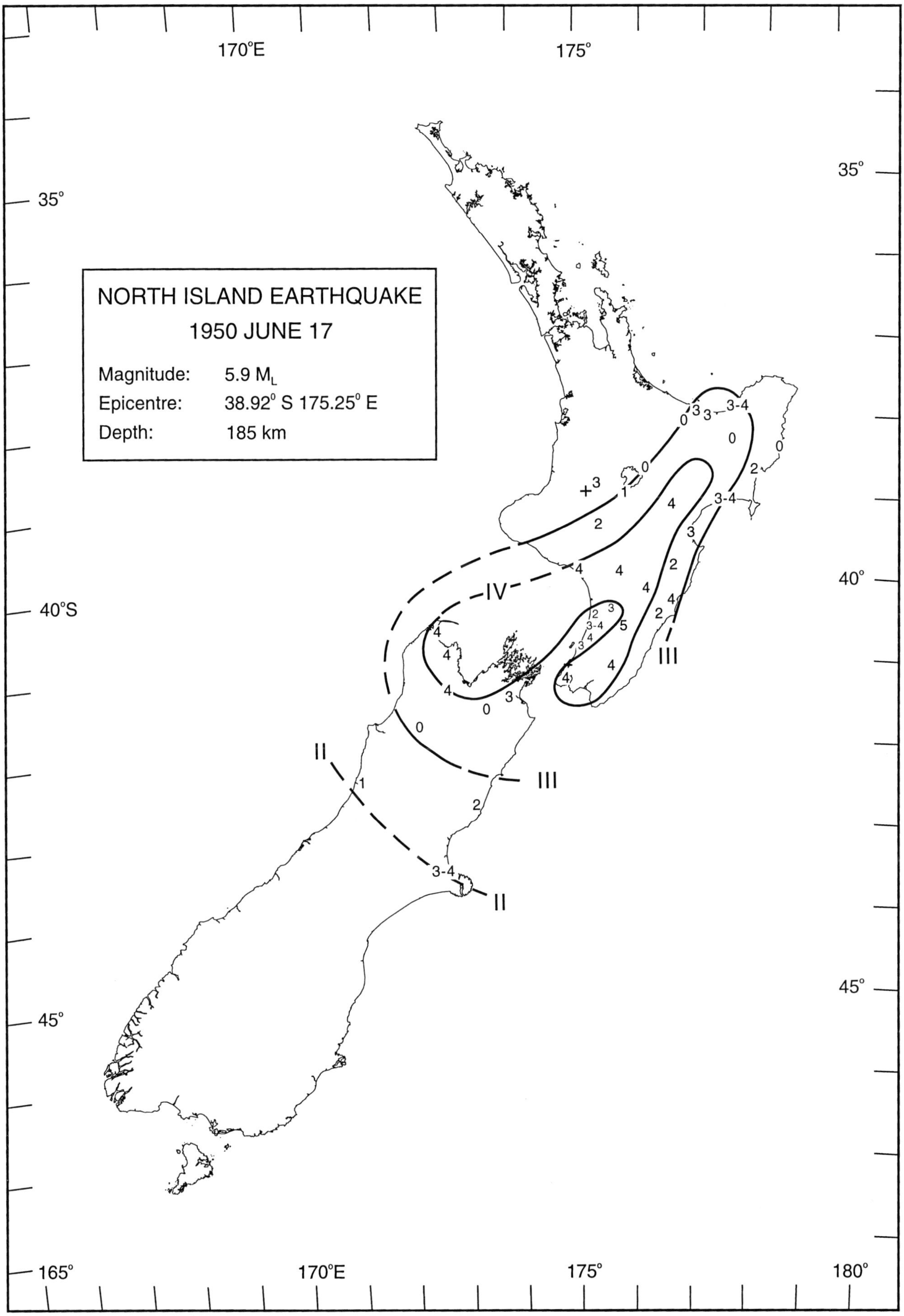

NORTH ISLAND EARTHQUAKE
1950 JUNE 17
Magnitude: 5.9 M$_L$
Epicentre: 38.92° S 175.25° E
Depth: 185 km
170°E
175°
165°
180°
35°
40°S
40°
45°
IV
III
II

38. ISOSEISMAL MAP OF THE CHEVIOT EARTHQUAKE - 1951 JANUARY 10

DATE:	1951 JANUARY 10
TIME:	19:15:18 UT
MAGNITUDE:	5.5 M_L*
EPICENTRE:	42.79°S 173.18°E (1)
DEPTH:	12 km (restricted)

The earthquake on 1951 January 10 was the largest of a swarm of earthquakes at approximately the same location, eight of which exceeded magnitude 5.0 M_L. They occurred between January 10 and January 27. The epicentral area included Cheviot, with the highest reported intensity of MM7, and Waiau and Culverden with a reported intensity of MM6. The felt area of the earthquake included Christchurch, the West Coast and there were isolated reports in the North Island from Wellington and Stratford.

The felt reports held at the Seismological Observatory have been reinterpreted by Eiby and a new map was drawn.

* Location and local magnitude have been revised since the publication of the Seismological Observatory bulletin S 97. Refer to introduction.

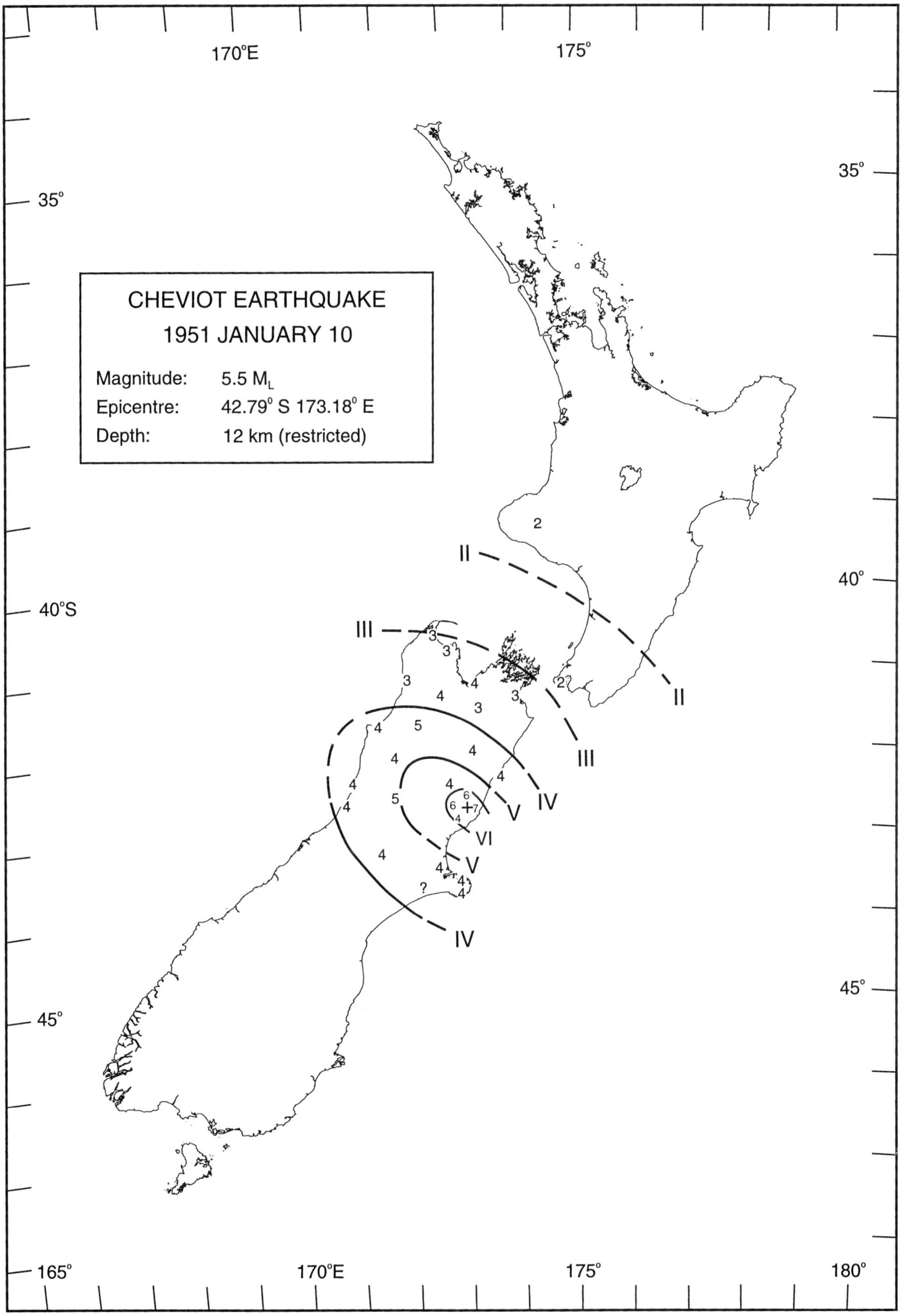
CHEVIOT EARTHQUAKE
1951 JANUARY 10
Magnitude: 5.5 M_L
Epicentre: 42.79° S 173.18° E
Depth: 12 km (restricted)
170°E
175°
35°
40°S
40°
45°
165°
180°
II
III
IV
V
VI

39. ISOSEISMAL MAP OF THE ARAMOANA EARTHQUAKE - 1951 FEBRUARY 10

DATE:	1951 FEBRUARY 10
TIME:	03:27:57.2 UT
MAGNITUDE:	6.1 M_L*
EPICENTRE:	40.21°S 177.04°E (1)
DEPTH:	33 km (restricted)

The earthquake on 1951 February 10 was felt over a large area of the North Island with maximum intensity of MM6 experienced at Waipawa and Porangahau.

The felt reports held at the Seismological Observatory have been reinterpreted by Eiby and a new map was drawn.

* Location and local magnitude have been revised since the publication of the Seismological Observatory bulletin S 97. Refer to introduction.

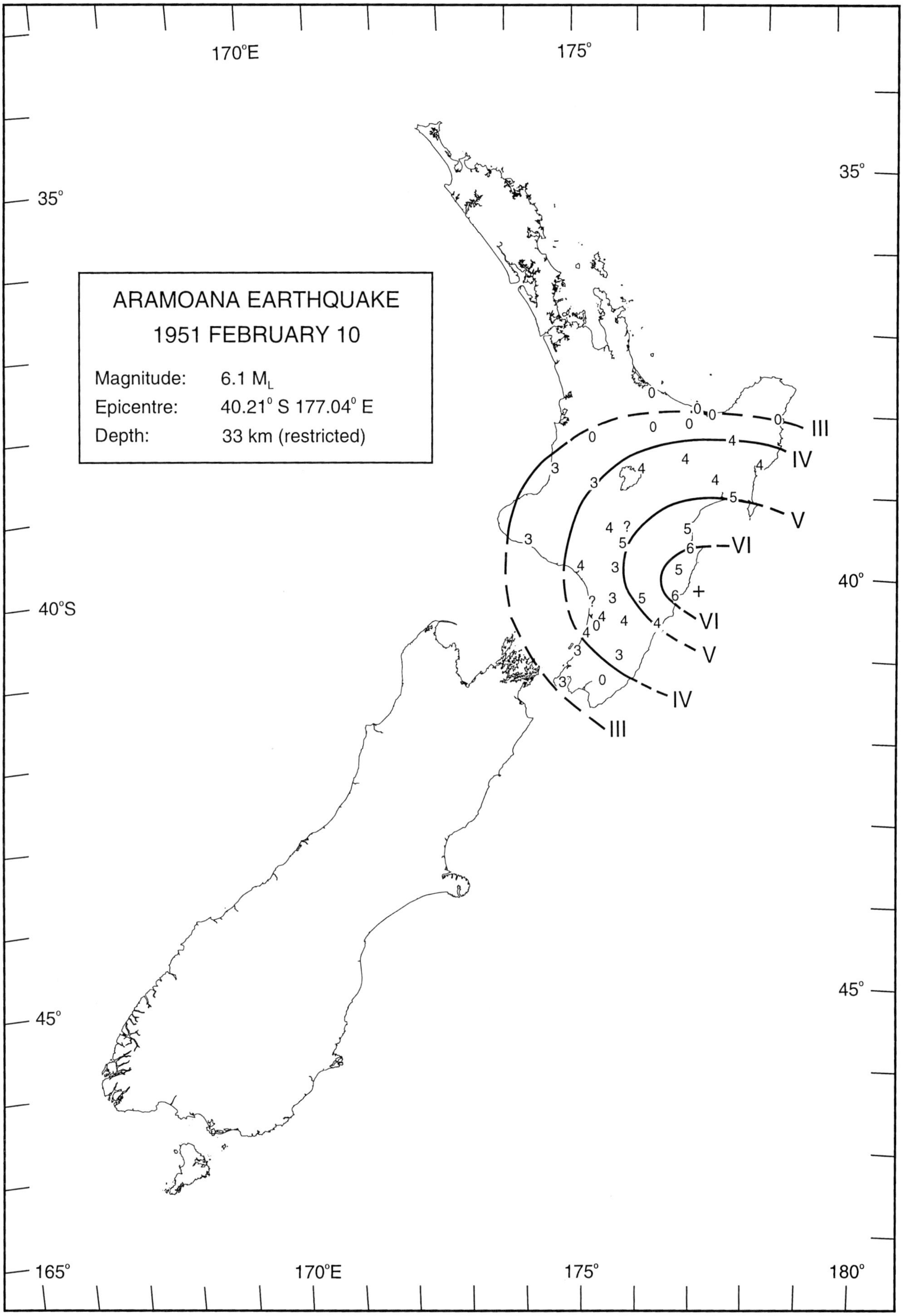

ARAMOANA EARTHQUAKE
1951 FEBRUARY 10
Magnitude: 6.1 M_L
Epicentre: 40.21° S 177.04° E
Depth: 33 km (restricted)
170°E
175°
180°
165°
35°
40°S
40°
45°
III
IV
V
VI

40. ISOSEISMAL MAP OF THE BAY OF PLENTY EARTHQUAKE - 1951 MARCH 28

DATE:	1951 MARCH 28
TIME:	01:55:15 UT
MAGNITUDE:	6.4 M_L*
EPICENTRE:	37.08°S 176.91°S (1)
DEPTH:	370 km (restricted)

A 370 km deep earthquake in the Bay of Plenty on 1951 March 28 was felt over most of the North Island east of a line through Wanganui, Taupo and Whakatane and in the north of the South Island with an intensity of MM3-4, with Opotiki being the only location reporting an intensity exceeding this.

The felt reports held at the Seismological Observatory have been reinterpreted by Eiby and a new map was drawn.

* Location and local magnitude have been revised since the publication of the Seismological Observatory bulletin S 97. Refer to introduction.

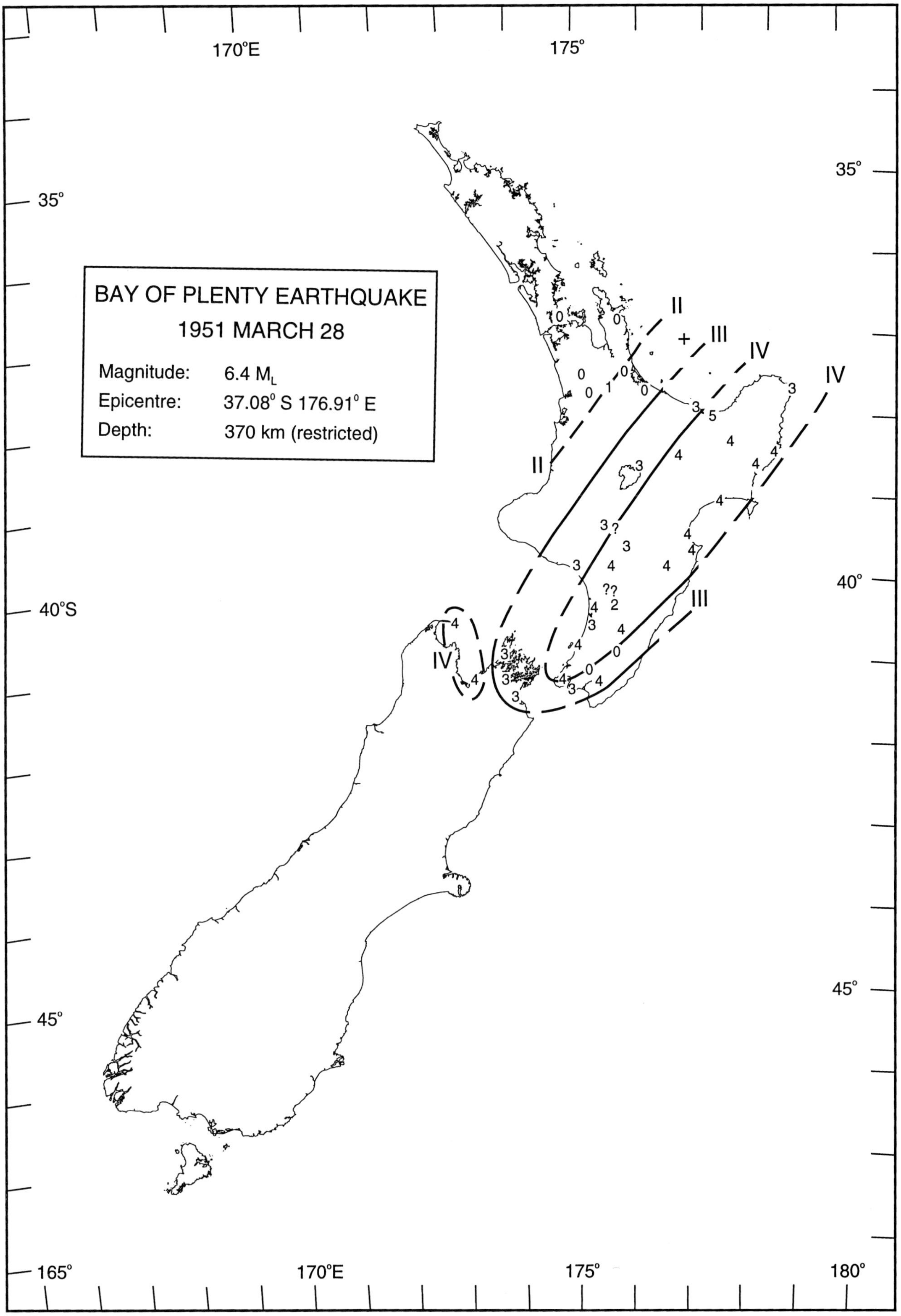

BAY OF PLENTY EARTHQUAKE
1951 MARCH 28
Magnitude: 6.4 M_L
Epicentre: 37.08° S 176.91° E
Depth: 370 km (restricted)
170°E
175°
35°
40°S
40°
45°
165°
180°
II
III
IV

41. ISOSEISMAL MAP OF THE CAPE RUNAWAY EARTHQUAKE - 1951 APRIL 23

DATE:	1951 APRIL 23
TIME:	06:50:20 UT
MAGNITUDE:	6.2 M_L*; 5.7 M_s (Dowrick & Smith 1990)
EPICENTRE:	37.56°S 177.76°E (1)
DEPTH:	125 km

The second of three earthquakes in 1951 to exceed a magnitude of 6.0 M_L, the earthquake on April 23 was felt strongly (MM5-6) over the East Cape peninsula, in eastern Bay of Plenty and northern Hawke's Bay and extensively in the North Island, excluding Taranaki and Northland. There were isolated reports in the South Island from Blenheim and Cheviot.

The felt reports held at the Seismological Observatory have been reinterpreted by Eiby and a new map was drawn.

---

* Location and local magnitude have been revised since the publication of the Seismological Observatory bulletin S 97. Refer to introduction.

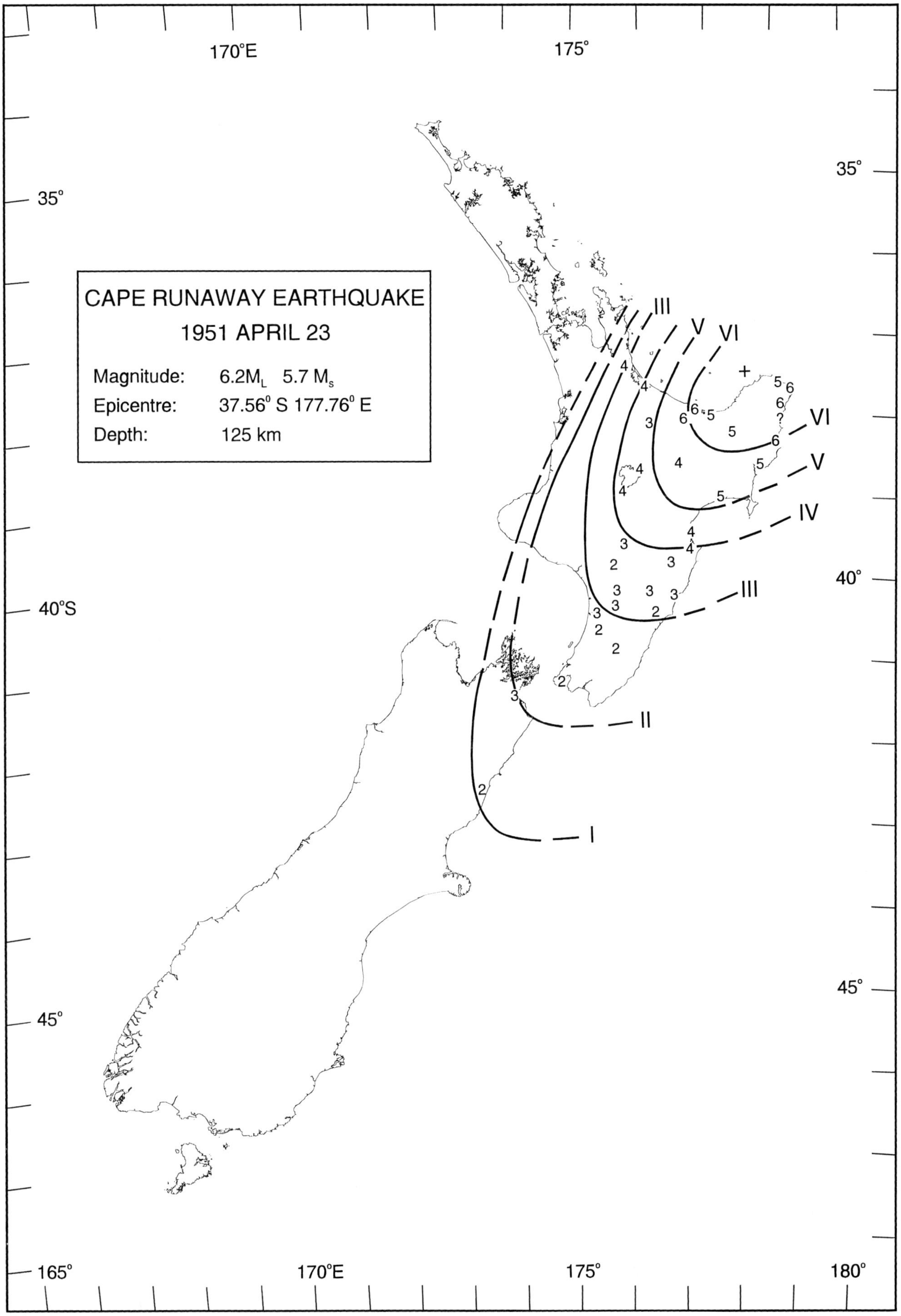

CAPE RUNAWAY EARTHQUAKE
1951 APRIL 23
Magnitude: 6.2M$_L$ 5.7 M$_s$
Epicentre: 37.56° S 177.76° E
Depth: 125 km
170°E
175°
165°
180°
35°
40°S
40°
45°
I
II
III
IV
V
VI

42. ISOSEISMAL MAP OF THE WESTERN HAWKE'S BAY EARTHQUAKE - 1951 JUNE 24

DATE:	1951 JUNE 24
TIME:	04:41:51 UT
MAGNITUDE:	6.3 M_L*
EPICENTRE:	39.46°S 176.20°E (1)
DEPTH:	33 km (restricted)

The earthquake on 1951 June 24 centred at the northern extremity of the Ruahine Range, western Hawke's Bay, was felt widely in the North Island south of a line through Kawhia and East Cape and in the northern part of the South Island, with isolated reports from Christchurch and Akaroa. The maximum intensity of MM6 was reported from Ohakune, Wairoa and Taihape. The earthquake was followed by aftershocks, the largest being of magnitude 5.0 M_L.

The felt reports held at the Seismological Observatory have been reinterpreted by Eiby and a new map was drawn.

* Location and local magnitude have been revised since the publication of the Seismological Observatory bulletin S 97. Refer to introduction.

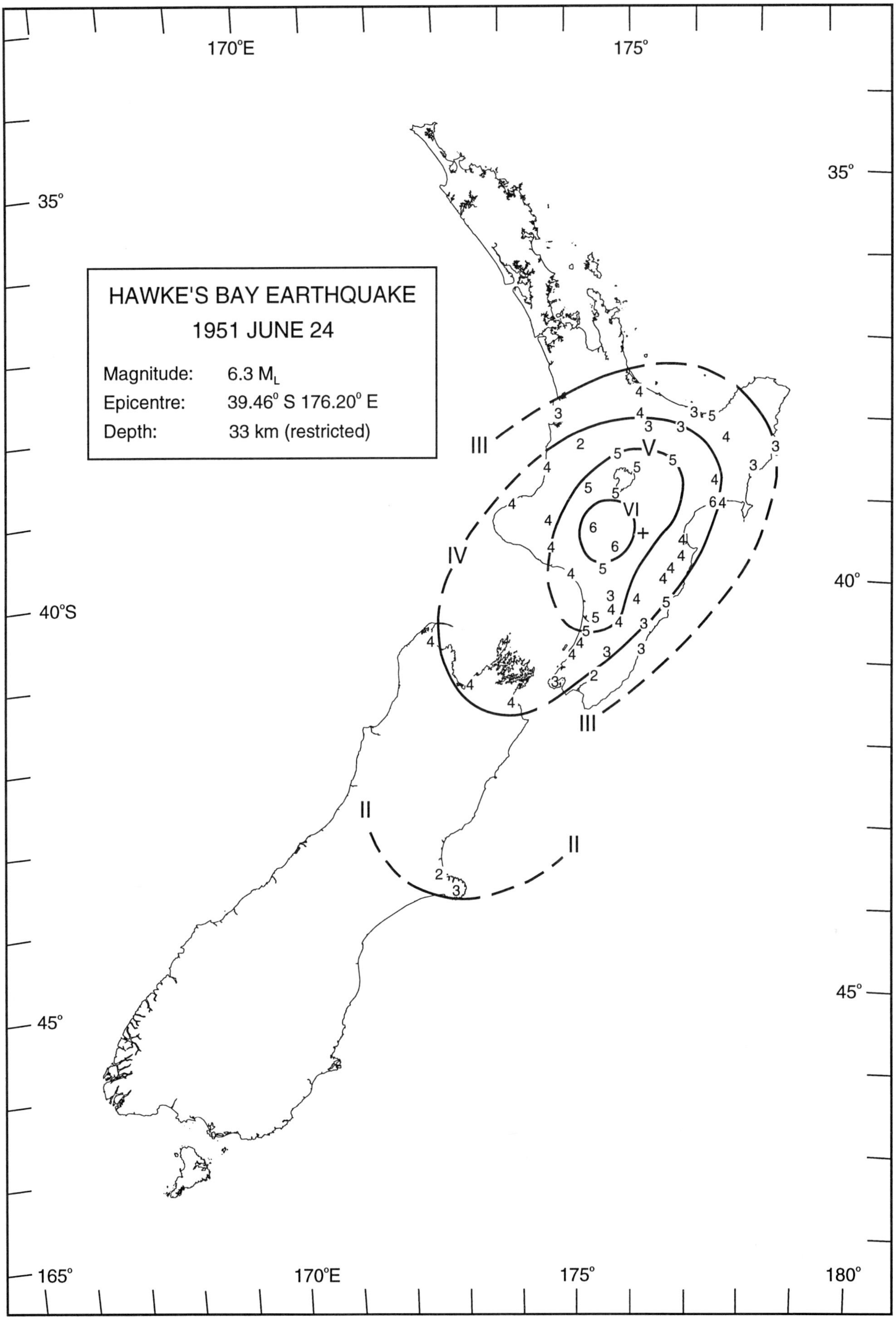

HAWKE'S BAY EARTHQUAKE
1951 JUNE 24
Magnitude: 6.3 M_L
Epicentre: 39.46° S 176.20° E
Depth: 33 km (restricted)
170°E
175°
35°
40°S
45°
165°
180°
40°
VI
V
IV
III
II

43. ISOSEISMAL MAP OF THE FIORDLAND EARTHQUAKE - 1951 JULY 07

DATE:	1951 JULY 07
TIME:	10:15:24 UT
MAGNITUDE:	5.6 M_L*
EPICENTRE:	44.87°S 167.35°E (1)
DEPTH:	33 km (restricted)

The earthquake on 1951 July 07 was felt widely in Otago and Southland. The maximum reported intensity was MM5 at Queenstown, some 100 km from the epicentre which was in an unpopulated mountainous area south of Milford Sound.

The felt reports held at the Seismological Observatory have been reinterpreted by Eiby and a new map was drawn.

---

* Location and local magnitude have been revised since the publication of the Seismological Observatory bulletin S 97. Refer to introduction.

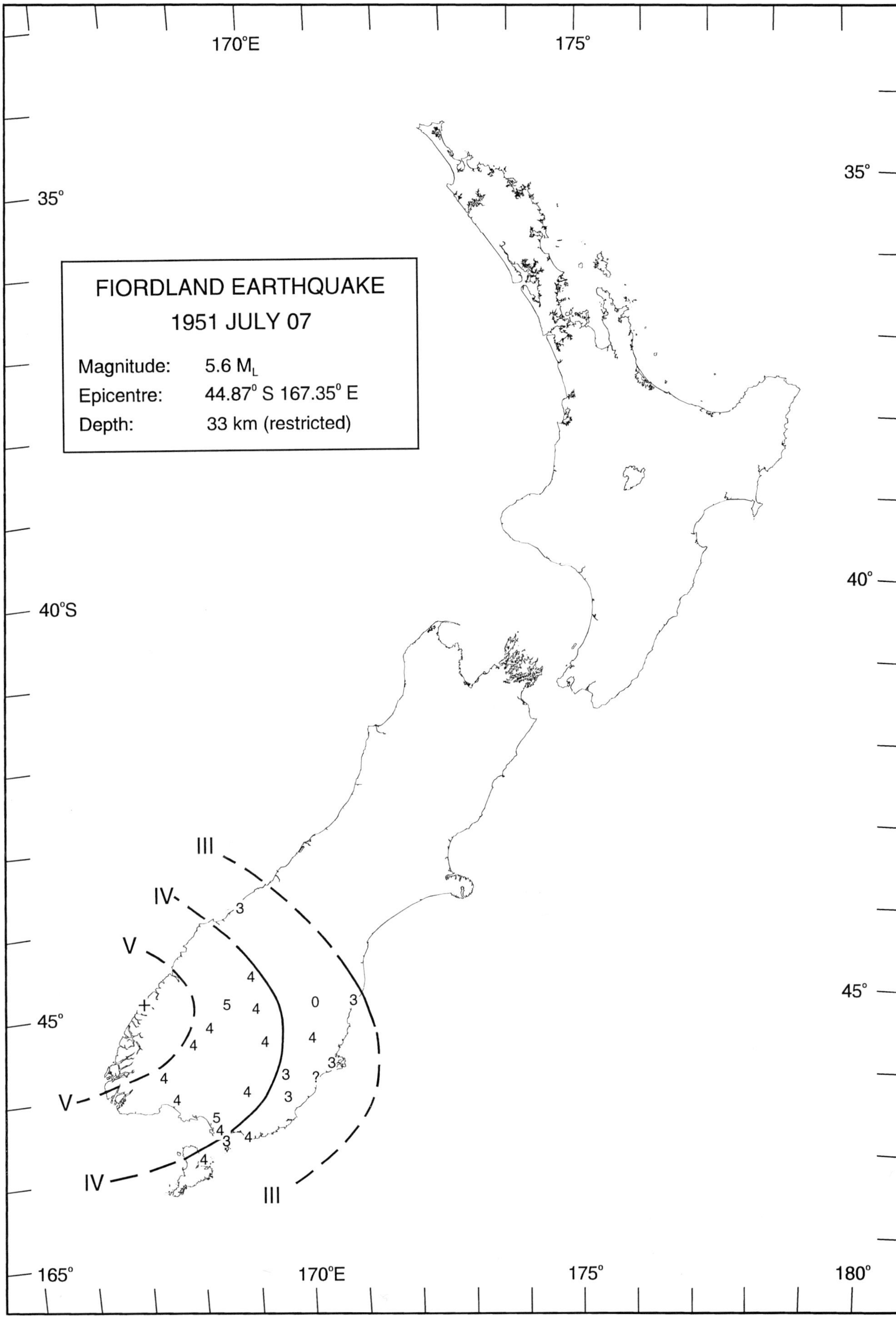

FIORDLAND EARTHQUAKE
1951 JULY 07
Magnitude: 5.6 M_L
Epicentre: 44.87° S 167.35° E
Depth: 33 km (restricted)
170°E
175°
35°
40°S
40°
45°
165°
180°
III
IV
V

44. ISOSEISMAL MAP OF THE NORTH WEST NELSON EARTHQUAKE - 1951 OCTOBER 03

DATE:	1951 OCTOBER 03
TIME:	17:38:48 UT
MAGNITUDE:	5.5 M_L*
EPICENTRE:	40.99°S 172.69°E (1)
DEPTH:	12 km (restricted)

'[The earthquake on 1951 October 03 was] felt in both islands of New Zealand, from Taumarunui to Hokitika; Max. MM V-VI.' (Seismological Observatory bulletin E 126.)

The felt reports held at the Seismological Observatory have been reinterpreted by Eiby and a new map was drawn.

* Location and local magnitude have been revised since the publication of the Seismological Observatory bulletin S 97. Refer to introduction.

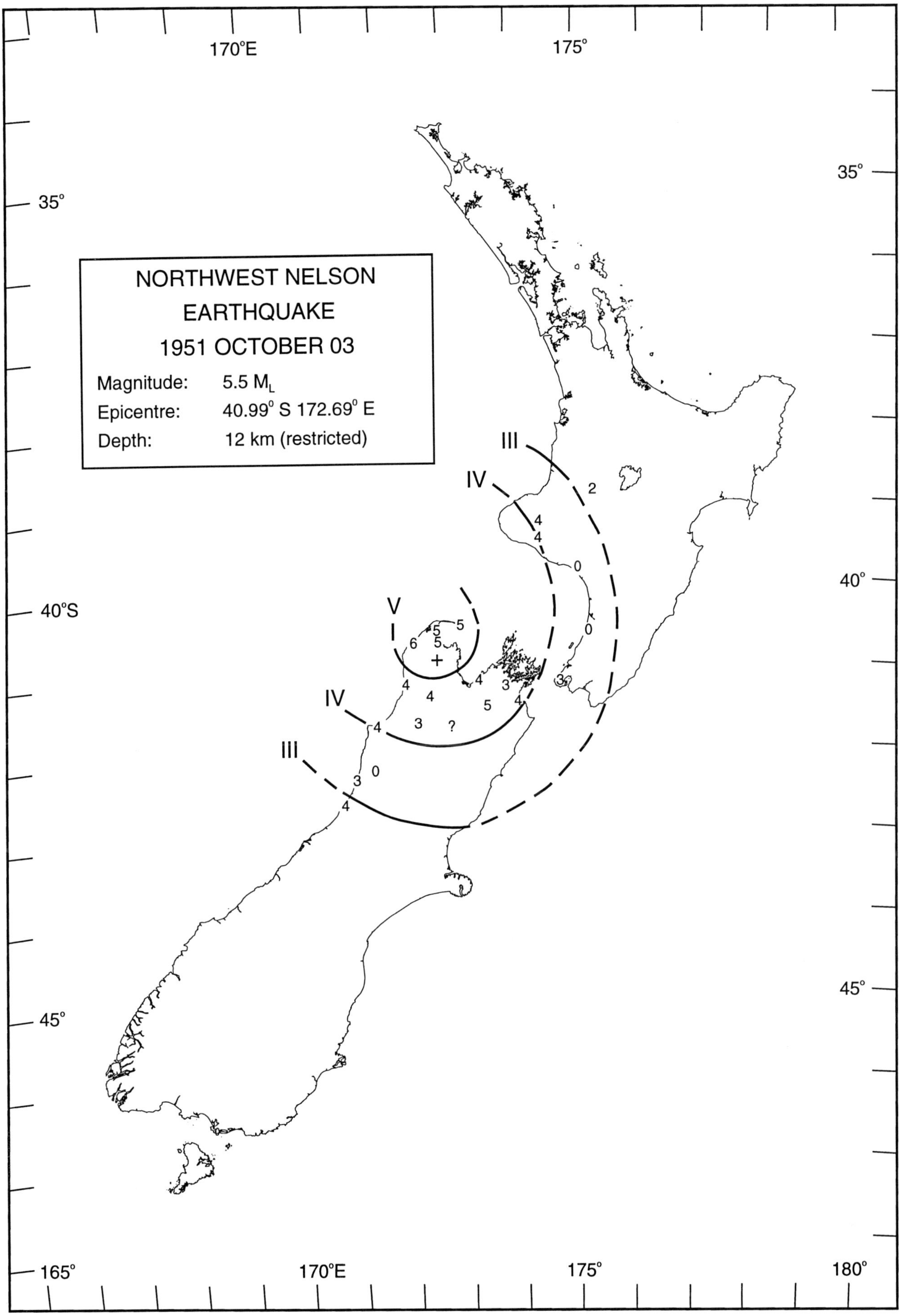

NORTHWEST NELSON
EARTHQUAKE
1951 OCTOBER 03
Magnitude: 5.5 M_L
Epicentre: 40.99° S 172.69° E
Depth: 12 km (restricted)
170°E
175°
35°
40°S
40°
45°
165°
180°
III
IV
V

45. ISOSEISMAL MAP OF THE SOUTHERN HAWKE'S BAY EARTHQUAKE - 1952 AUGUST 28

DATE:	1952 AUGUST 28
TIME:	10:40:15 UT
MAGNITUDE:	5.8 M_L*
EPICENTRE:	39.99°S 176.96°E (1)
DEPTH:	12 km (restricted)

'[The earthquake on 1952 August 28 was] felt extensively in Hawke's Bay and southern part of the North Island, Max MM VI in southern Hawke's Bay.' (Seismological Observatory bulletin E 129.)

The felt reports held at the Seismological Observatory have been reinterpreted by Eiby and a new map was drawn.

* Location and local magnitude have been revised since the publication of the Seismological Observatory bulletin S 98. Refer to introduction.

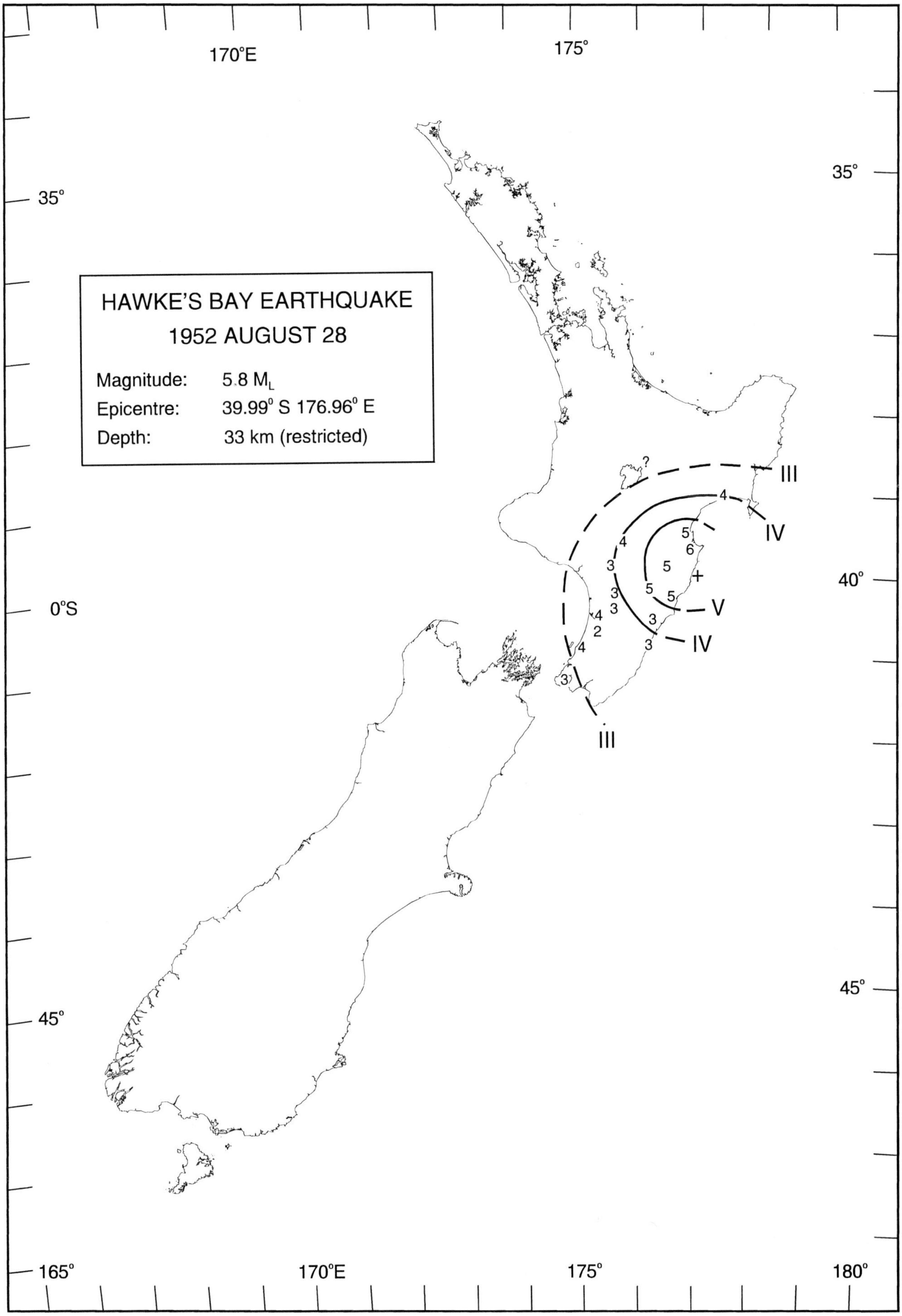

HAWKE'S BAY EARTHQUAKE
1952 AUGUST 28
Magnitude: 5.8 M_L
Epicentre: 39.99° S 176.96° E
Depth: 33 km (restricted)
170°E
175°
165°
180°
35°
40°
45°
0°S
III
IV
V

46. ISOSEISMAL MAP OF THE FOXTON EARTHQUAKE - 1953 APRIL 11

DATE:	1953 APRIL 11
TIME:	10:27:53 UT
MAGNITUDE:	5.4 M_L*
EPICENTRE:	40.45°S 174.78°E (1)
DEPTH:	106 km

The earthquake on 1953 April 11 was widely felt in Taranaki and along the southeast coast of the North Island and in the north of the South Island. The maximum intensity was MM5 at Otaki and Farewell Spit.

The felt reports held at the Seismological Observatory have been reinterpreted by Eiby and a new map was drawn.

* Location and local magnitude have been revised since the publication of the Seismological Observatory bulletin S 99. Refer to introduction.

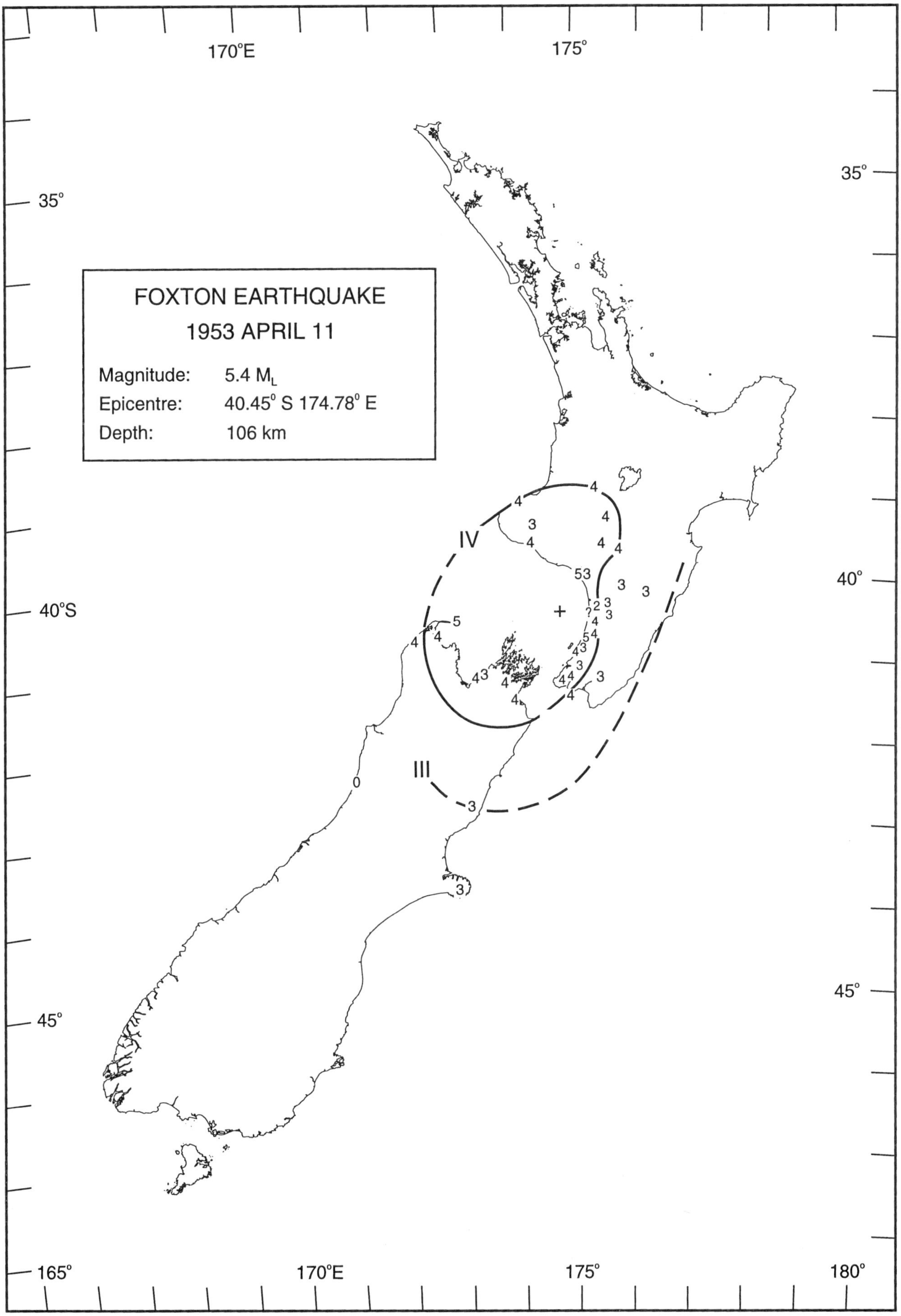

FOXTON EARTHQUAKE
1953 APRIL 11
Magnitude: 5.4 M_L
Epicentre: 40.45° S 174.78° E
Depth: 106 km
170°E
175°
35°
40°S
40°
45°
165°
180°
IV
III

47. ISOSEISMAL MAP OF THE TOKAANU EARTHQUAKE - 1953 JULY 04

DATE:	1953 JULY 04
TIME:	02:07:28 UT
MAGNITUDE:	5.5 M_L*
EPICENTRE:	38.86°S 175.68°E (1)
DEPTH:	12 km (restricted)

'A shock of shallow origin on July 4 reached intensity MM7 in the Tokaanu region, causing some minor damage. The perceptible area was comparatively small, being confined to the central and western parts of the North Island.' (Seismological Observatory bulletin S 99.)

The felt reports held at the Seismological Observatory have been reinterpreted by Eiby and a new map was drawn.

* Location and local magnitude have been revised since the publication of the Seismological Observatory bulletin S 99. Refer to introduction.

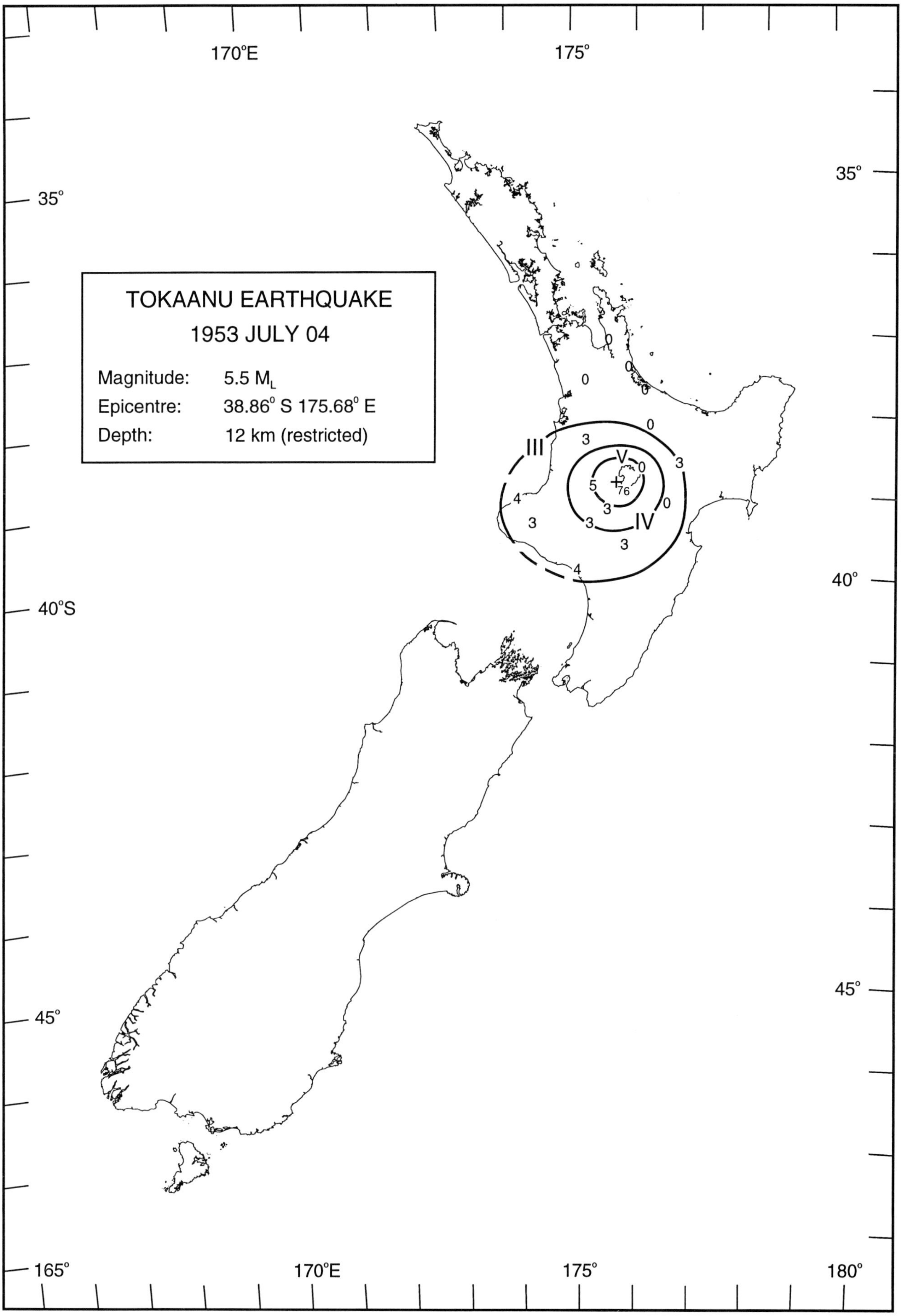
TOKAANU EARTHQUAKE
1953 JULY 04
Magnitude: 5.5 M_L
Epicentre: 38.86° S 175.68° E
Depth: 12 km (restricted)
170°E
175°
165°
180°
35°
40°S
40°
45°
III
IV
V

48. ISOSEISMAL MAP OF THE BAY OF PLENTY EARTHQUAKE - 1953 SEPTEMBER 29

DATE:	1953 SEPTEMBER 29
TIME:	01:36:51 UT
MAGNITUDE:	7.2 M_L*
EPICENTRE:	37.59°S 176.48°E (1)
DEPTH:	273 km

'On 29 September a strong shock originated in the Bay of Plenty ... The instrumental magnitude (7-7¼) was abnormally high for shocks at that depth, and consequently it was felt over a very large area with irregular distribution of intensity. The maximum felt intensity (MM7) occurred in the eastern Bay of Plenty and East Cape Peninsula, where considerable minor damage resulted.' (Seismological Observatory bulletin S 99.)

The felt reports held at the Seismological Observatory have been reinterpreted by Eiby and a new map was drawn.

* Location and local magnitude have been revised since the publication of the Seismological Observatory bulletin S 99. Refer to introduction.

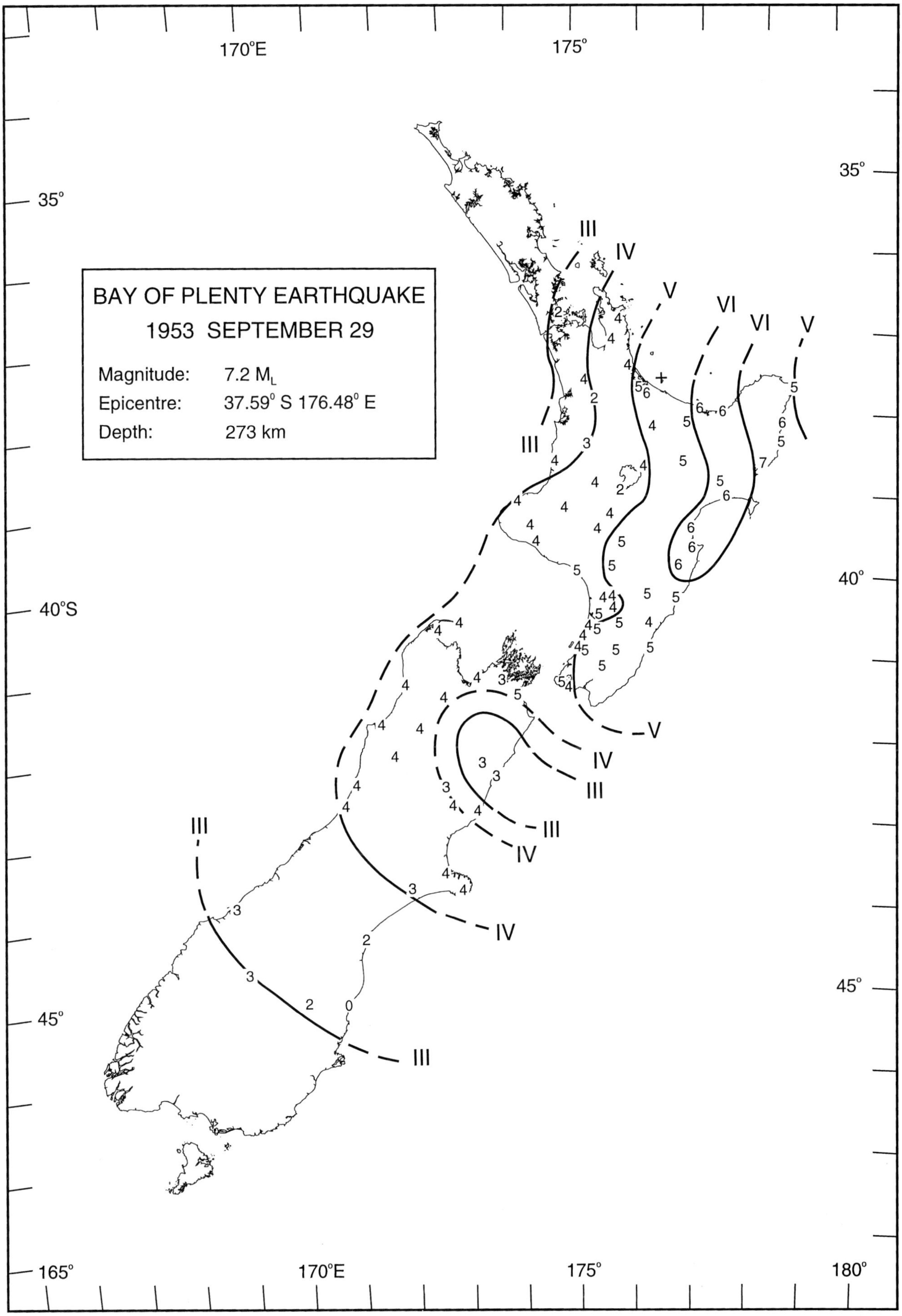
BAY OF PLENTY EARTHQUAKE
1953 SEPTEMBER 29
Magnitude: 7.2 M_L
Epicentre: 37.59° S 176.48° E
Depth: 273 km
170°E
175°
35°
40°S
45°
165°
180°
III
IV
V
VI

49. ISOSEISMAL MAP OF THE NORTH TARANAKI BIGHT EARTHQUAKE - 1953 OCTOBER 18

DATE:	1953 OCTOBER 18
TIME:	03:38:06 UT
MAGNITUDE:	5.3 M_L*
EPICENTRE:	38.64°S 174.03°E (1)
DEPTH:	12 km (restricted)

'On 18 October an earthquake originating north of Taranaki, with epicentre in lat. 38.6°S, long. 173.9°E [now revised*], affected a considerable area of the North Island from near Helensville to Wanganui. The maximum intensity reported was MM5 in the coastal regions from about Kawhia to New Plymouth.' (Seismological Observatory bulletin S 99.)

The felt reports held at the Seismological Observatory have been reinterpreted by Eiby and a new map was drawn.

* Location and local magnitude have been revised since the publication of the Seismological Observatory bulletin S 99. Refer to introduction.

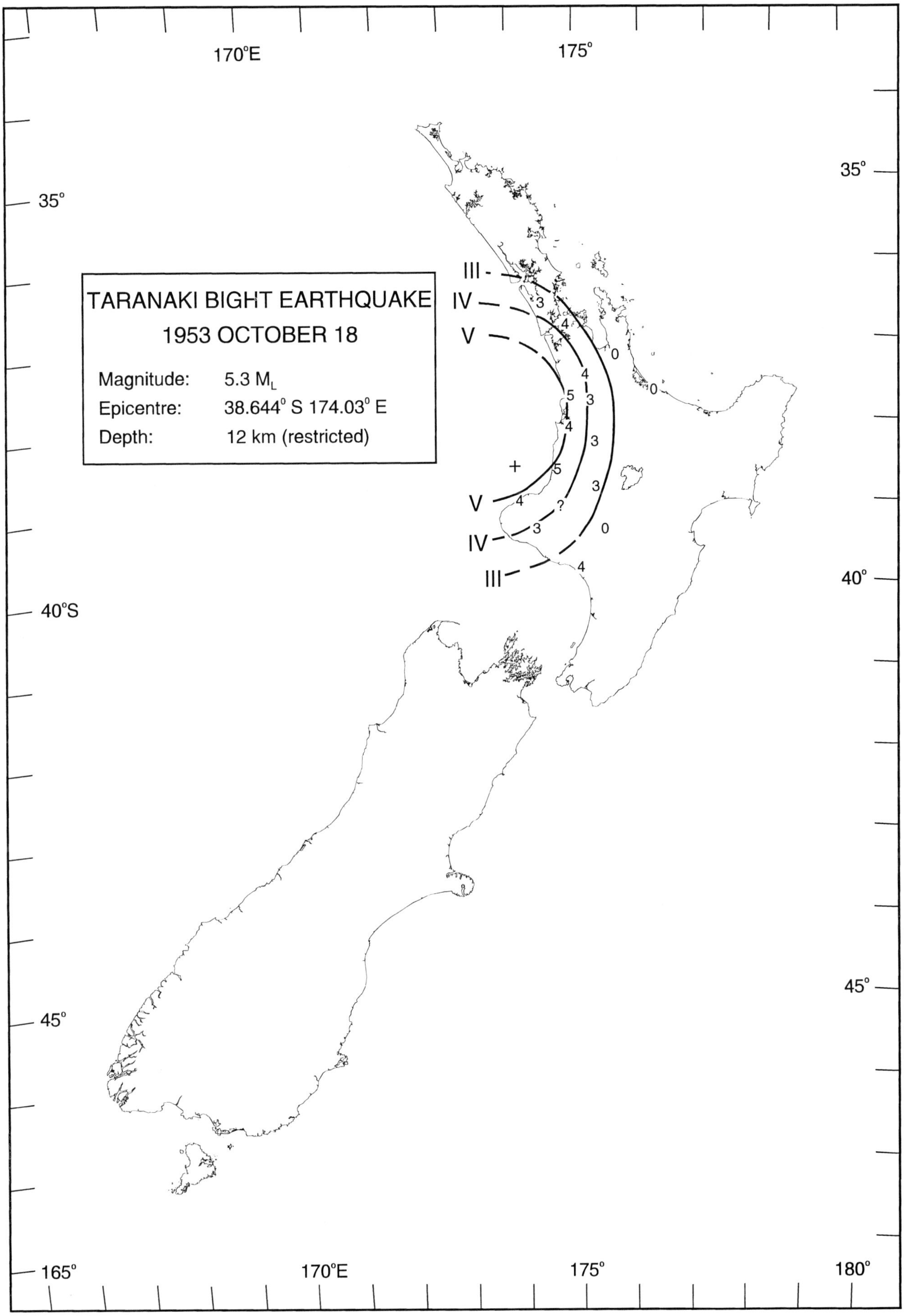

TARANAKI BIGHT EARTHQUAKE
1953 OCTOBER 18
Magnitude: 5.3 M_L
Epicentre: 38.644° S 174.03° E
Depth: 12 km (restricted)
170°E
175°
35°
40°S
45°
165°
180°
40°
III
IV
V

50. ISOSEISMAL MAP OF THE FIORDLAND EARTHQUAKE - 1954 MAY 04

DATE:	1954 MAY 04
TIME:	02:41:58 UT
MAGNITUDE:	5.4 M_L *
EPICENTRE:	44.92°S 167.60°E (1)
DEPTH:	12 km (restricted)

'Another shock [following that on 1954 February 18] was felt extensively in the southern part of the South Island on 4 May ... The magnitude was near 6, and the maximum felt intensity MM4-5.' (Seismological Observatory bulletin S 106.)

The felt reports held at the Seismological Observatory have been reinterpreted by Eiby and a new map was drawn.

---

* Location and local magnitude have been revised since the publication of the Seismological Observatory bulletin S 106. Refer to introduction.

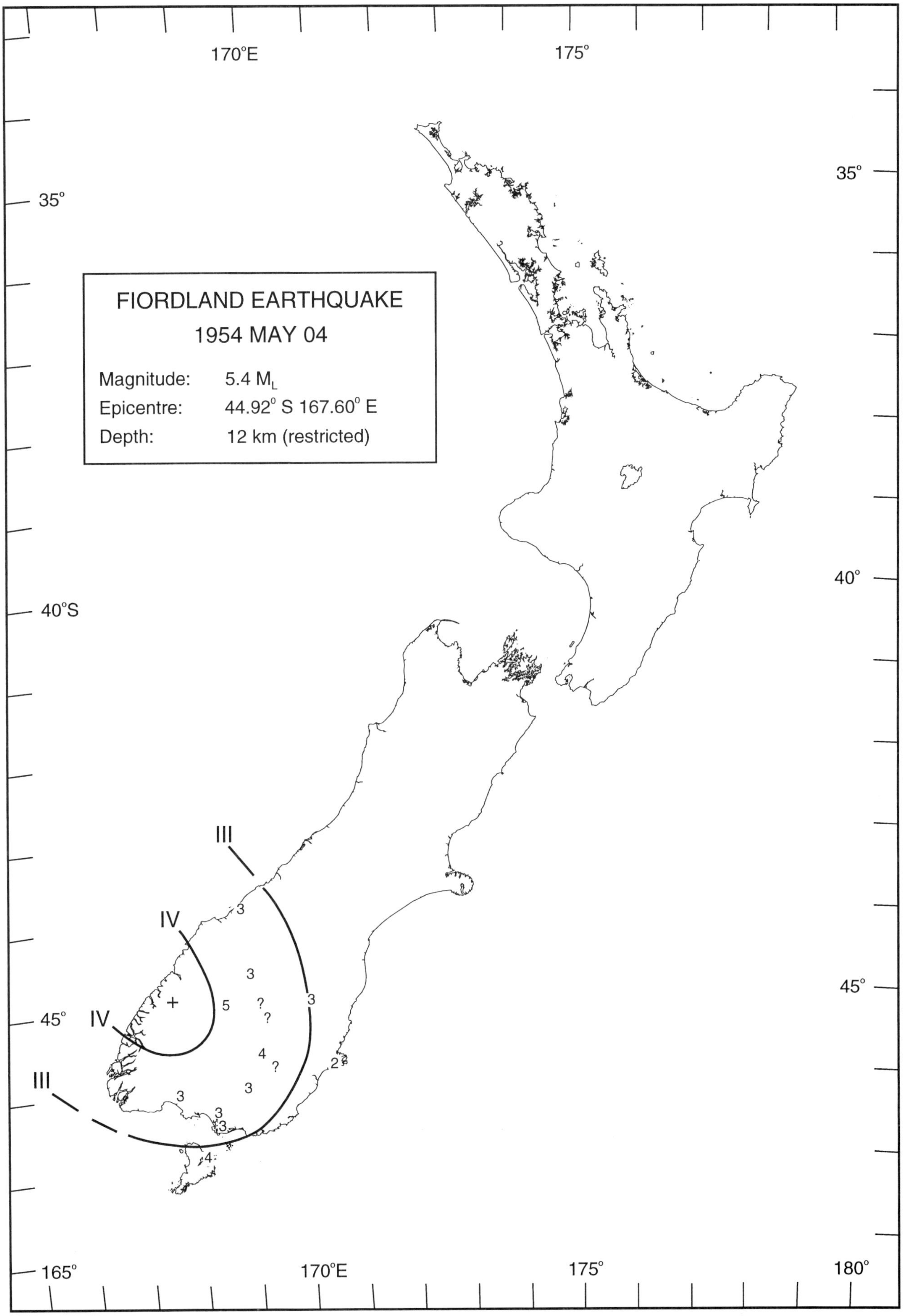

FIORDLAND EARTHQUAKE
1954 MAY 04
Magnitude: 5.4 M_L
Epicentre: 44.92° S 167.60° E
Depth: 12 km (restricted)
170°E
175°
35°
40°S
40°
45°
165°
180°
III
IV
3
5
?
4
2

51. ISOSEISMAL MAP OF THE WAIPURU EARTHQUAKE - 1955 FEBRUARY 08

DATE:	1955 FEBRUARY 08
TIME:	10:38:57 UT
MAGNITUDE:	5.1 M_L*
EPICENTRE:	39.90°S 175.80°E (1)
DEPTH:	33 km (restricted)

The earthquake, of magnitude 5.1 M_L, on 1955 February 8 was felt over a wide area in the Rangitikei and Manawatu districts from Chateau Tongariro to Foxton with a maximum intensity of MM4. It was felt as far away as Wellington and New Plymouth.

The felt reports held at the Seismological Observatory have been reinterpreted by Eiby and a new map was drawn.

* Location and local magnitude have been revised since the publication of the Seismological Observatory bulletin E 136. Refer to introduction.

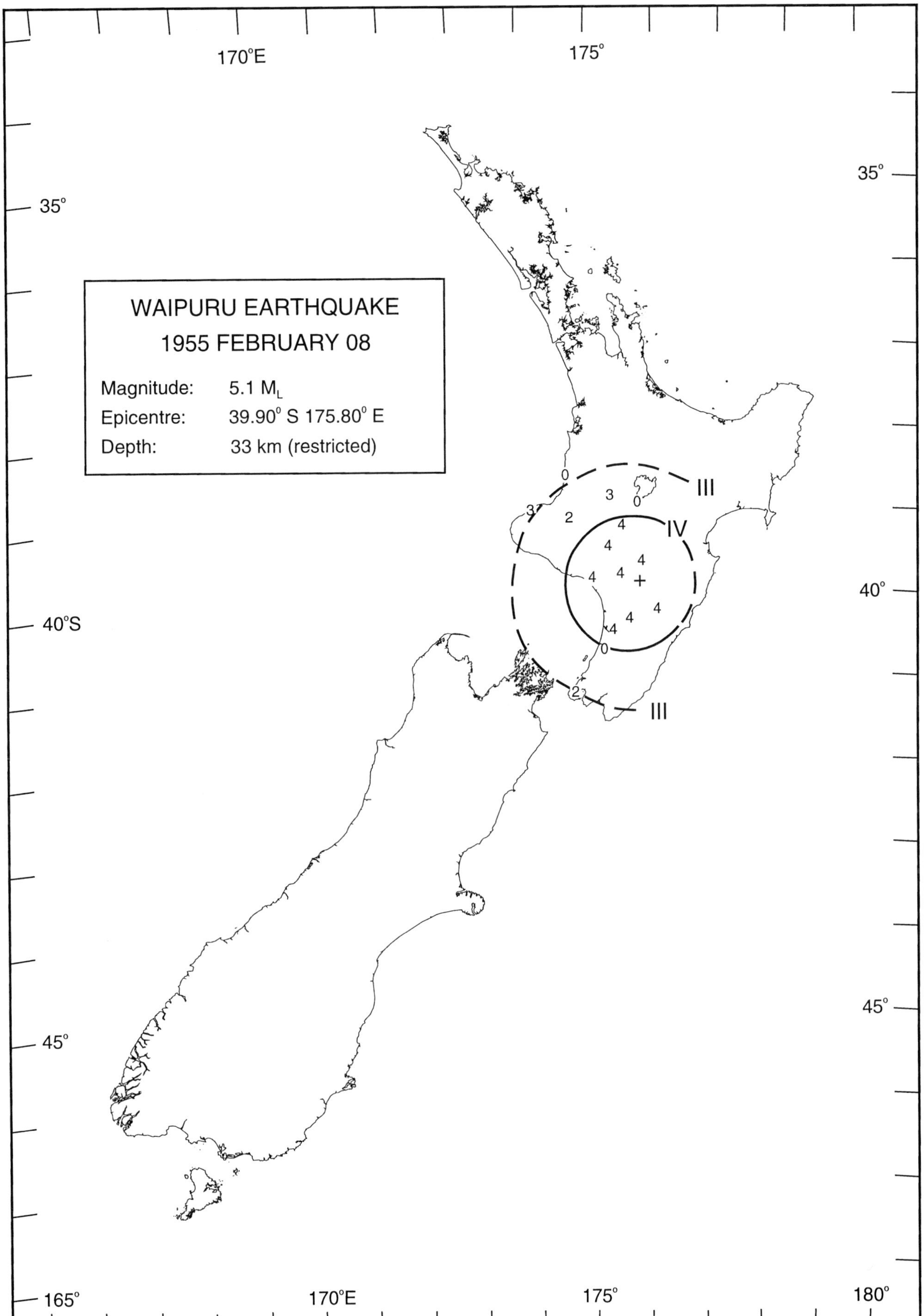

WAIPURU EARTHQUAKE
1955 FEBRUARY 08
Magnitude: 5.1 M_L
Epicentre: 39.90° S 175.80° E
Depth: 33 km (restricted)
170°E
175°
35°
40°S
40°
45°
165°
180°
III
IV

52. ISOSEISMAL MAP OF THE BAY OF PLENTY EARTHQUAKE - 1956 JANUARY 30

DATE:	1956 JANUARY 30
TIME:	08:43:02 UT
MAGNITUDE:	5.8 M_L^*; 6.4 M_s (Dowrick & Smith 1990)
EPICENTRE:	37.10°S 177.42°E (1)
DEPTH:	12 km (restricted)

'On 1956 January 30, a magnitude 6 [5.8 M_L^*] earthquake centred in the Bay of Plenty some 70 miles [110 km] to the north-east of Tauranga, was felt from Auckland city to the East Cape region.

'A special questionnaire was issued concerning the effects of this earthquake. Owing to the large number of answers received, only the maximum intensity for each locality is listed. The actual observations are shown on the isoseismal map appended to the bulletin. One figure in many cases represents a number of independent and accordant observations from the same settlement. It is rare for two such observations to differ by more than one degree of the MM intensity scale and the local irregularities revealed by the map are considered to represent real variations in intensity.' (Seismological Observatory bulletin E 137)

The isoseismal map is reproduced from the Seismological Observatory bulletin E 137.

* Location and local magnitude have been revised since the publication of the Seismological Observatory bulletin E 137.

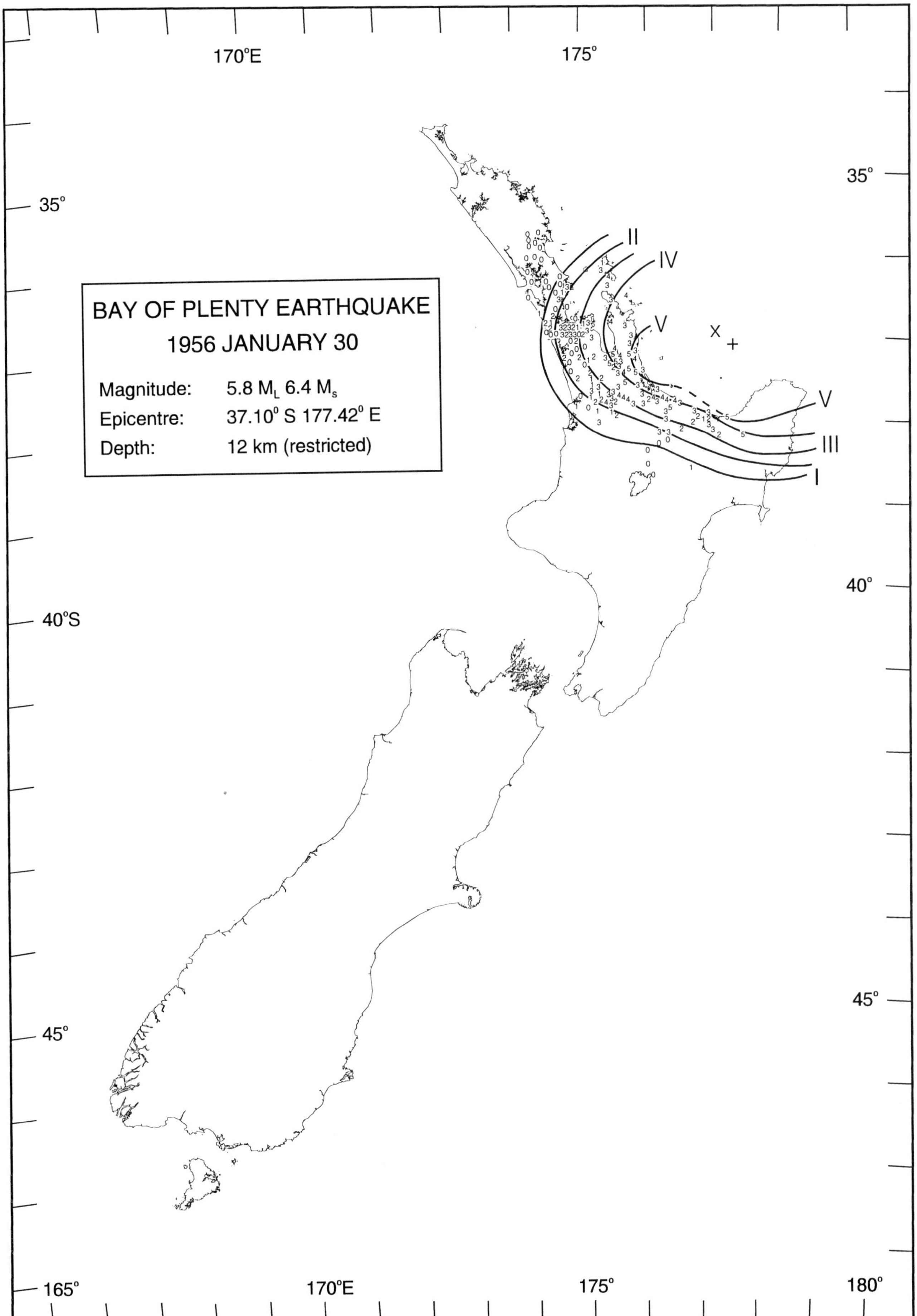
BAY OF PLENTY EARTHQUAKE
1956 JANUARY 30
Magnitude: 5.8 M_L 6.4 M_s
Epicentre: 37.10° S 177.42° E
Depth: 12 km (restricted)
170°E
175°
35°
40°S
45°
165°
180°
40°
II
IV
V
V
III
I

53. ISOSEISMAL MAP OF THE TOKAANU EARTHQUAKE - 1956 MARCH 02

DATE:	1956 MARCH 02
TIME:	22:43:51 UT
MAGNITUDE:	5.4 M_L*
EPICENTRE:	38.90°S 175.80°E (1)
DEPTH:	12 km (restricted)

'Slight damage to chimneys and stacked goods in shops occurred in Tokaanu on 1956 March 2 as the result of a shock of magnitude 5.3 centred some 10 miles [15 km] to the north-west of that settlement. Felt intensities reached MM6-7. Owing to the small focal depth, the shock was not felt at distances greater than about 60 miles [100 km] from the epicentre. Numerous aftershocks followed. Officers of the N. Z. Geological Survey who visited the area reported that gas was "boiling" from the bed of Lake Taupo near the village of Waihi. They attributed this to the liberation of methane which had been trapped in the sediment of the lake bed. A wave 3 ft [1 m] high was reported at the delta of the Tongariro River. No new displacements of any of the many known faults in the area were detected.' (Seismological Observatory bulletin E 137)

The isoseismal map is reproduced from the Seismological Observatory bulletin E 137.

* Location and local magnitude have been revised since the publication of the Seismological Observatory bulletin E 137.

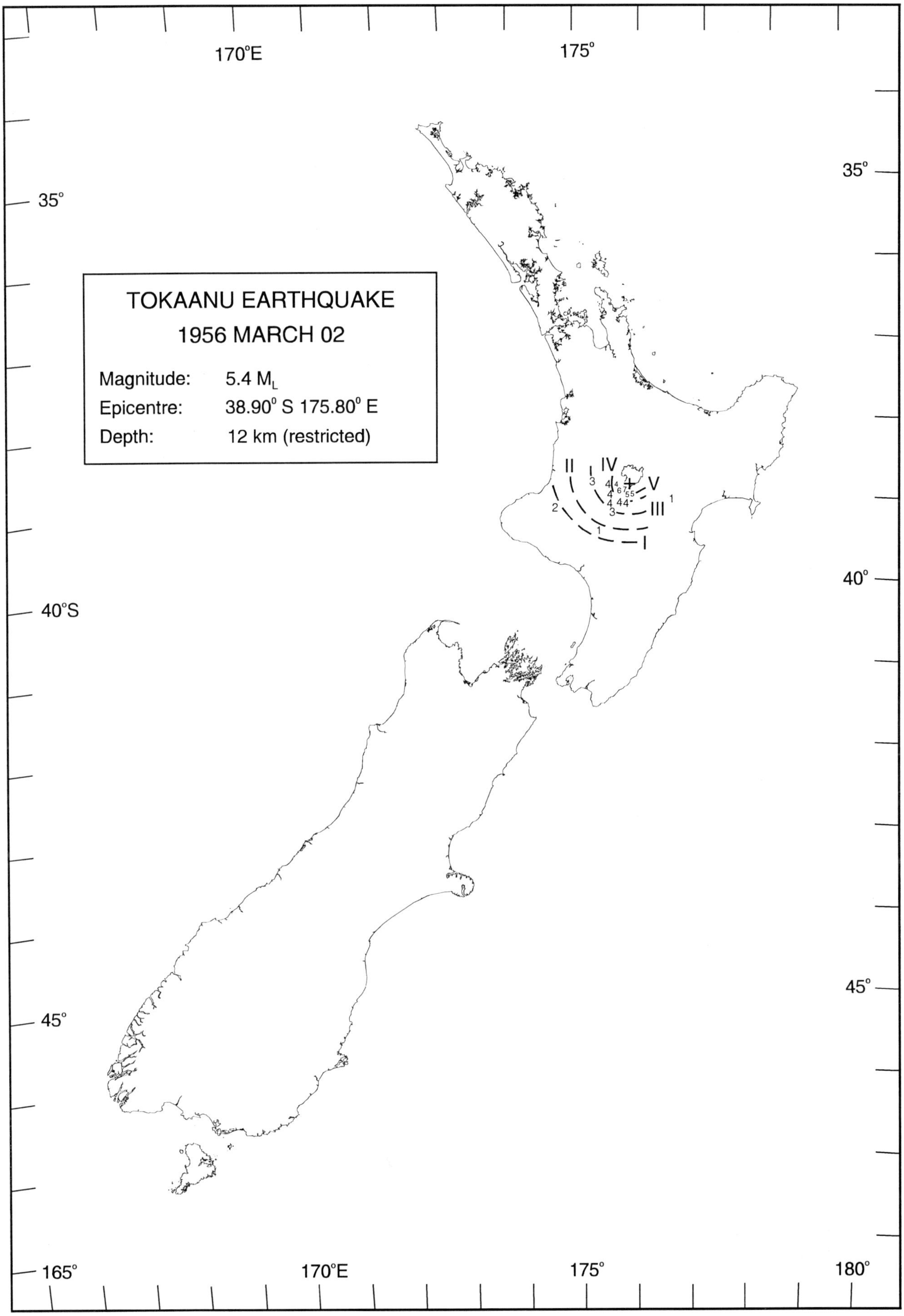
TOKAANU EARTHQUAKE
1956 MARCH 02
Magnitude: 5.4 M_L
Epicentre: 38.90° S 175.80° E
Depth: 12 km (restricted)
170°E
175°
35°
40°S
45°
165°
180°
40°
II
IV
V
III
I

54. ISOSEISMAL MAP OF THE EAST CAPE EARTHQUAKE - 1956 DECEMBER 28

DATE:	1956 DECEMBER 28	
TIME:	14:24:33 UT	14:24:23.9 UT
MAGNITUDE:	6.3 M_L^*	
EPICENTRE:	38.3°S 177.5°E C (ISS)	38.10°S 178.44°E (1)
DEPTH:	33 km (restricted)	12 km (restricted)

'The most severe earthquake [in 1956] took place on 1956 December 28. Its magnitude was 6.4 [6.3 M_L^*], and in the sparsely populated area between Opotiki and Tolaga Bay felt intensities of at least MM7 were reached, with some minor damage and landslides. The area of perceptibility extended from western Bay of Plenty to the northern Manawatu, and there were many aftershocks, several of magnitude 5 or more.' (Seismological Observatory bulletin E 137)

There is some difficulty assigning an epicentre for this event, the International Seismological Summary for 1956 (ISS) epicentre from teleseismic data differing markedly from the generally preferable, locally derived solution. However a better fit to the isoseismal data is obtained with the ISS epicentre. The ISS comments that "The consistently greater pP-P interval recorded by stations than should be found for a focus of this depth suggests a double shock. Accordingly an aftershock has been worked using readings which fit this hypothesis some of which have been attributed to the main shock." The ISS indicate a second shock 39 seconds after the main event. The presence of a second shock has not been recognised in the analysis of the New Zealand data and it is possible that the locally derived epicentre results from incorrectly labelled phases.

The isoseismal map is reproduced from the Seismological Observatory bulletin E 137. The preferred ISS epicentre is indicated.

---

* Location and local magnitude have been revised since the publication of the Seismological Observatory bulletin E 137.

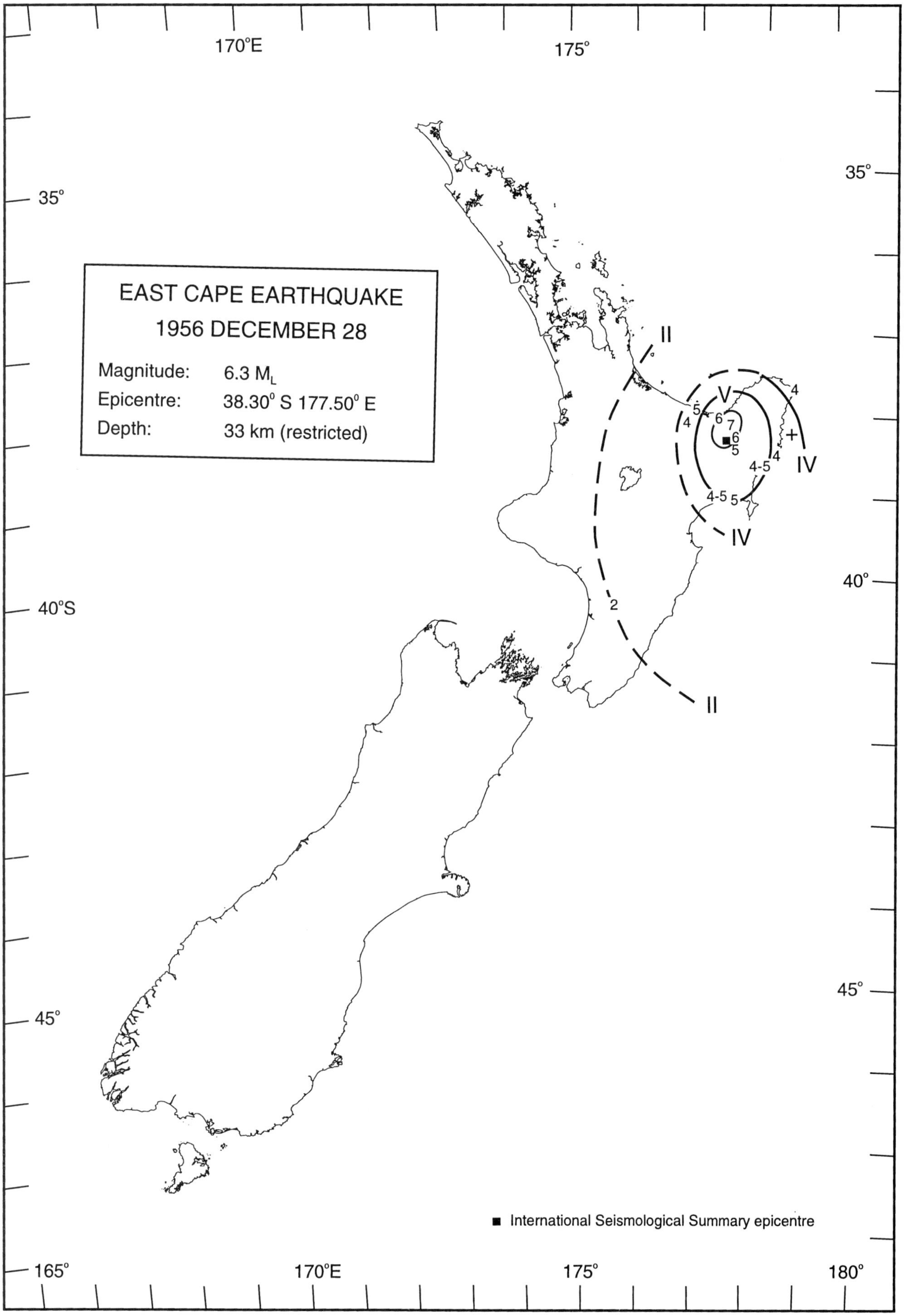

EAST CAPE EARTHQUAKE
1956 DECEMBER 28
Magnitude: 6.3 M_L
Epicentre: 38.30° S 177.50° E
Depth: 33 km (restricted)
170°E
175°
35°
40°S
45°
165°
180°
40°
II
V
IV
4-5
2
■ International Seismological Summary epicentre

55. ISOSEISMAL MAP OF THE SOUTHERN KERMEDEC RIDGE EARTHQUAKE - 1957 FEBRUARY 09

DATE:	1957 FEBRUARY 09
TIME:	13:29:20 UT
MAGNITUDE:	6.7 M_L*
EPICENTRE:	34.25°S 179.51 W (1)
DEPTH:	314 km

'The [earthquake] on Feb. 9 originated at sea, 275 miles [440 km] north of East Cape. It had a focal depth of about 85 miles [314 km*] and a magnitude of 6¾, and the area of perceptibility extended as far south as Greymouth. Intensities do not seem to have greatly exceeded MM4, and the earthquake does not appear to have been felt far west of a line from Foxton to Whakatane. Isolated minor damage covers most of the eastern North Island.' (Seismological Observatory bulletin E 138)

The isoseismal map is reproduced from the Seismological Observatory bulletin E 138 and has been included in this Atlas to indicate the intensities from an earthquake whose epicentre is well to the northeast of East Cape.

* Location and local magnitude have been revised since the publication of the Seismological Observatory bulletin E 138.

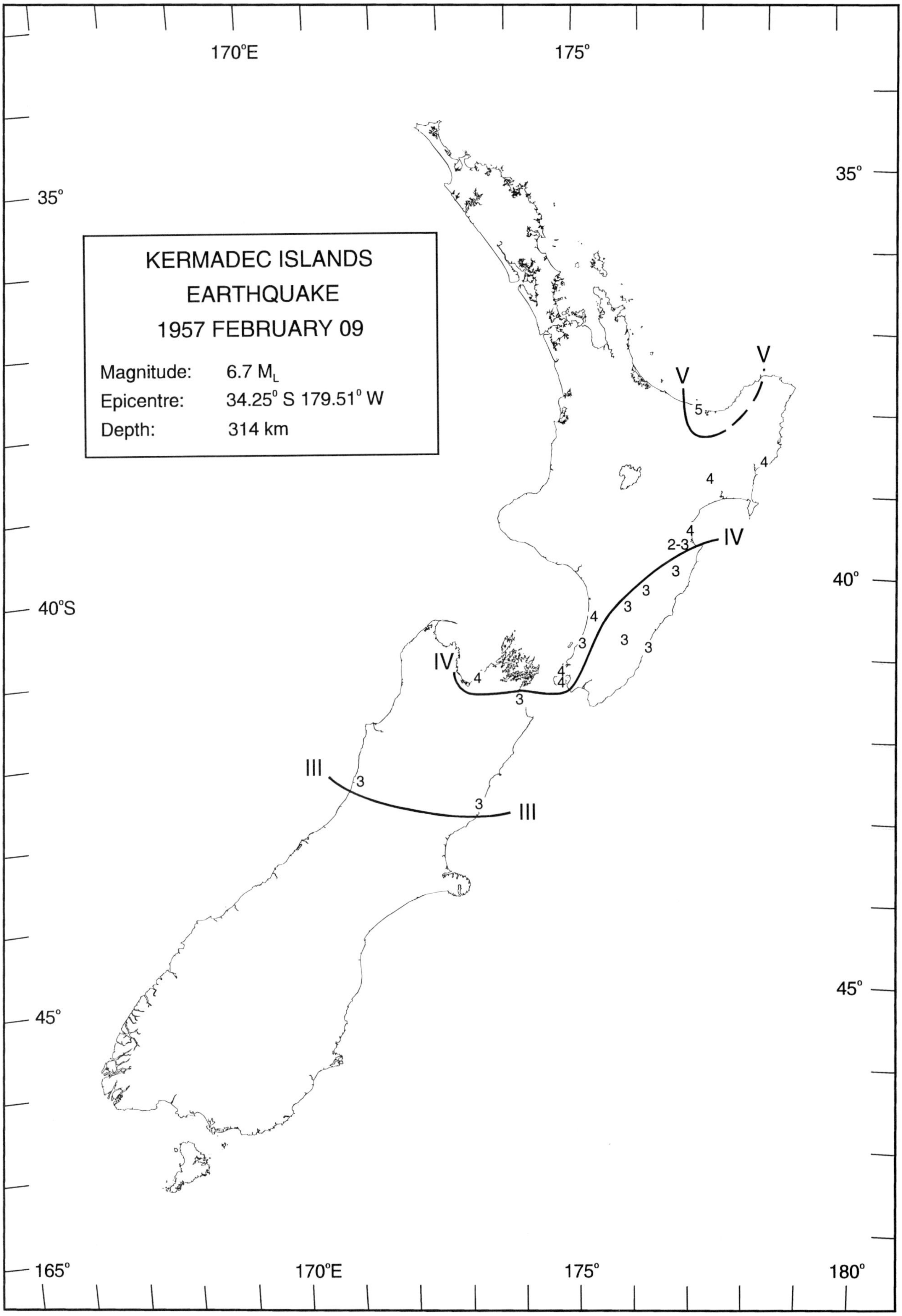
KERMADEC ISLANDS
EARTHQUAKE
1957 FEBRUARY 09
Magnitude: 6.7 M_L
Epicentre: 34.25° S 179.51° W
Depth: 314 km
170°E
175°
35°
35°
40°
40°S
45°
45°
165°
170°E
175°
180°
V
V
5
4
4
IV
4
2-3
3
3
3
4
3
3
3
IV
4
4
4
3
III
3
3
III

56. ISOSEISMAL MAP OF THE TONGARIRO EARTHQUAKE - 1957 FEBRUARY 22

DATE:	1957 FEBRUARY 22
TIME:	00:30:10.7 UT
MAGNITUDE:	5.6 M_L* ; 4.9 M_s (Dowrick & Smith 1990)
EPICENTRE:	39.19°S 175.14°E (1)
DEPTH:	5 km (restricted)

'[The earthquake on 1957 February 22] originated to the west of Tongariro National Park, and was felt over the central and southern parts of the North Island. It was of shallow focal depth, and had a magnitude of 5.5 [5.6 M_L*]. Considered in terms of damage to property, this was an important shock. Nearly 100 insurance claims were filed by property owners in the Ohakune-Raetihi district, and north to Taumarunui. In most cases, however, the effects were limited to cracked plaster, the fall of objects from shelves, and damage to sanitary fittings. A few instances of fallen chimneys and lesser chimney damage were also reported to the Earthquake and War Damage Commission.' (Seismological Observatory bulletin E 138)

The isoseismal map is reproduced from the Seismological Observatory bulletin E 138.

* Location and local magnitude have been revised since the publication of the Seismological Observatory bulletin E 138. Refer to introduction.

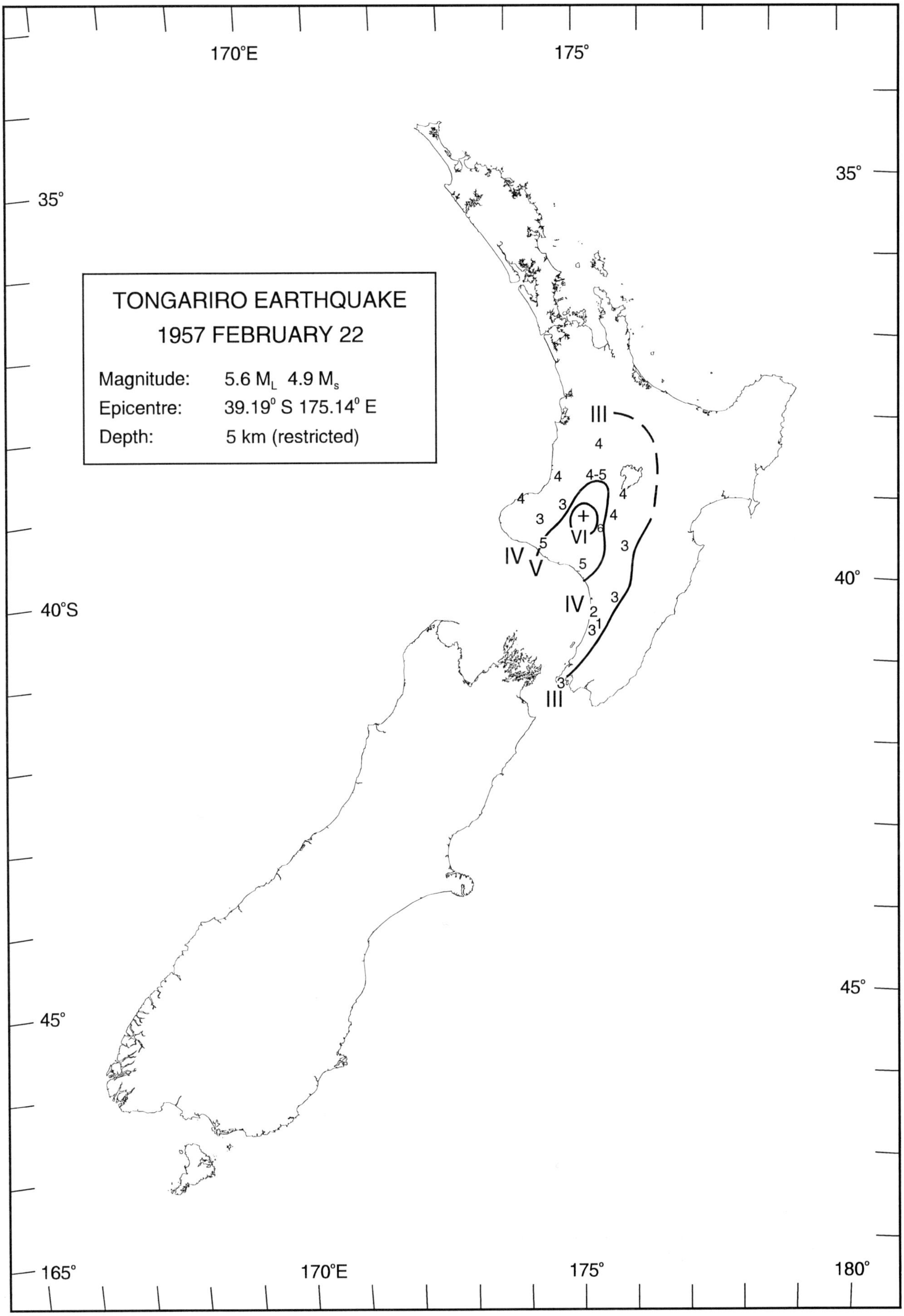

TONGARIRO EARTHQUAKE
1957 FEBRUARY 22
Magnitude: 5.6 M_L 4.9 M_s
Epicentre: 39.19° S 175.14° E
Depth: 5 km (restricted)
170°E
175°
35°
40°S
40°
45°
165°
180°
III
IV
V
VI

57. ISOSEISMAL MAP OF THE LAKE TAUPO EARTHQUAKE - 1957 MARCH 13

DATE:	1957 MARCH 13
TIME:	09:11:30 UT
MAGNITUDE:	6.0 M_L*
EPICENTRE:	38.68°S 175.33°E (1)
DEPTH:	273 km

'As might be expected from its deep origin (170 miles [270 km]) and its magnitude of 6.5 [6.0 M_L*], the shock of March 13 was felt over most of the country between the Bay of Plenty and Greymouth. The centre has been located a few miles north-west of Lake Taupo. As [the map] indicates, intensities were fairly uniform over a large area, and although no serious damage occurred, minor insurance claims were received from both Auckland and Christchurch.' (Seismological Observatory bulletin E 138)

The revised magnitude of 6.0 M_L must be considered as a minimum as it is derived from WEL data (Wood Anderson) only and the maximum trace amplitude could not be read. The isoseismal map is reproduced from the Seismological Observatory bulletin E 138. Whether there is justification for drawing isoseismals on the distribution of the given data is debatable.

* Location and local magnitude have been revised since the publication of the Seismological Observatory bulletin E 138. Refer to introduction.

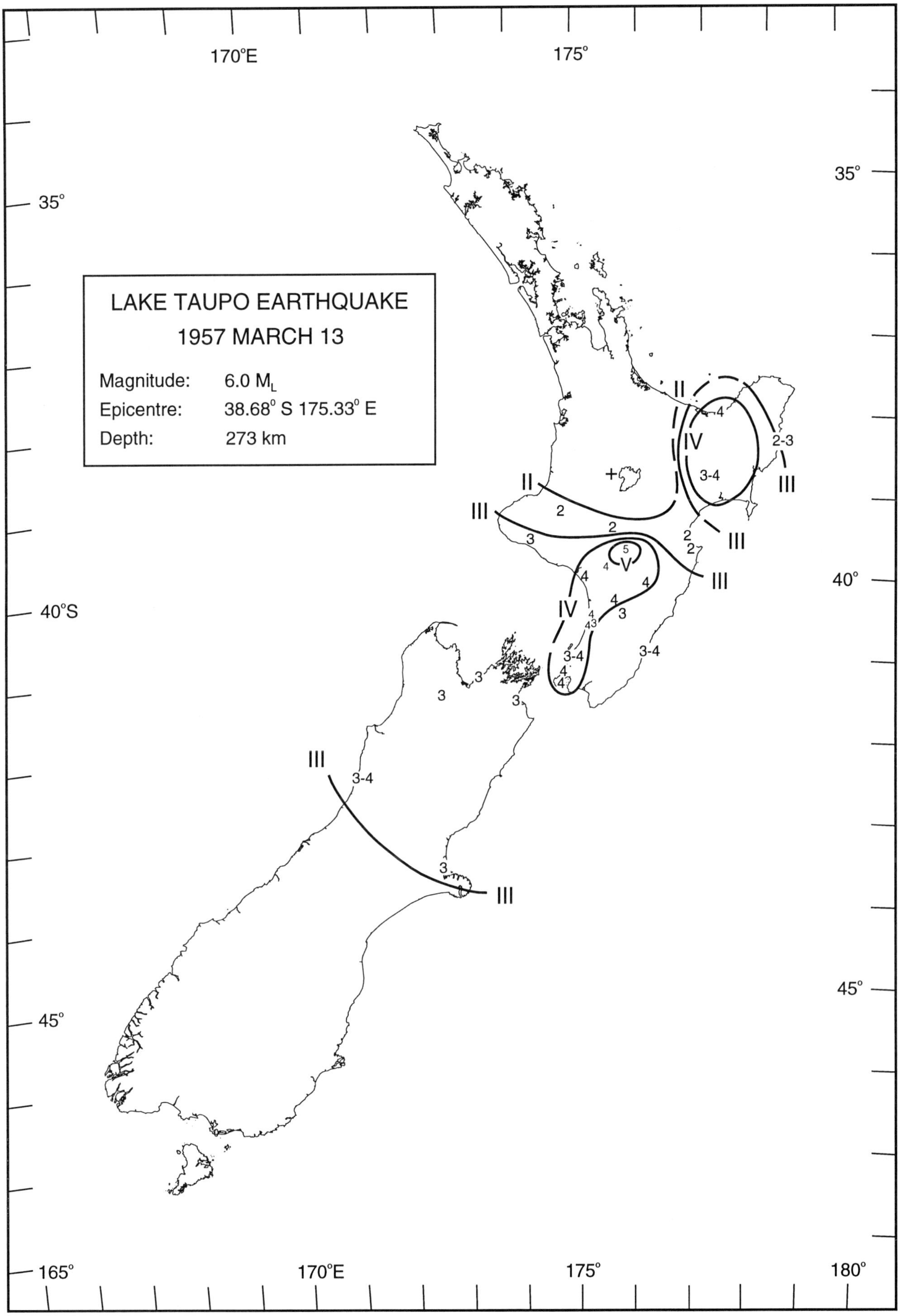
LAKE TAUPO EARTHQUAKE
1957 MARCH 13
Magnitude: 6.0 M_L
Epicentre: 38.68° S 175.33° E
Depth: 273 km
170°E
175°
35°
40°S
45°
165°
180°
40°
II
III
IV
V
2-3
3-4

58. ISOSEISMAL MAP OF THE CASTLEPOINT EARTHQUAKE - 1957 AUGUST 21

DATE:	1957 AUGUST 21
TIME:	05:48:02 UT
MAGNITUDE:	5.6 M_L*
EPICENTRE:	41.03°S 176.00°E (1)
DEPTH:	33 km (restricted)

'The epicentre of the shock on Aug. 21 lies in thinly populated country west of Castlepoint. The shock was of shallow origin, and its magnitude was 5.6. The felt area extended as far north as Te Kuiti, but there were no reports from across Cook Strait. A chimney at Masterton was damaged, ceiling plaster cracked in Wellington, and goods displaced from shelves in Levin, but these were isolated occurrences.' (Seismological Observatory bulletin E 138)

The felt information appearing in the Seismological Observatory bulletin E 138 has been plotted and the isoseismal map drawn by Eiby.

* Location and local magnitude have been revised since the publication of the Seismological Observatory bulletin E 138. Refer to introduction.

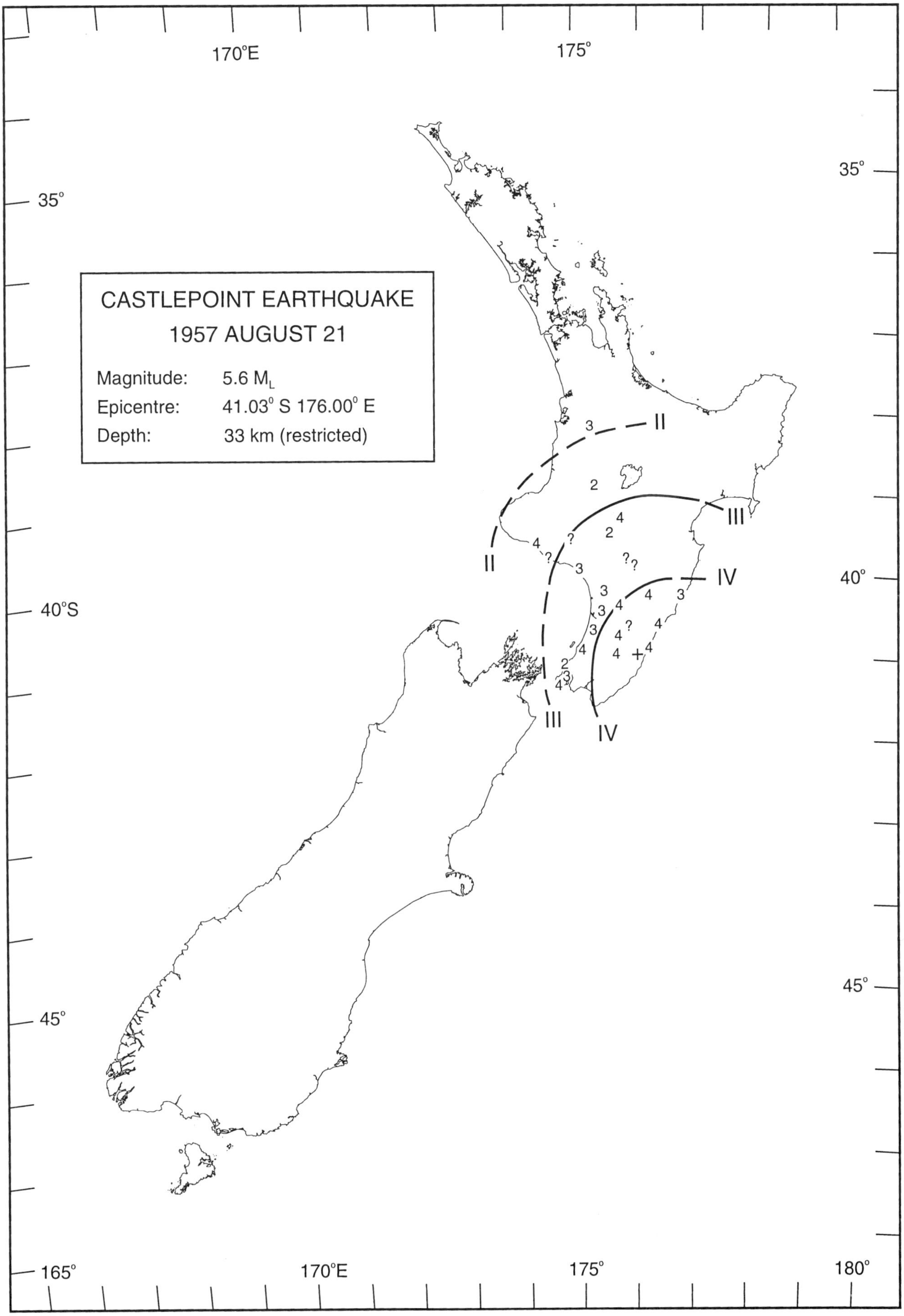
CASTLEPOINT EARTHQUAKE
1957 AUGUST 21
Magnitude: 5.6 M_L
Epicentre: 41.03° S 176.00° E
Depth: 33 km (restricted)
170°E
175°
165°
180°
35°
40°S
40°
45°
II
III
IV

59. ISOSEISMAL MAP OF THE SOUTH TARANAKI BIGHT EARTHQUAKE - 1957 SEPTEMBER 26

DATE:	1957 SEPTEMBER 26
TIME:	12:03:06 UT
MAGNITUDE:	5.7 M_L*
EPICENTRE:	40.21°S 173.80°E (1)
DEPTH:	145 km

'The shock on Sept. 26, with an epicentre in the South Taranaki Bight, had a magnitude of 6.0 [5.7 M_L*] and a depth of 70 miles [145 km*]. Damage was again minor, and seems to have been confined to the North Island coast, between New Plymouth and Wellington; but the felt area extended from Te Kuiti to Banks Peninsula.' (Seismological Observatory bulletin E 138)

The revised magnitude of 5.7 M_L must be considered as a minimum as it is derived from WEL data (Wood Anderson seismograph) only and the maximum trace amplitude could not be read.

The map has been reproduced from the Seismological Observatory bulletin E 138. However the isoseismals are poorly constrained.

* Location and local magnitude have been revised since the publication of the Seismological Observatory bulletin E 138. Refer to introduction.

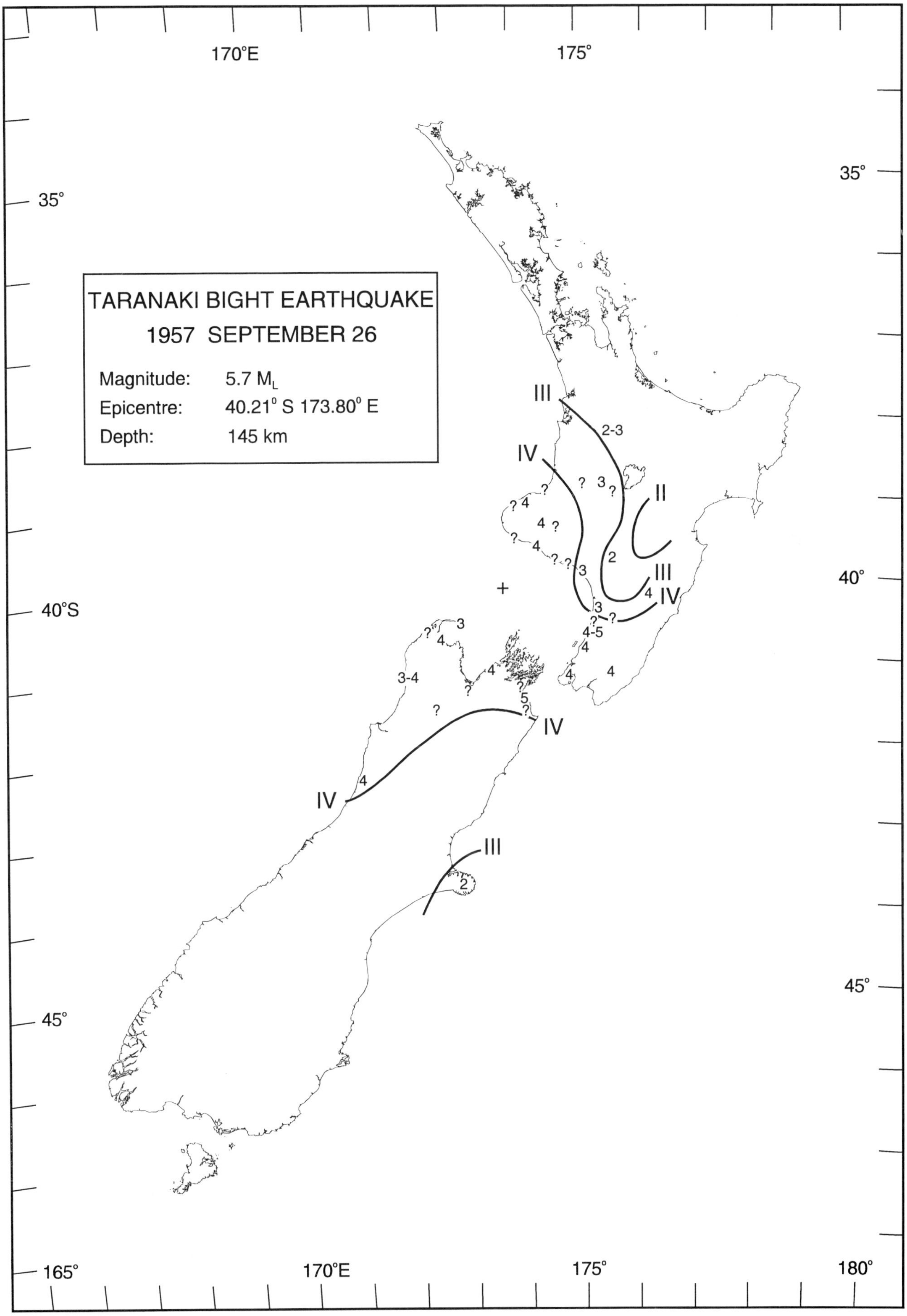
TARANAKI BIGHT EARTHQUAKE
1957 SEPTEMBER 26
Magnitude: 5.7 M_L
Epicentre: 40.21° S 173.80° E
Depth: 145 km
170°E
175°
35°
40°S
45°
165°
180°
40°
II
III
IV
2-3
3-4
4-5
2
3
4
5
?

60a. ISOSEISMAL MAP OF THE ASHLEY CLINTON EARTHQUAKE - 1958 JANUARY 31

DATE:	1958 JANUARY 31
TIME:	06:32:42 UT
MAGNITUDE:	6.1 M_L^* ; 5.2 M_s (Dowrick & Smith 1990)
EPICENTRE:	39.95°S 176.53°E (1)
DEPTH:	33 km (restricted)

'The shallow shock of January 31 had a magnitude of 5.9 [6.1 M_L*], and was centred to the east of the Ruahine Range, near Ashley Clinton, some 25 miles [40 km] north of Dannevirke [*]. Although in no sense a major earthquake, it resulted in a troublesome amount of minor damage. Officers of the Seismological Observatory and the Geological Survey visited the area, but no evidence of geological movements was found. Between Dannevirke and Raumati, some four or five miles [6 to 8 km] to the east, there is an area of typical 'slump' or 'landslide' topography which did not appear to have been affected.

'At the time of the visit, no instrumental epicentre was available, and the party did not travel as far north as Ashley Clinton, where it was later found that a large proportion of the chimneys was damaged, and that plaster had cracked or fallen. When the number of chimneys affected is related to the size of the settlement, and the nature of the other damage considered, it seems clear that the intensity in Ashley Clinton was a little above that in Dannevirke, and should probably be described as MM7.

'An interesting feature of this earthquake is the absence of identifiable aftershocks; although a shock of magnitude 4.3 on February 28, with an epicentre some 20 miles [30 km] to the south-east and one with a magnitude of 3.9 40 miles [65 km] to the east, and five hours later than the main shock on January 31 might be regarded as associated events.' (Seismological Observatory bulletin E 139)

The map has been reproduced from the Seismological Observatory bulletin E 139 and it is based on felt reports received by the Seismological Observatory. A second more detailed map of the intensities experienced in southern Hawke's Bay, based on a more extensive data set also appeared in the Seismological Observatory bulletin E 139. It is reproduced on the following pages. There are differences between the two maps.

---

* Location and local magnitude have been revised since the publication of the Seismological Observatory bulletin E 139. Refer to introduction. The revised epicentre, 30 km due east of the earlier graphically determined epicentre, possibly renders the inner isoseismal lines inappropriate. The observed intensities, however, are not necessarily at variance with the revised location. A reappraisal of the data is probably required.

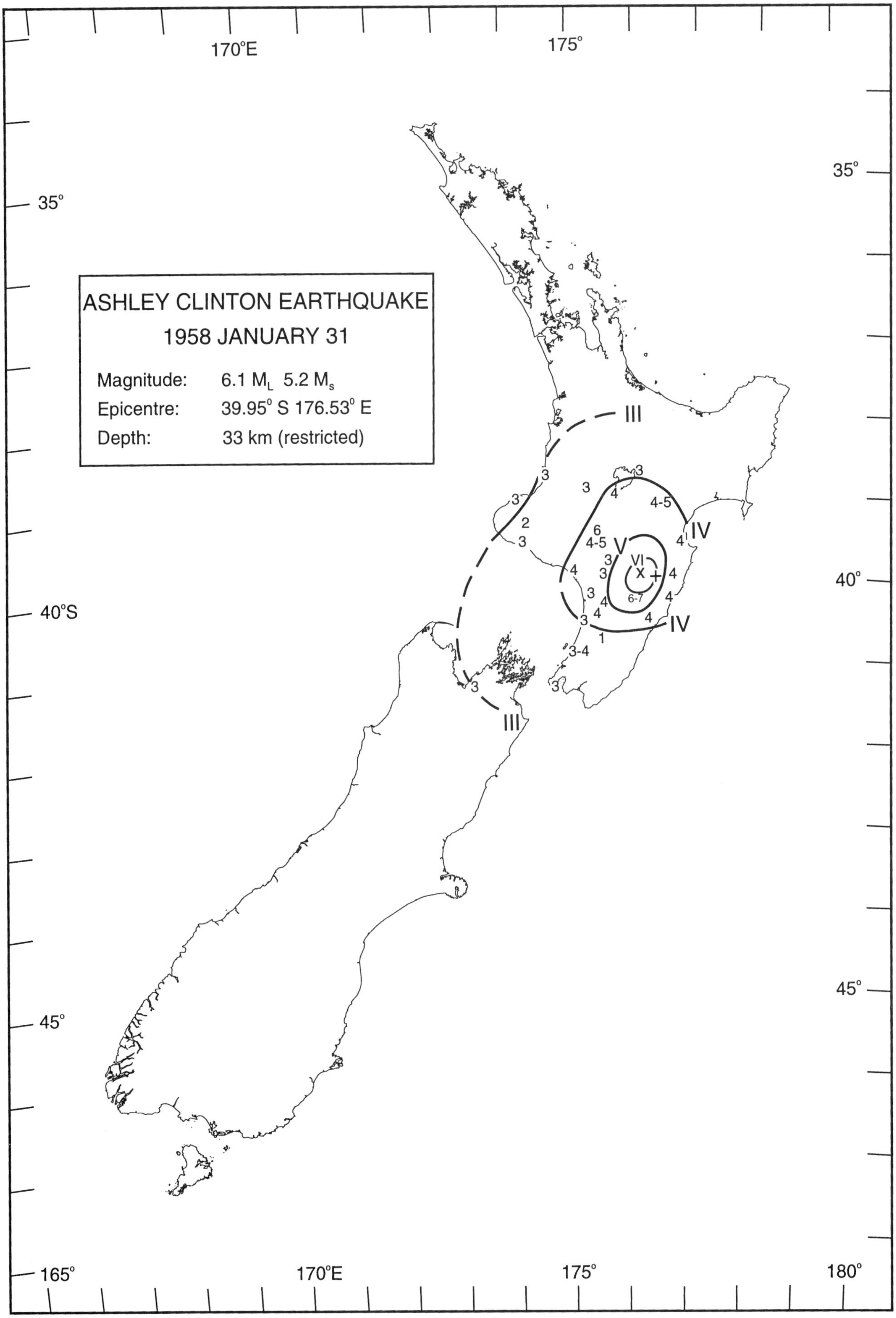

ASHLEY CLINTON EARTHQUAKE
1958 JANUARY 31
Magnitude: 6.1 M_L 5.2 M_s
Epicentre: 39.95° S 176.53° E
Depth: 33 km (restricted)
170°E
175°
35°
40°S
45°
165°
180°
40°
III
IV
V
VI
6-7
4-5
3-4

60b. ISOSEISMAL MAP OF THE ASHLEY CLINTON EARTHQUAKE - 1958 JANUARY 31

DATE:	1958 JANUARY 31
TIME:	06:32:42 UT
MAGNITUDE:	6.1 M_L*; 5.2 M_s (Dowrick & Smith 1990)
EPICENTRE:	39.95°S 176.53°E (1)
DEPTH:	33 km (restricted)

'The nature of the damage in Dannevirke itself suggested an epicentre in the immediate vicinity. Cracks in the walls and other structural damage affected the Post Office, the Courthouse, the Bank of New Zealand, and the Regent Cinema, all of which lie within a radius of a few hundred feet [100 metres]. Along the main street, the dis-arrangement of goods in the shops became progressively less in both directions. The same applied to the cracking of plate glass windows, and the glazier confirmed that he had received no calls beyond the town area. There is no obvious difference in foundation conditions over the area.

'Outside the business area, there was a region of cracked chimneys (all poorly constructed) extending east and south-east to parts of Tiratu, about 3 miles [5 km] from the Post Office. In the north-westerly direction, damage of a similar kind did not extend more than half a mile [300 m]. A large number of defective chimneys, which could easily have been brought down, were found to be still standing, allowing an outer limit to be assigned to the region of maximum intensity. It was originally considered that the epicentre must lie within the region, and the nature of the cracks in the damaged buildings and the direction of overthrow of objects was consistent with that view. It now seems clear that there was an area of reduced intensity between Dannevirke and Norsewood, the real centre of damage being close to Ashley Clinton, in closer accord with the instrumental epicentre.

'The large-scale map shown [opposite], which has been based on the field observations and upon the reports of the assessors to the Earthquake and War Damage Commission, indicates a slightly larger region of intensity MM5 and above than do the felt reports sent to the Observatory, which form the basis of [the map on the previous page]. The differences are largely accounted for by the fact that minor damage outside the home of the observers would not come to their notice until after their reports had been made. In all, about 170 insurance claims were lodged, mostly from southern Hawke's Bay, although isolated cases of damage to structures in poor condition occurred in Napier, Wanganui, and even Wellington.' (Seismological Observatory bulletin E 139)

The map has been reproduced from the Seismological Observatory bulletin E 139, the origin of the data being given above.

---

* Location and local magnitude have been revised since the publication of the Seismological Observatory bulletin E 139. Refer to introduction. The revised epicentre, 30 km due east of the earlier graphically determined epicentre, possibly renders the inner isoseismal lines inappropriate. The observed intensities, however, are not necessarily at variance with the revised location. A reappraisal of the data is probably required.

ASHLEY CLINTON EARTHQUAKE
1958 JANUARY 31

Magnitude: 6.1 M_L 5.2 M_s
Epicentre: 39.95° S 176.53° E
Depth: 33 km (restricted)

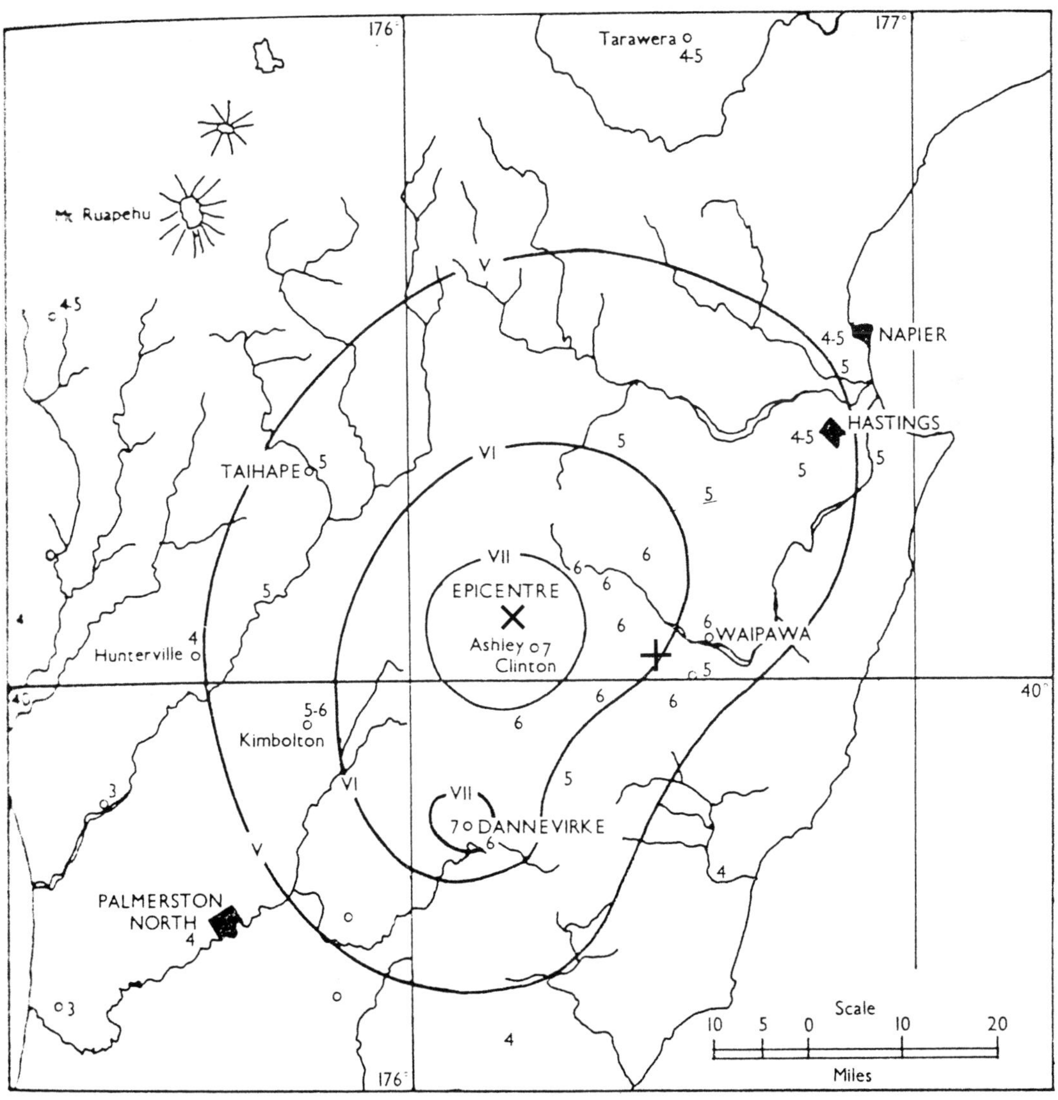

61. ISOSEISMAL MAP OF THE BAY OF PLENTY EARTHQUAKE - 1958 DECEMBER 10

DATE:	1958 DECEMBER 10
TIME:	07:03:03 UT
MAGNITUDE:	6.3 M_L*
EPICENTRE:	37.13°S 176.95°E (1)
DEPTH:	284 km

'An even more extensively felt shock [than that on 1958 January 31] was that on December 10, at a depth of 200 miles [284 km*] under the Bay of Plenty. This shock, of magnitude 6¾ [6.3 M_L*], was felt with intensities ranging from MM3 to MM6 over the whole of the area between Coromandel Peninsula, East Cape, and Banks Peninsula, with an isolated report from as far south as Queenstown. The intensities produced by a shock at so great a depth would be expected to vary only slowly with epicentral distance. The rather irregular form of the isoseismals must in consequence be largely attributed to variations in ground conditions.

'Approximately 100 insurance claims resulted, none of them serious, although isolated chimneys suffered as far south as Blenheim and Christchurch. Apart from these, there was only a small amount of cracked plaster, and the dis-arrangement of displayed crockery and goods in shops.' (Seismological Observatory bulletin E 139)

The map has been reproduced from the Seismological Observatory bulletin E 139.

* Location and local magnitude have been revised since the publication of the Seismological Observatory bulletin E 139. Refer to introduction.

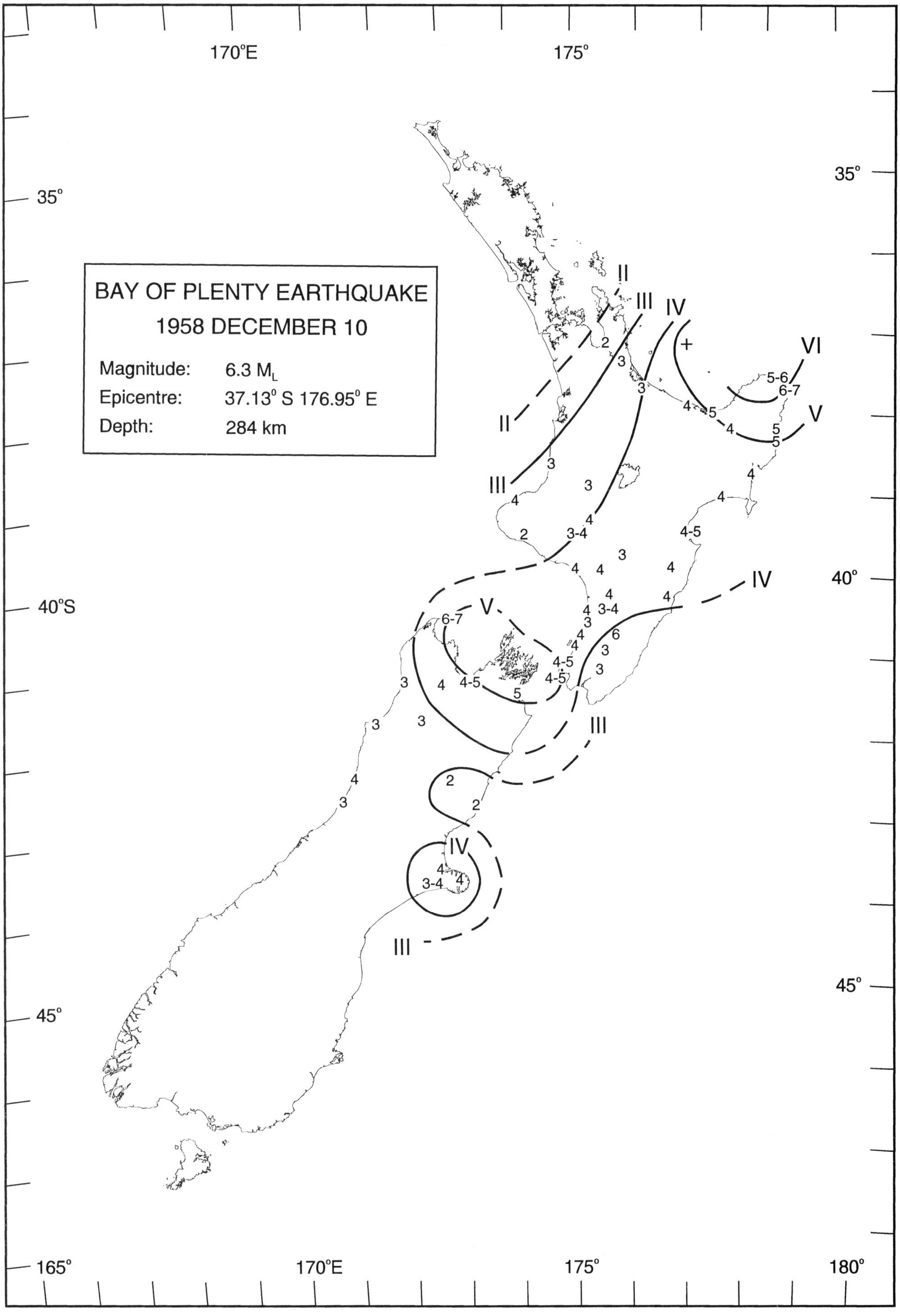

BAY OF PLENTY EARTHQUAKE
1958 DECEMBER 10
Magnitude: 6.3 M_L
Epicentre: 37.13° S 176.95° E
Depth: 284 km
170°E
175°
35°
40°S
40°
45°
165°
180°
II
III
IV
V
VI

62. ISOSEISMAL MAP OF THE MARLBOROUGH SOUNDS EARTHQUAKE - 1959 MAY 22

DATE:	1959 MAY 22
TIME:	06:57:09 UT
MAGNITUDE:	5.7 M_L* ; 4.9 M_s (Dowrick & Smith 1990)
EPICENTRE:	41.22°S 174.58°E (1)
DEPTH:	86 km

'The largest shallow shock [in 1959], on May 22 had an instrumental magnitude of 6.0 [5.7 M_L*], and an epicentre in the Marlborough Sounds region. The felt area extended from Taumarunui to Banks Peninsula ... It was also reported to have been felt aboard fishing vessels in Cook Strait. In terms of property damage, this was probably the most severe shock since the Wairarapa earthquakes of 1942. Picton suffered most severely, but structural damage was confined to ageing buildings dating from the first decade of the century. Chimneys, plaster and lavatory pans in Blenheim and Wellington were also affected. In all some 460 insurance claims were lodged with the Earthquake and War Damage Commission.' (Seismological Observatory bulletin E 140)

The isoseismal map has been reproduced from the Seismological Observatory bulletin E 140.

* Location and local magnitude have been revised since the publication of the Seismological Observatory bulletin E 140. Refer to introduction.

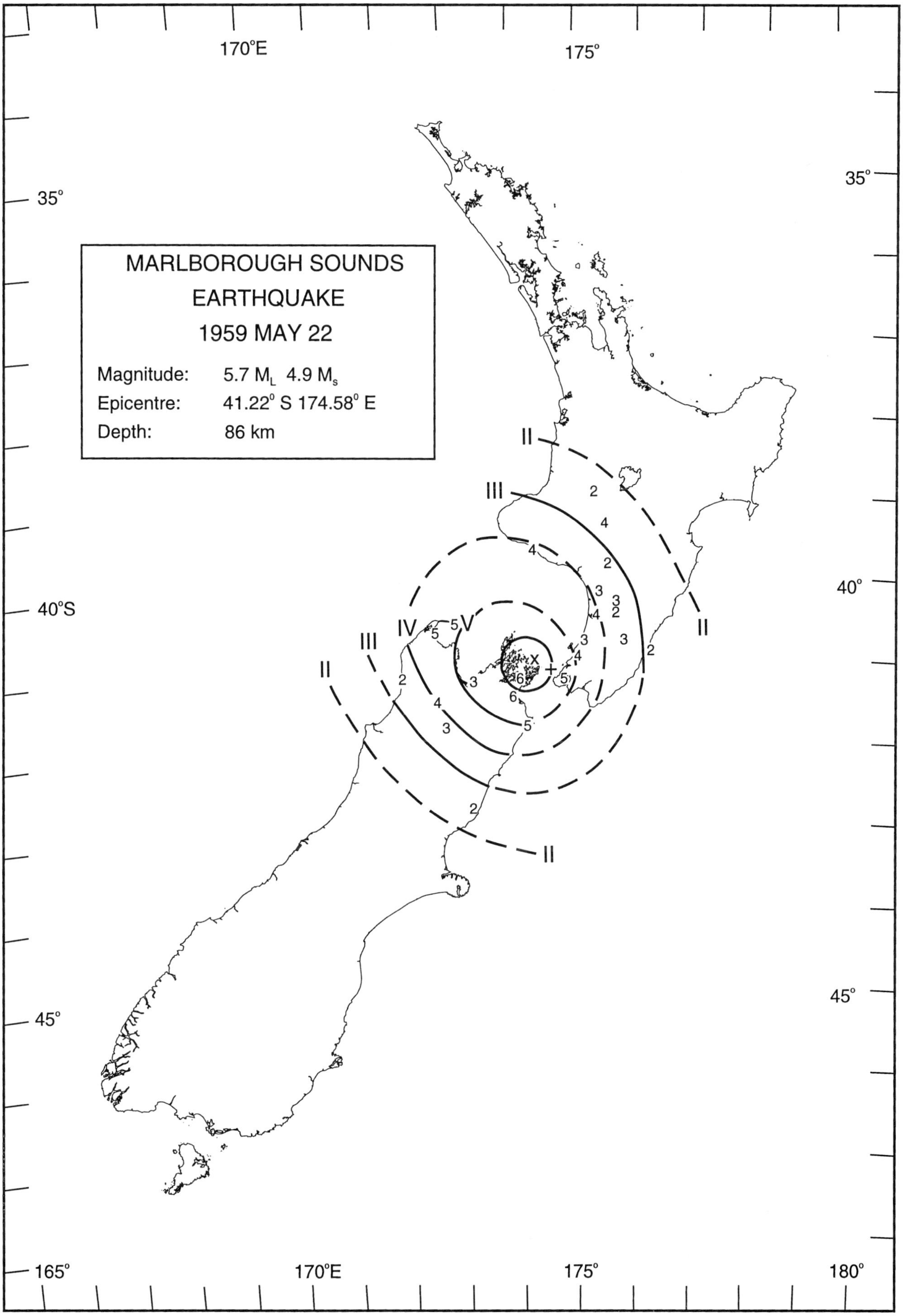

170°E
175°
35°
35°
40°S
40°
45°
45°
165°
170°E
175°
180°
MARLBOROUGH SOUNDS
EARTHQUAKE
1959 MAY 22
Magnitude: 5.7 M_L 4.9 M_s
Epicentre: 41.22° S 174.58° E
Depth: 86 km
II
III
IV
V

63. ISOSEISMAL MAP OF THE EAST CAPE EARTHQUAKE - 1960 FEBRUARY 03

DATE:	1960 FEBRUARY 03
TIME:	02:21:08.0 UT
MAGNITUDE:	6.0 M_L^* ; 5.1 M_s (Dowrick & Smith 1990)
EPICENTRE:	37.56°S 177.98°E (1)
DEPTH:	33 km (restricted)

'On Feb. 3, a shallow earthquake of magnitude 6.4 [6.0 M_L^*] occurred near East Cape, and was felt over the whole of the North Island, an isolated observation being received from as far north as Kaitaia.

'Felt intensities close to the epicentre appear somewhat low (MM5), and abnormal focal depth was suspected. Readings of PKP from Scandinavian stations exclude this possibility. The absence of large settlements may account for lack of observations suggesting a higher intensity. Only one insurance claim was filed.' (Seismological Observatory bulletin E 141)

The isoseismal map has been reproduced from the Seismological Observatory bulletin E 141.

---

* Location and local magnitude have been revised since the publication of the Seismological Observatory bulletin E 141. Refer to introduction.

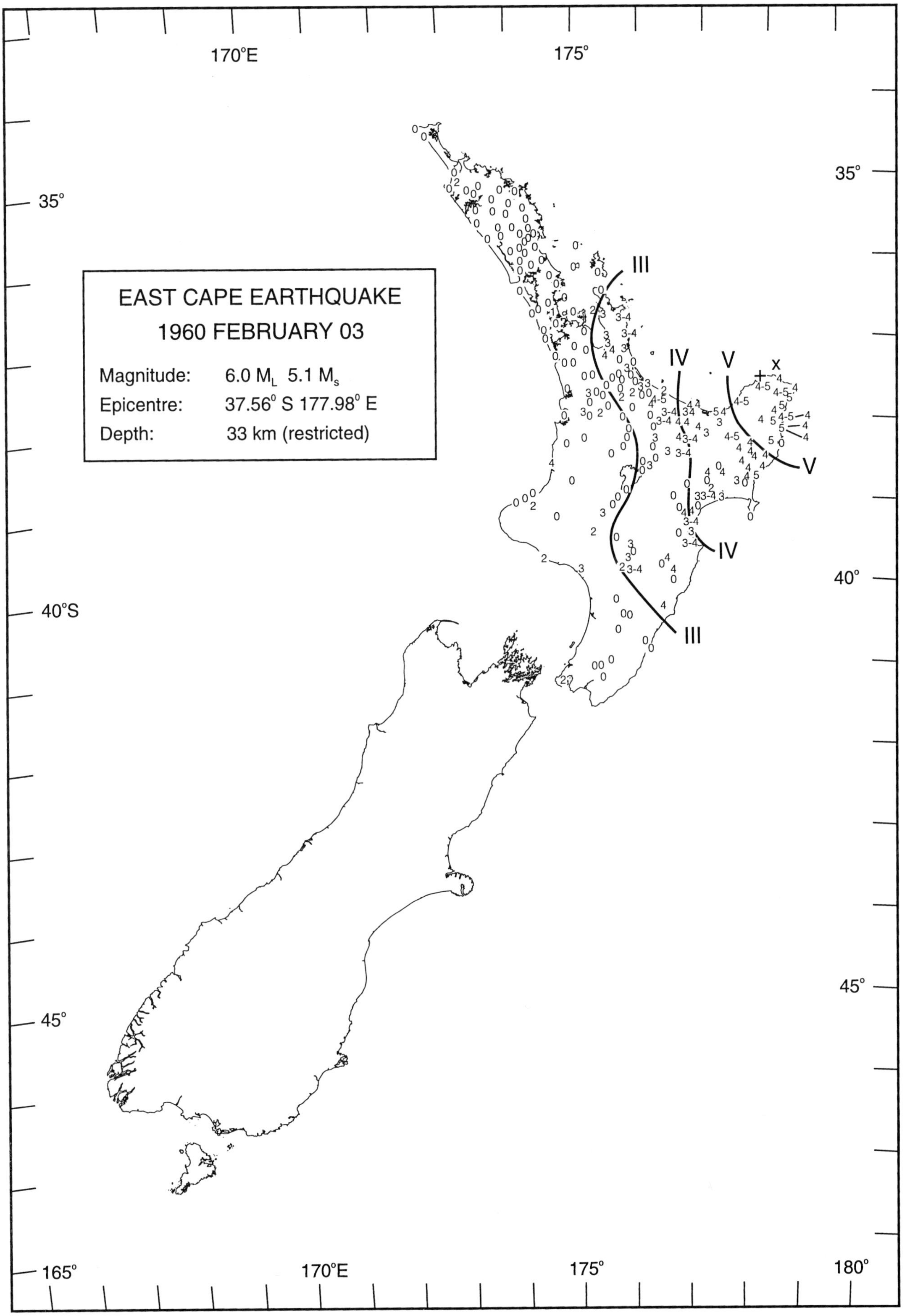
EAST CAPE EARTHQUAKE
1960 FEBRUARY 03
Magnitude: 6.0 M_L 5.1 M_s
Epicentre: 37.56° S 177.98° E
Depth: 33 km (restricted)
170°E
175°
35°
40°S
45°
165°
180°
40°
III
IV
V
x

64. ISOSEISMAL MAP OF THE ACHERON RIVER EARTHQUAKE - 1960 FEBRUARY 21

DATE:	1960 FEBRUARY 21
TIME:	00:46:54.8 UT
MAGNITUDE:	5.8 M_L*; 5.4 M_s (Dowrick & Smith 1990)
EPICENTRE:	42.28°S 173.10°E (1)
DEPTH:	50 km (restricted)*

'Another shallow [*] shock [the first being on 1960 February 03] of magnitude 6.4 [5.8 M_L*] took place of February 21.

'The felt area extended from Waitara, Taumarunui and Dannevirke in the North Island to Fairlie and Franz Josef in the South. Minor damage was reported from Murchison and Kaikoura, where intensities reached MM6. The surrounding observations indicate that the region of consistently high intensity lay to the east of Murchison and near Lake Rotoroa, and that the reports of MM6 at Kaikoura and at Karamea on the opposite coast were probably the result of local ground peculiarities. An hour after the shock, water flowing from Lake Rotoroa was reported to be turbid. It was also observed that the water in a small bay between Kaikoura and Clarence Bridge was discoloured until the following morning, but the evidence linking this to the earthquake is slight. A noteworthy feature of this shock is the large number of observers, both men and women, who reported nausea. This appeared to be experienced at all intensities from MM2 to MM6, and is particularly surprising as the questionnaire issued makes no reference to an effect of this kind. The effect was sufficiently pronounced to call for special remark under the heading 'Further Comments'. The swaying of trees and power poles was also widely remarked upon. Taken together, these observations may indicate movements of rather longer period than normal. Neither this shock nor that near East Cape on Feb. 3 was accompanied by appreciable foreshock or aftershock activity.' (Seismological Observatory bulletin E 141)

The map has been reproduced from the Seismological Observatory bulletin E 141.

---

* Location and local magnitude have been revised since the publication of the New Zealand Seismological Report 1960. Refer to introduction. However the depth of the epicentre was not well constrained and depths of 33 km to over 100 km fitted equally well. A restricted depth of 50 km achieved consistency in the magnitude data and was considered appropriate for the location and felt data.

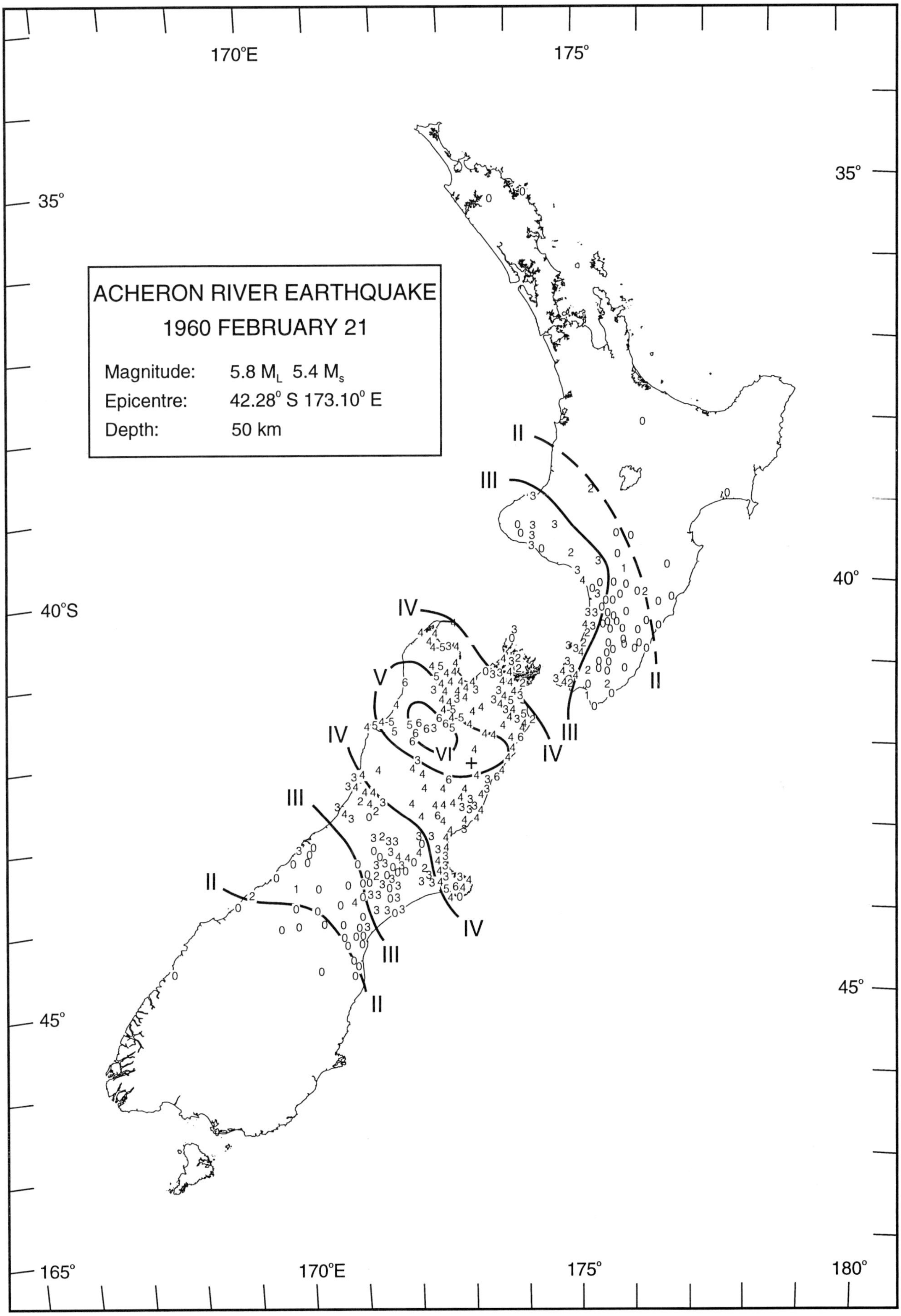

ACHERON RIVER EARTHQUAKE
1960 FEBRUARY 21
Magnitude: 5.8 M_L 5.4 M_s
Epicentre: 42.28° S 173.10° E
Depth: 50 km
170°E
175°
35°
40°S
40°
45°
165°
180°
II
III
IV
V
VI

65. ISOSEISMAL MAP OF THE FIORDLAND EARTHQUAKE - 1960 MAY 24

DATE:	1960 MAY 24
TIME:	14:46:37.9 UT
MAGNITUDE:	6.5 M_L*; 6.5 M_s (Dowrick & Smith 1990)
EPICENTRE:	44.19°S 167.73°E (1)
DEPTH:	12 km (restricted)

'The shallow earthquake in the northern part of Fiordland on May 24 had a magnitude of 7.0 [6.5 M_L*], and it must be reckoned fortunate that the shock was centred at sea, some 20 miles [30 km] off a sparsely populated part of the coast. The felt area included most of the South Island (except Marlborough) and Stewart Island, but no reports were received from north of Cook Strait. A questionnaire was issued, and the observations have been plotted on an isoseismal map. The maximum intensity reported was MM6, at several places near the southern end of Lake Wanaka, goods being thrown from shelves, and plaster cracked at Cardrona and Luggate. The lack of observers and buildings in the region between this area and the epicentre leaves the boundary of the MM6 isoseismal very uncertain. Isolated cases of minor damage occurred throughout Otago and Southland, more than 80 claims being filed with the Earthquake and War Damage Commission. Because the shock occurred in the early hours of the morning, the boundary of the felt area appears to be marked by the MM3 isoseismal, lesser intensities being insufficient to wake even light sleepers....

The main shock was preceded by a foreshock of magnitude 5.3 on February 13, and at least 12 other foreshocks took place between then and the main event on May 24. Aftershocks were numerous, continuing spasmodically until at least the end of September. More than 100 of them had magnitudes of 4 or greater. '(Seismological Observatory bulletin E 141)

The isoseismal map has been reproduced from the Seismological Observatory bulletin E 141.

* Location and local magnitude have been revised since the publication of the Seismological Observatory bulletin E 141. Refer to introduction.

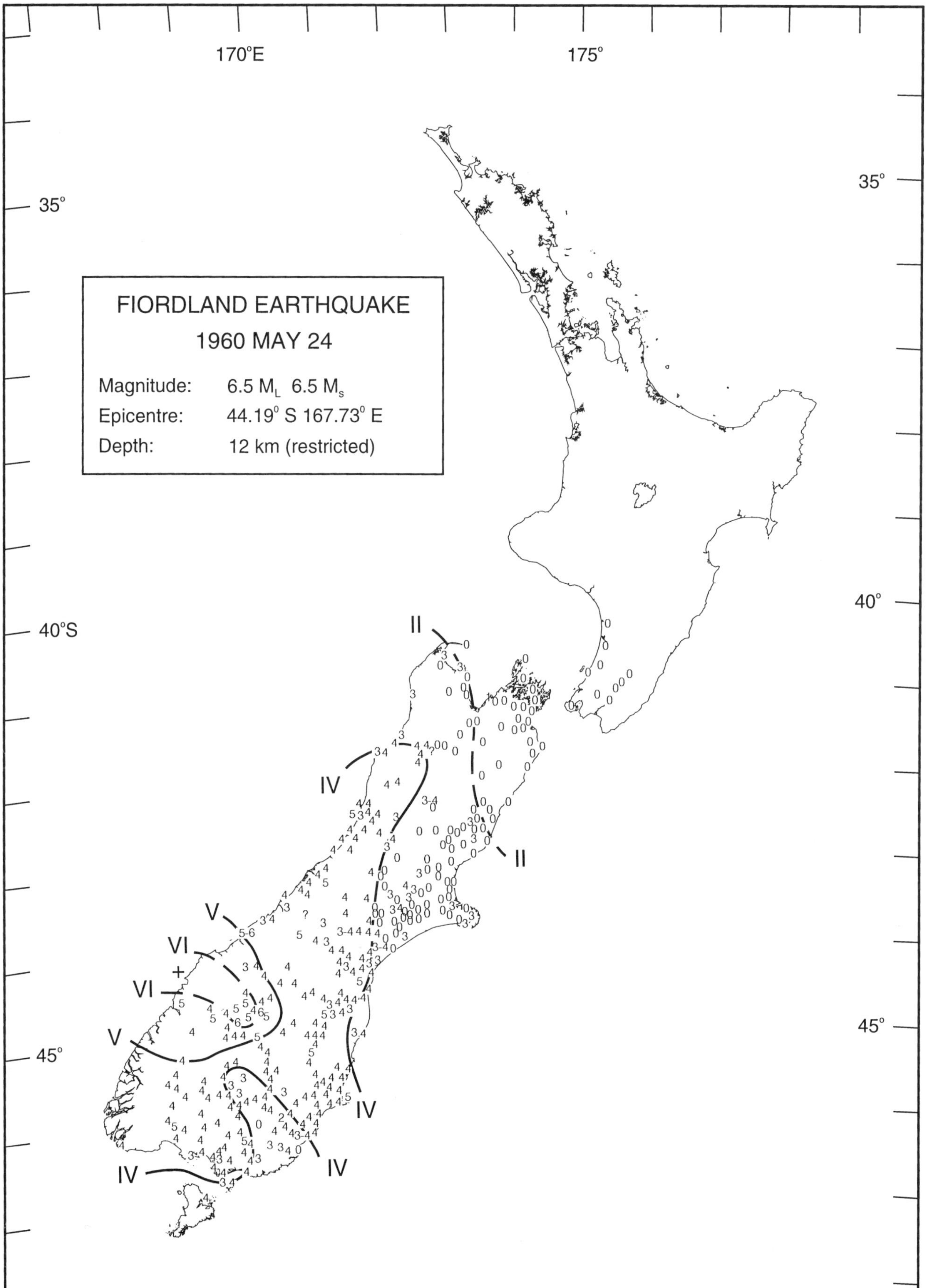

FIORDLAND EARTHQUAKE
1960 MAY 24
Magnitude: 6.5 M$_L$ 6.5 M$_s$
Epicentre: 44.19° S 167.73° E
Depth: 12 km (restricted)
170°E
175°
35°
40°S
40°
45°
165°
180°
II
IV
V
VI

66. ISOSEISMAL MAP OF THE NORTH ISLAND EARTHQUAKE - 1961 FEBRUARY 03

DATE:	1961 FEBRUARY 03
TIME:	12:33:29.5 UT
MAGNITUDE:	5.9 M_L*
EPICENTRE:	37.66°S 175.97°E (1)
DEPTH:	308 km

'A deep earthquake on February 3, with a focal depth of 320 km [revised depth 308 km*] showed a similarly irregular pattern of felt intensities [as the large deep earthquake on 1961 July 26], but the felt area was more restricted, as might be expected from the greater depth and smaller magnitude ... This was the deepest shock of the year. Its epicentre is near Te Aroha.' (Seismological Observatory bulletin E 142)

The isoseismal map has been reproduced from the Seismological Observatory bulletin E 142.

* Location and local magnitude have been revised since the publication of the Seismological Observatory bulletin E 142. Refer to introduction.

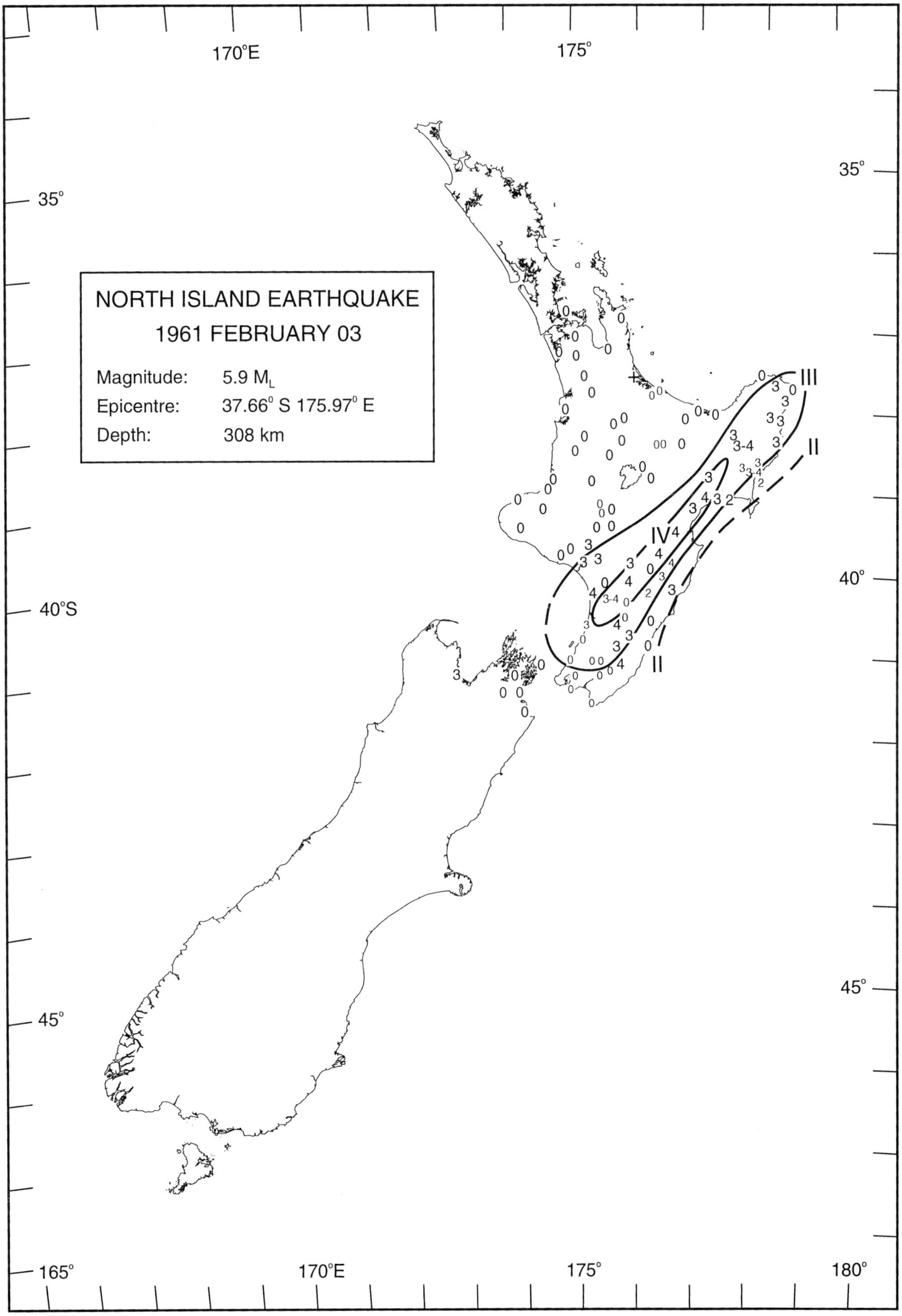

NORTH ISLAND EARTHQUAKE
1961 FEBRUARY 03
Magnitude: 5.9 M_L
Epicentre: 37.66° S 175.97° E
Depth: 308 km
170°E
175°
35°
40°S
40°
45°
165°
180°
III
II
IV

67. ISOSEISMAL MAP OF THE DANNEVIRKE EARTHQUAKE - 1961 MAY 14

DATE:	1961 MAY 14
TIME:	00:12:38.2 UT
MAGNITUDE:	5.3 M_L*
EPICENTRE:	40.38°S 176.20°E (1)
DEPTH:	33 km (restricted)

'The shallow earthquake near Dannevirke of May 14 had a magnitude of 5.4 [5.3 M_L*] and a felt area that includes most of the central and southern parts of the North Island, the maximum reported intensity (in Dannevirke itself) being MM5.' (Seismological Observatory bulletin E 142)

The isoseismal map has been reproduced from the Seismological Observatory bulletin E 142.

* Location and local magnitude have been revised since the publication of the Seismological Observatory bulletin E 142. Refer to introduction.

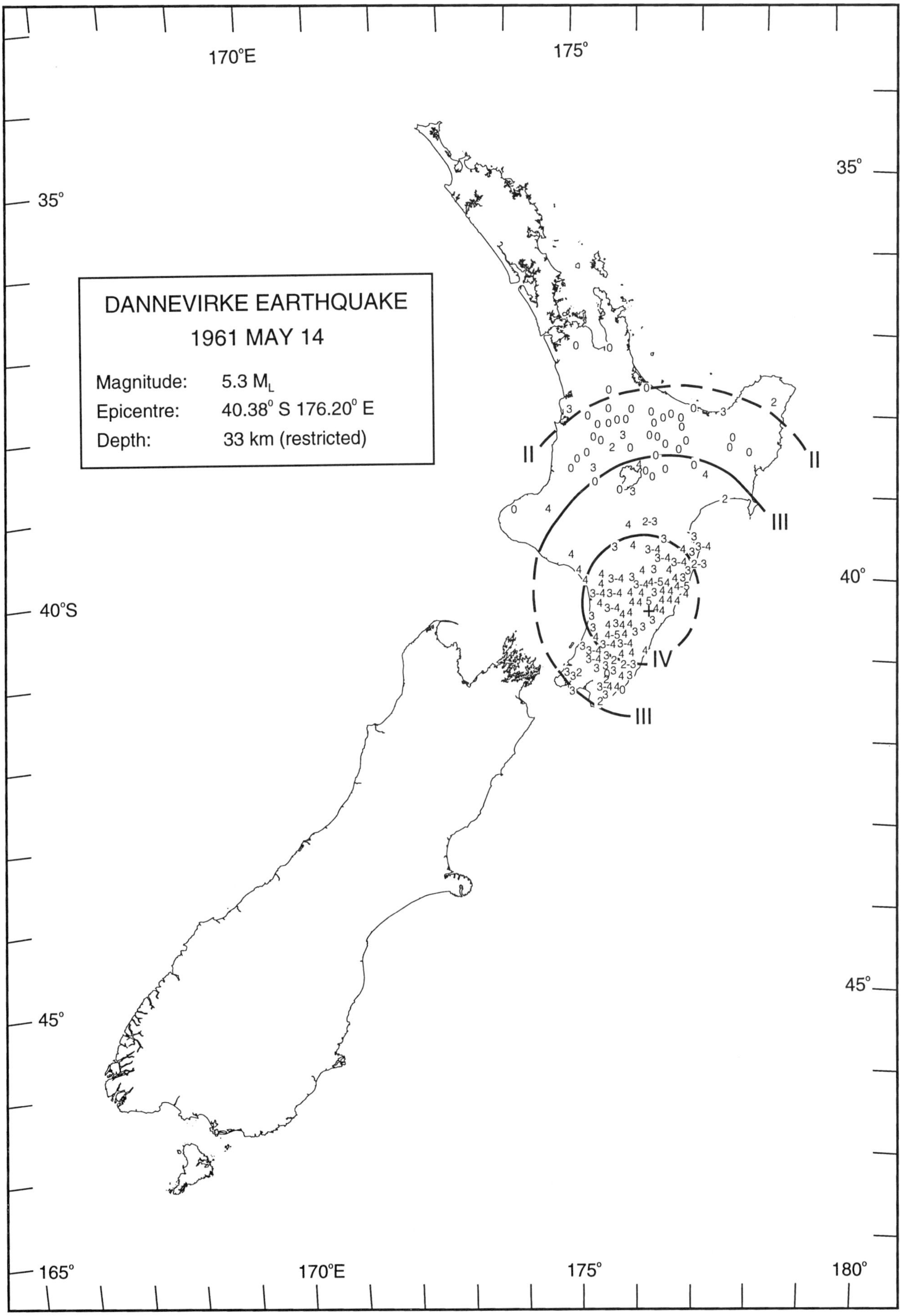

DANNEVIRKE EARTHQUAKE
1961 MAY 14
Magnitude: 5.3 M_L
Epicentre: 40.38° S 176.20° E
Depth: 33 km (restricted)
170°E
175°
35°
40°S
40°
45°
165°
180°
II
III
IV

68. ISOSEISMAL MAP OF THE BARRIER RANGE EARTHQUAKE - 1961 JULY 04

DATE:	1961 JULY 04
TIME:	08:23:31.2 UT
MAGNITUDE:	5.1 M_L^*
EPICENTRE:	44.45°S 168.08°E (1)
DEPTH:	12 km (restricted)

'Another widely felt shallow shock [in 1961] on July 4 had an epicentre near the northern end of Lake Wakatipu and a felt area extending from Bruce Bay to Stewart Island. The epicentral region is sparsely populated, the maximum intensity reported being MM4, at Milford Sound.' (Seismological Observatory bulletin E 142)

The isoseismal map has been reproduced from the Seismological Observatory bulletin E 142.

* Location and local magnitude have been revised since the publication of the Seismological Observatory bulletin E 142. Refer to introduction.

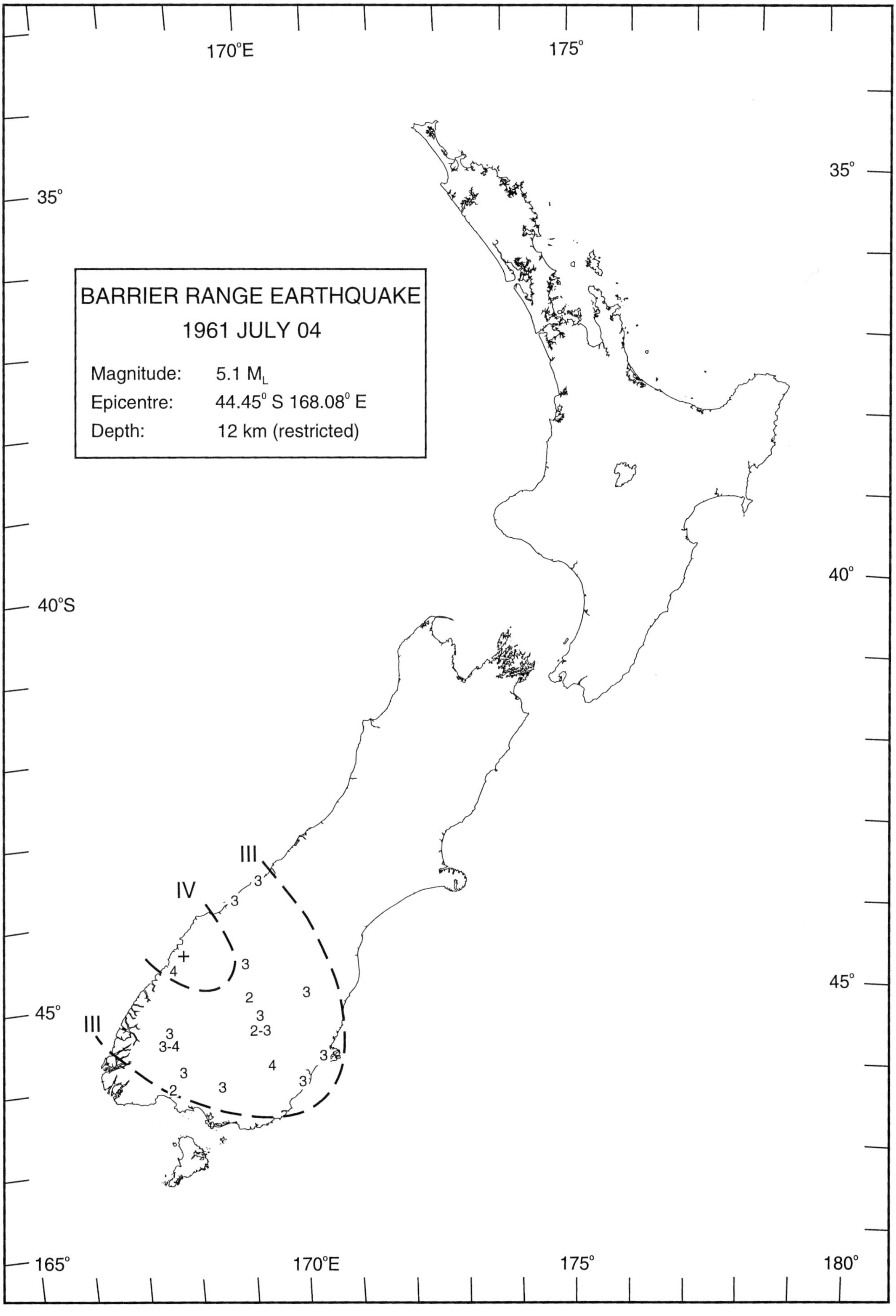

BARRIER RANGE EARTHQUAKE
1961 JULY 04
Magnitude: 5.1 M_L
Epicentre: 44.45° S 168.08° E
Depth: 12 km (restricted)
170°E
175°
180°
165°
35°
40°S
40°
45°
III
IV
3
4
2
2-3
3-4

69. ISOSEISMAL MAP OF THE BAY OF PLENTY EARTHQUAKE - 1961 JULY 26

DATE:	1961 JULY 26
TIME:	09:19:04.4 UT
MAGNITUDE:	5.8 M_L*
EPICENTRE:	37.68°S 176.84°E (1)
DEPTH:	204 km

'The largest earthquake [in 1961] was the deep-focus shock of July 26 which was centred in the Bay of Plenty and had a magnitude of 6¾ [5.8 M_L*]. Its focal depth was 230 km [revised to 204 km*], and the felt area included most of the North Island (with the exception of the Northland peninsula) and several places in northern Marlborough. The usual questionnaire was issued. Most of the reports received indicate an intensity of MM3 or MM4. Although the higher values predominated in reports from Gisborne-East Cape region, it is difficult to suggest isoseismals, and it seems likely that the local ground conditions are largely responsible for the irregular distribution apparent.' (Seismological Observatory bulletin E 142)

The map has been reproduced from the Seismological Observatory bulletin E 142.

* Location and local magnitude have been revised since the publication of the Seismological Observatory bulletin E 142. Refer to introduction.

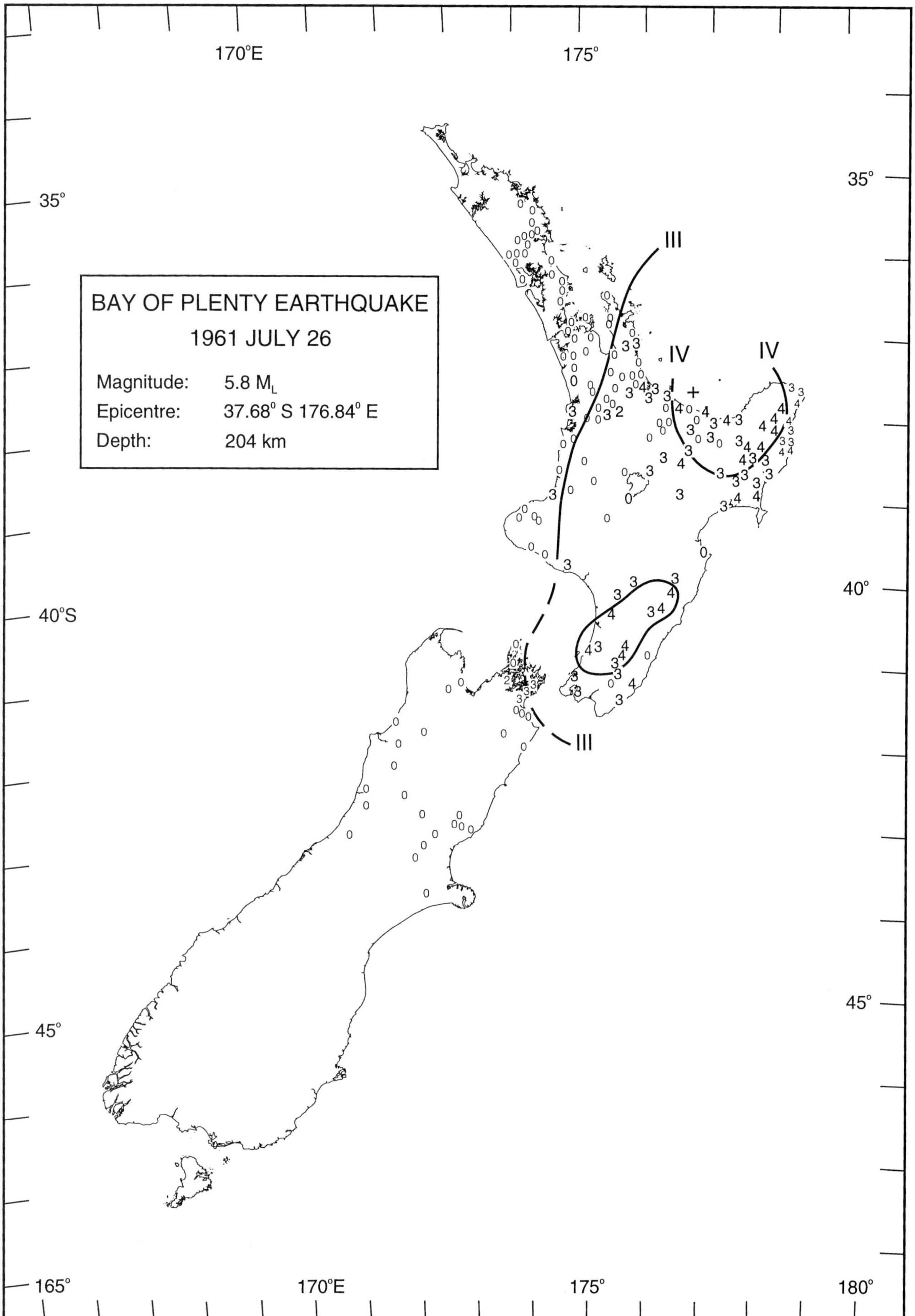
BAY OF PLENTY EARTHQUAKE
1961 JULY 26
Magnitude: 5.8 M_L
Epicentre: 37.68° S 176.84° E
Depth: 204 km
170°E
175°
35°
40°S
45°
165°
180°
40°
III
IV

70. ISOSEISMAL MAP OF THE HIKURANGI TRENCH EARTHQUAKE - 1961 DECEMBER 27

DATE:	1961 DECEMBER 27
TIME:	23:47:53.2 UT
MAGNITUDE:	6.3 M_L* ; 6.5 M_s (Dowrick & Smith, 1990)
EPICENTRE:	41.51°S 176.11°E (1)
DEPTH:	12 km (restricted)

'The largest shallow shock [in 1961], on December 27 had a magnitude of 6.3, and originated at sea about 100 km [revised to 70 km*] south east of Masterton, near the southern end of the Hikurangi Trench. It was followed by a series of aftershocks extending for some 200 km in a belt 20-30 km wide, at right angles to the coast, and some 40 km to the south of the main shock... The principal shock was responsible for some minor damage in the Wairarapa, 77 insurance claims being lodged with the Earthquake and War damage Commission; but replies to the Observatory questionnaire provide little evidence of intensities greater than MM5. The felt area extended from a little south of the Firth of Thames to southern Nelson and Kaikoura.' (Seismological Observatory bulletin E 142)

The map has been reproduced from the Seismological Observatory bulletin E 142.

---

* Location only has been revised since the publication of the Seismological Observatory bulletin E 142. As the original amplitude data has been misplaced the local magnitude has not been revised. Refer to introduction.

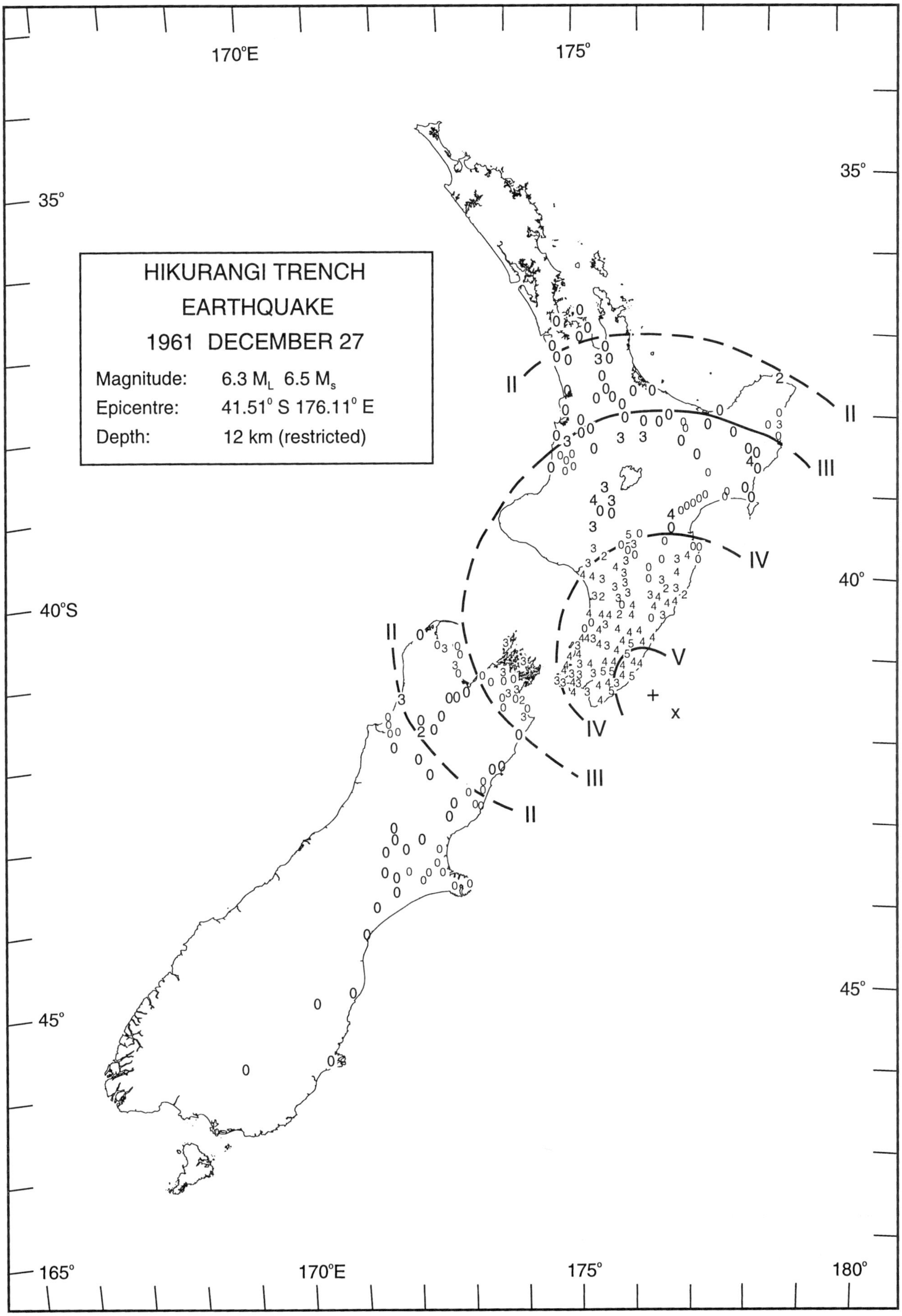
HIKURANGI TRENCH
EARTHQUAKE
1961 DECEMBER 27
Magnitude: 6.3 M_L 6.5 M_s
Epicentre: 41.51° S 176.11° E
Depth: 12 km (restricted)
170°E
175°
35°
40°S
40°
45°
165°
180°
II
III
IV
V
x

71. ISOSEISMAL MAP OF THE ARIA EARTHQUAKE - 1962 JANUARY 23

DATE:	1962 JANUARY 23
TIME:	06:49:41.5 UT
MAGNITUDE:	5.5 M_L*; 4.5 M_s (Dowrick & Smith 1990)
EPICENTRE:	38.58°S 174.60°E (1)
DEPTH:	12 km (restricted)

'The earthquake near Aria on January 23 ... caused some damage. This shallow earthquake of magnitude 5.5 is of interest because it lies further to the north than the majority of shallow shocks west of the N. Z. Sub-Crustal Rift. The felt area included Auckland city and extended southwards to Palmerston North and Shannon; but only isolated reports were received from places east of a line through Tauranga, Taupo, and Taihape. Chimneys fell at Aria and Mahoenui, and a chimney at Benneydale was cracked.' (Seismological Observatory bulletin E 143)

The map from the Seismological Observatory bulletin E 143 has been redrawn using the listed data. Close examination of the map in this publication revealed not only that the overlay of MM data and isoseismal lines onto the map of New Zealand had been skewed by nearly half a degree, but also that some intensities differed from those listed. Examination of the original felt report forms indicated that the first assessment of assigned intensities had been altered at a later date, presumably by a more experienced analyst, and the correct list but incorrect map had been printed.

* Location and local magnitude have been revised since the publication of the Seismological Observatory bulletin E 143. Refer to introduction.

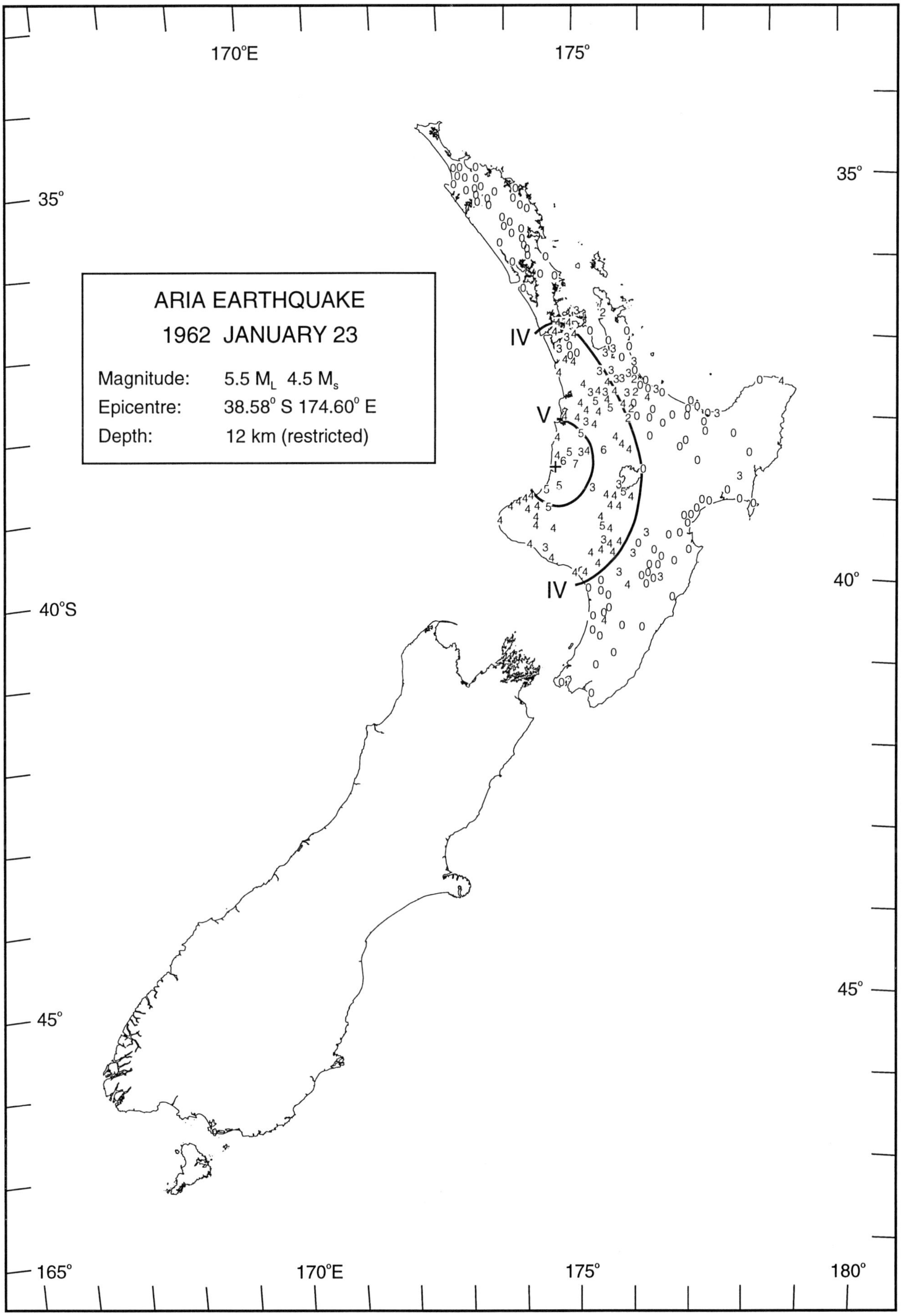
ARIA EARTHQUAKE
1962 JANUARY 23
Magnitude: 5.5 M_L 4.5 M_s
Epicentre: 38.58° S 174.60° E
Depth: 12 km (restricted)
170°E
175°
180°
165°
35°
40°
40°S
45°
IV
V

72. ISOSEISMAL MAP OF THE WESTPORT EARTHQUAKE - 1962 MAY 10

DATE:	1962 MAY 10
TIME:	00:27:14.7 UT
MAGNITUDE:	5.7 M_L^*; 5.9 M_s (Dowrick & Smith 1990)
EPICENTRE:	41.67°S 171.44°E (1)
DEPTH:	12 km (restricted)

'A sequence of over 80 earthquakes occurred in 1962 off the coast near Westport, New Zealand. The first and largest shock took place on May 10 and had a magnitude of 5.9 [5.7 M_L^*]; the next largest, on May 17, had a magnitude of 5.6 [not revised*]. The epicentres of these two earthquakes have been accurately determined from readings at local seismograph stations; the strong crustal phases, felt effects, and arrival times of phases at overseas stations all show that the foci were shallow.

'The earthquake was most strongly felt at Westport, where the intensity was 7 on the Modified Mercalli Scale; reports from Greymouth indicate intensity MM6-7. The felt area covered all the South Island north of Otago, and there were a few isolated reports from the south of the North Island.

'More damage was caused by the Westport sequence than by any other New Zealand earthquake since those in the Wairarapa in 1942. This may be attributed to the closeness of the epicentres (25 km) to Westport, and the shallowness of the foci; for the magnitudes were not great. According to preliminary estimates from the Earthquake and War Damage Commission, the total value of the damage was approximately £125,000, which included damage to about 2,500 chimneys. In Westport brick- and plaster-work in buildings was damaged, and electrical and water services were disrupted. Chimneys were damaged over a distance of 140 km, from Granity (25 km NE of Westport) to Hokitika (35 km SW of Greymouth). There were no casualties.' (Adams & Le Fort 1963)

The map from the Seismological Observatory bulletin E 143 has been redrawn using the listed data as some intensities on the map differed from those listed. Examination of the original felt report forms indicated that the first assessment of assigned intensities had been altered at a later date, presumably by a more experienced analyst, and the correct list but incorrect map had been printed.

* Location and local magnitude have been revised since the publication of the Seismological Observatory bulletin E 143. Refer to introduction.

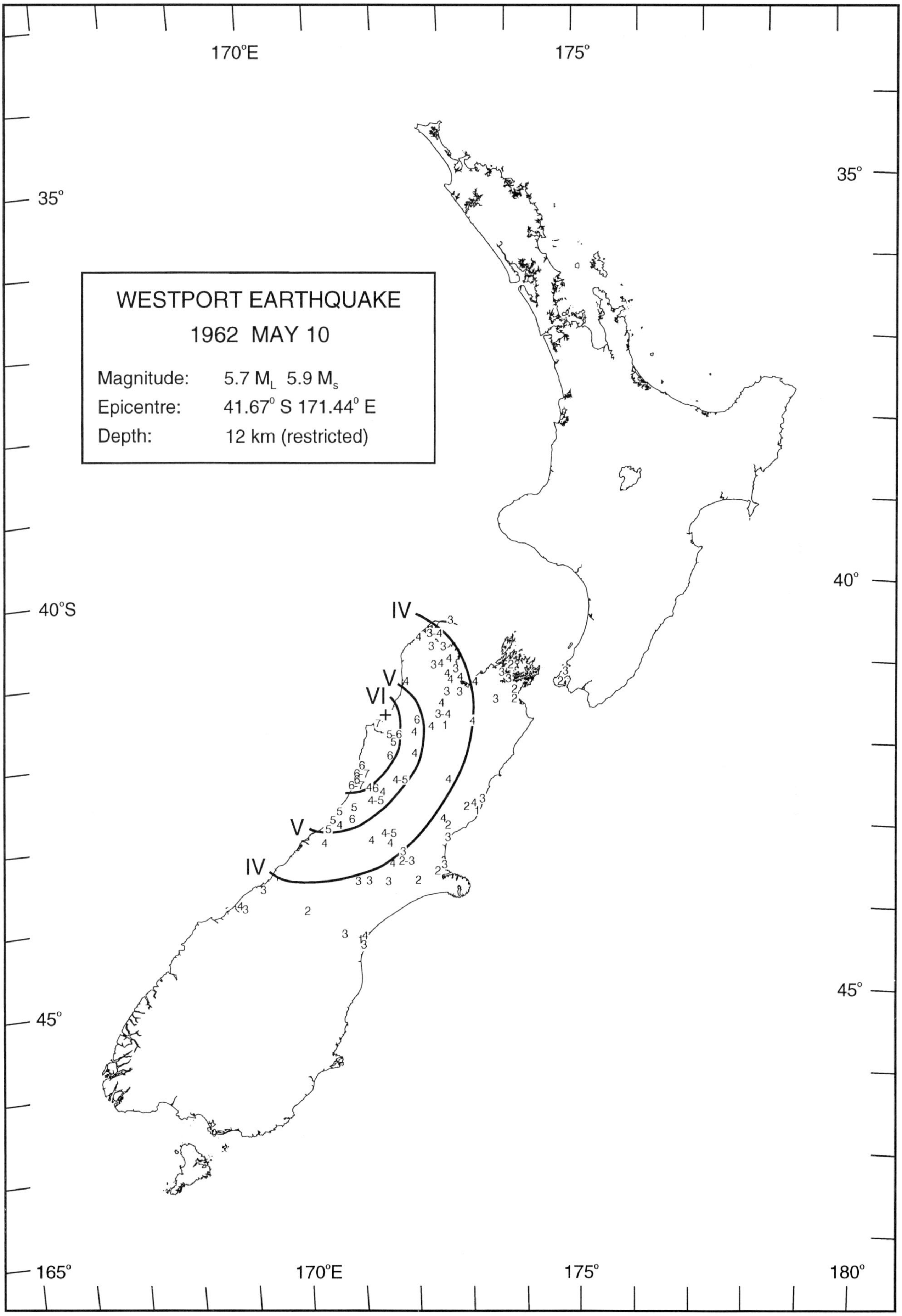

WESTPORT EARTHQUAKE
1962 MAY 10
Magnitude: 5.7 M_L 5.9 M_s
Epicentre: 41.67° S 171.44° E
Depth: 12 km (restricted)
170°E
175°
165°
180°
35°
40°S
40°
45°
IV
V
VI

73. ISOSEISMAL MAP OF THE MARLBOROUGH EARTHQUAKE - 1962 JULY 29

DATE:	1962 JULY 29
TIME:	18:19:49.7 UT
MAGNITUDE:	5.7 M_L*
EPICENTRE:	41.45°S 173.47°E (1)
DEPTH:	109 km

The earthquake on 1962 July 29, of magnitude of 5.7 M_L, occurred at a depth of 109 km beneath Marlborough and was widely felt over the southern part of the North Island and the northern part of the South Island. Only one report, at Rotomanu, indicated MM5. MM4 was widely reported in northwest Nelson and Marlborough.

The map has been reproduced from the Seismological Observatory bulletin E 143.

* Location and local magnitude have been revised since the publication of the Seismological Observatory bulletin E 143. Refer to introduction.

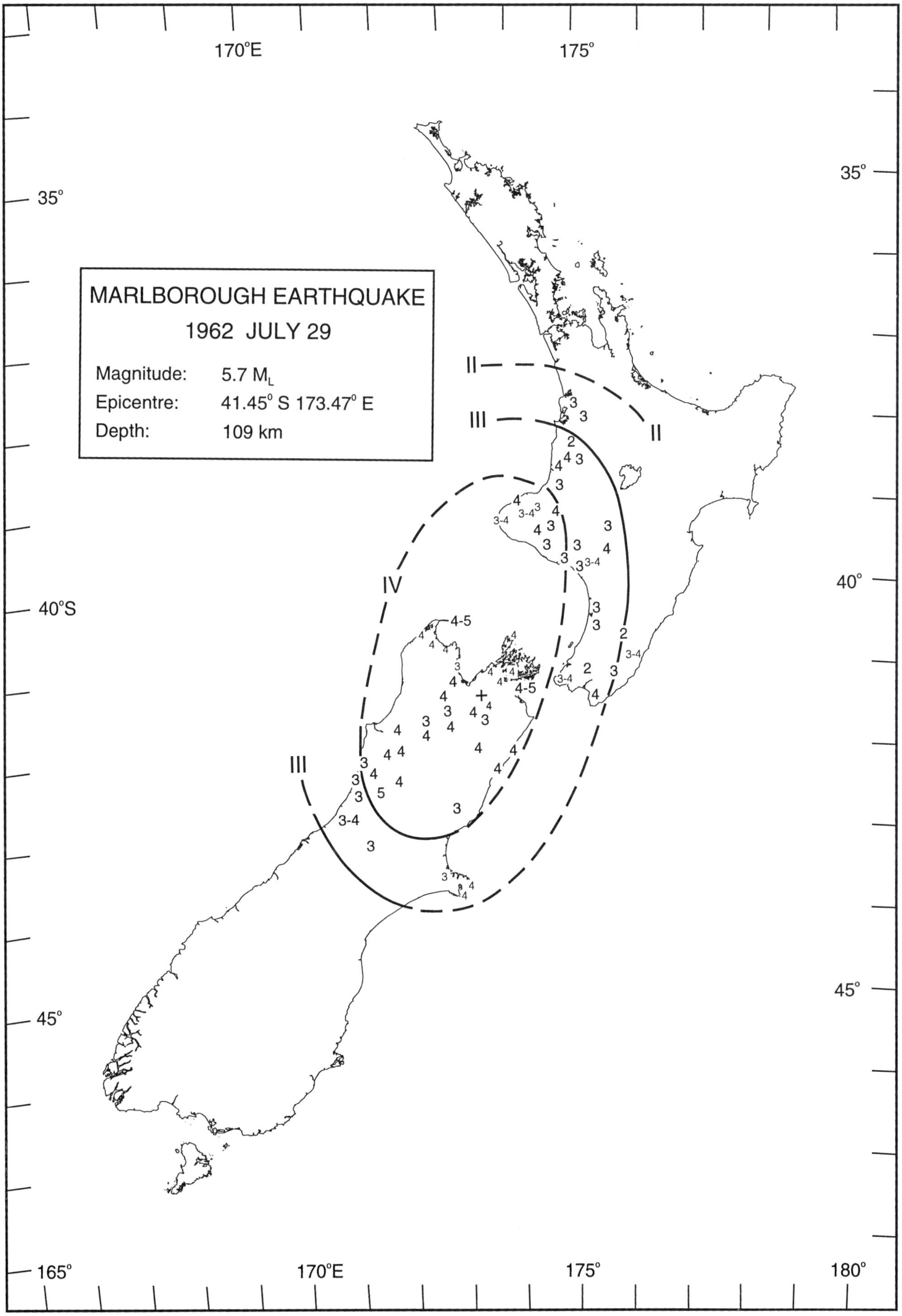

MARLBOROUGH EARTHQUAKE
1962 JULY 29
Magnitude: 5.7 M_L
Epicentre: 41.45° S 173.47° E
Depth: 109 km
170°E
175°
35°
40°S
45°
165°
180°
40°
II
III
IV

74. ISOSEISMAL MAP OF THE BRUCE BAY EARTHQUAKE - 1962 OCTOBER 15

DATE:	1962 OCTOBER 15
TIME:	23:36:31.5 UT
MAGNITUDE:	5.7 M_L* ; 5.4 M_s (Dowrick & Smith 1990)
EPICENTRE:	43.54°S 169.77°E (1)
DEPTH:	12 km (restricted)

'The largest earthquake [in 1962] occurred on October 15. It had a magnitude of 6.1 [5.7 M_L*], and a shallow origin off the coast of southern Westland to the north of Bruce Bay. Damage was confined to the fall of isolated chimney pots in the Fox Glacier - Bruce Bay area, and of goods from shelves at Fox Glacier, Lake Tekapo and Mount Cook. More spectacular effects were to be seen in the Southern Alps. The Chief Ranger at Mt. Cook, Mr. M. Burke, reports that "Magnificent avalanches fell from all glaciers and ice shelves in the Hooker and Tasman valleys, and the faces of both Mt. Cook and Mt. Sefton were almost completely obscured by billowing clouds of ice-dust and snow. Tourists on the Tasman Glacier trip were treated to a most impressive sight, as large avalanches fell from every peak on the main divide from Mt. Cook to the Minarets. I was a member of a party of four rangers returning from the second Hooker swing bridge, who had a first class view of the avalanches on the slope of Sefton and of numerous rockfalls on the slopes of Mt. Wakefield above the track. Some rock fell quite close to the track in places." This account is confirmed by Mr. M. Barrie, a climber who was in the vicinity of the Ball Hut. He reports that his party was greatly delayed on its return journey to Husky Camp by the rock that had shaken down on to the road. He further points out that falls were not restricted to moraine and other loose material, but also affected "virgin rock". It appears likely that felt intensities approached MM7.' (Seismological Observatory bulletin E 143)

The map from the Seismological Observatory bulletin E 143 has been redrawn using the listed data. Close examination of the map in this publication revealed not only that the overlay of MM data and isoseismal lines onto the map of New Zealand had been skewed by nearly half a degree, but also that some intensities differed from those listed. Examination of the original felt report forms indicated that the first assessment of assigned intensities had been altered at a later date, presumably by a more experienced analyst, and the correct list but incorrect map had been printed.

* Location and local magnitude have been revised since the publication of the Seismological Observatory bulletin E 143. Refer to introduction.

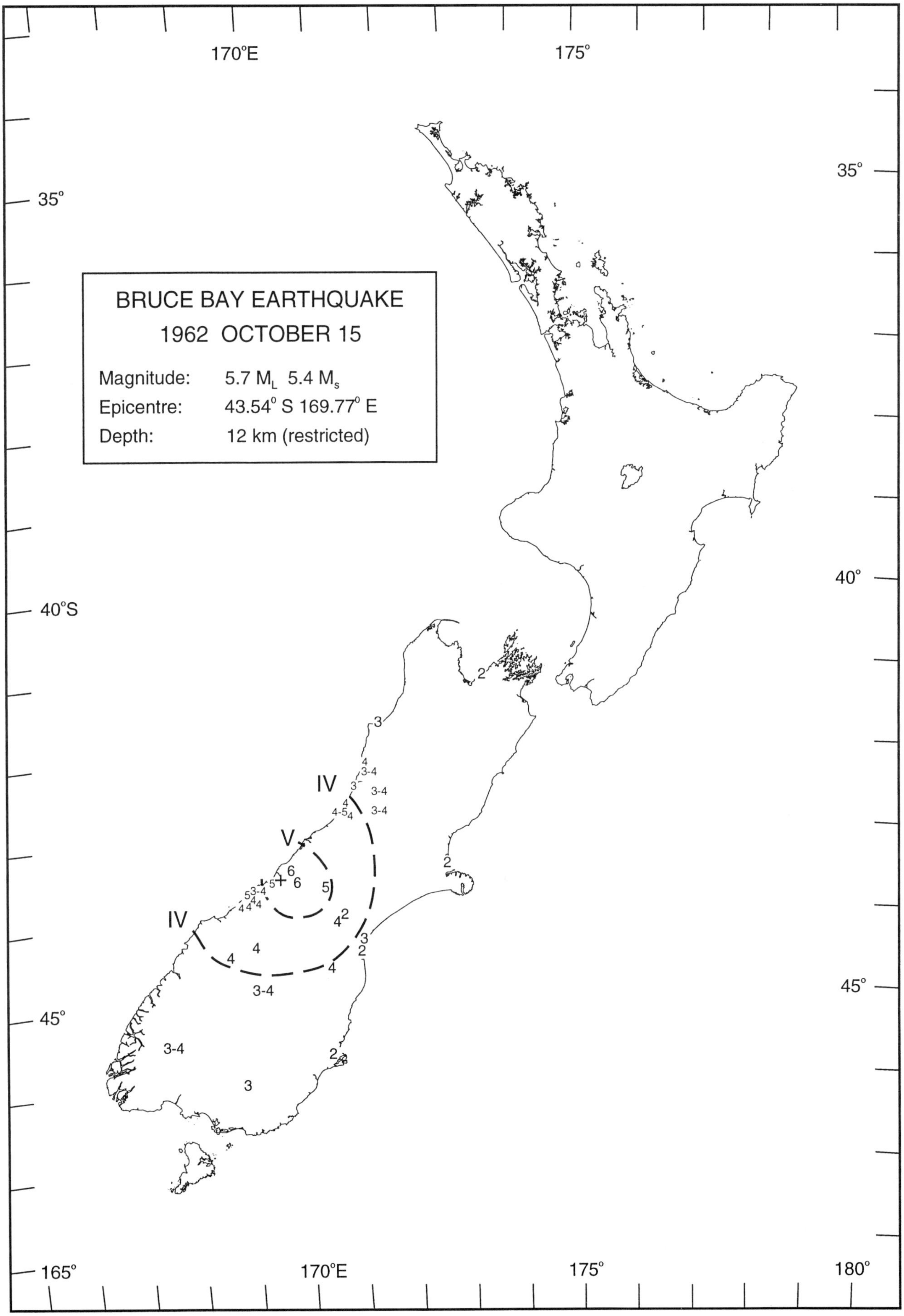
BRUCE BAY EARTHQUAKE
1962 OCTOBER 15
Magnitude: 5.7 M_L 5.4 M_s
Epicentre: 43.54° S 169.77° E
Depth: 12 km (restricted)
170°E
175°
35°
40°S
40°
45°
165°
180°
IV
V
IV
6
6
5
4
3-4
3
2

75. ISOSEISMAL MAP OF THE HUIARAU RANGE EARTHQUAKE - 1963 APRIL 12

DATE:	1963 APRIL 12
TIME:	08:41:43.9 UT
MAGNITUDE:	5.5 M_L* ; 5.7 M_s (Dowrick & Smith 1990)
EPICENTRE:	38.66°S 176.80°E (1)
DEPTH:	20 km (restricted)

'The largest New Zealand earthquake during the year [1963] occurred on April 12, and was centred in the Huiarau Range between Lake Taupo and Waikaremoana. It had a magnitude of 6.0 [5.5 M_L*] and was widely felt in the central parts of the North Island. There was slight damage at Minginui and at Tarawera. A few aftershocks followed, all of them small.' (Seismological Observatory bulletin E 144)

The isoseismal map has been reproduced from the Seismological Observatory bulletin E 144.

* Location and local magnitude have been revised since the publication of the Seismological Observatory bulletin E 144. Refer to introduction.

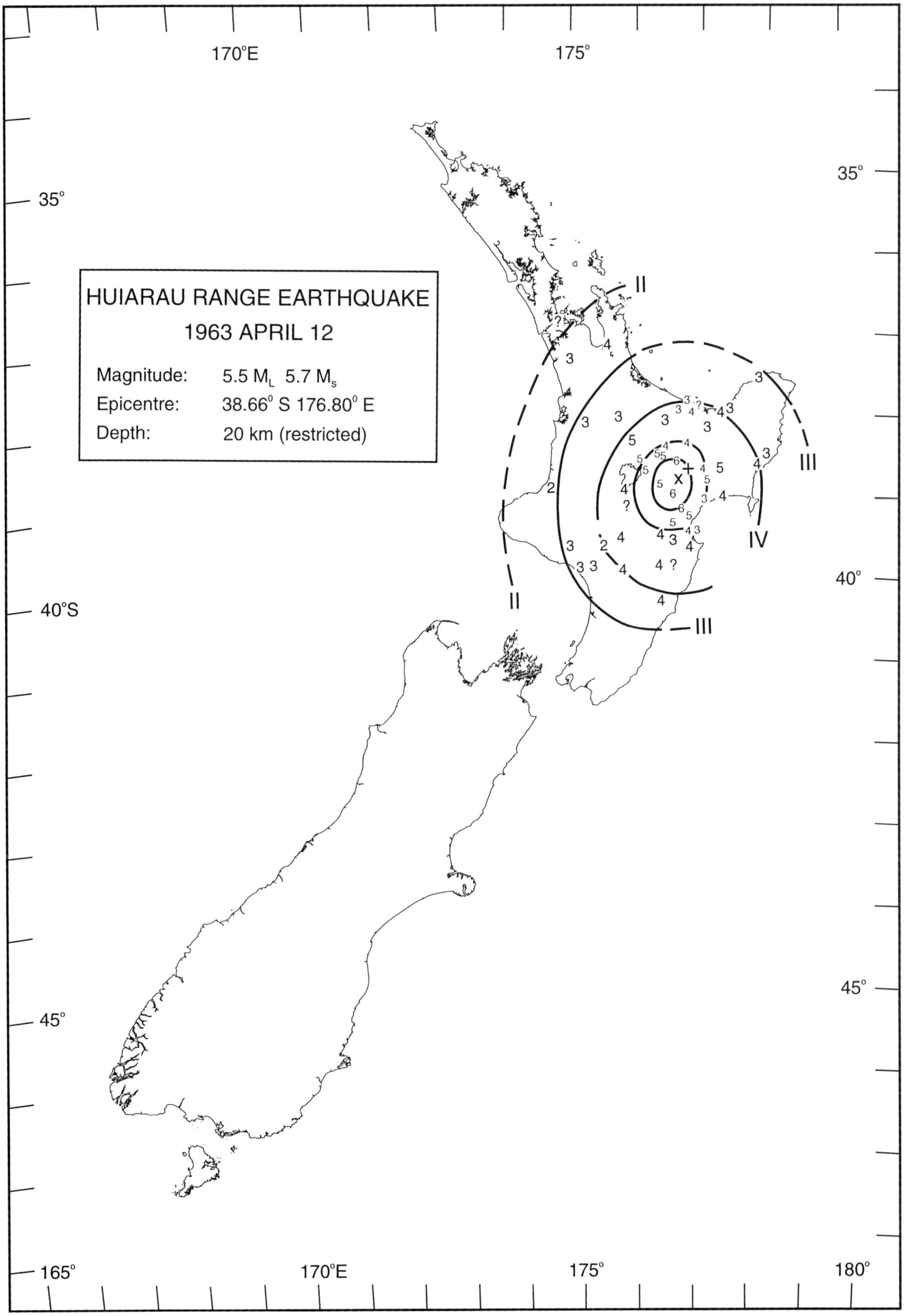

HUIARAU RANGE EARTHQUAKE
1963 APRIL 12
Magnitude: 5.5 ML 5.7 Ms
Epicentre: 38.66° S 176.80° E
Depth: 20 km (restricted)
170°E
175°
180°
165°
35°
40°S
40°
45°
II
III
IV

76. ISOSEISMAL MAP OF THE SOUTH TARANAKI EARTHQUAKE - 1963 JULY 14

DATE:	1963 JULY 14
TIME:	17:06:32.1 UT
MAGNITUDE:	5.5 M_L^*
EPICENTRE:	39.52°S 174.72°E (1)
DEPTH:	166 km

'The shock on July 14 of magnitude 5.9 [5.5 M_L^*] and at a depth of 135 kilometres [166 km*] was felt in the southern part of the North Island and the northern part of the South Island, from Gisborne to Christchurch. The origin was beneath southern Taranaki.' (Seismological Observatory bulletin E 144)

The isoseismal map has been reproduced from the Seismological Observatory bulletin E 144.

* Location and local magnitude have been revised since the publication of the Seismological Observatory bulletin E 144. Refer to introduction.

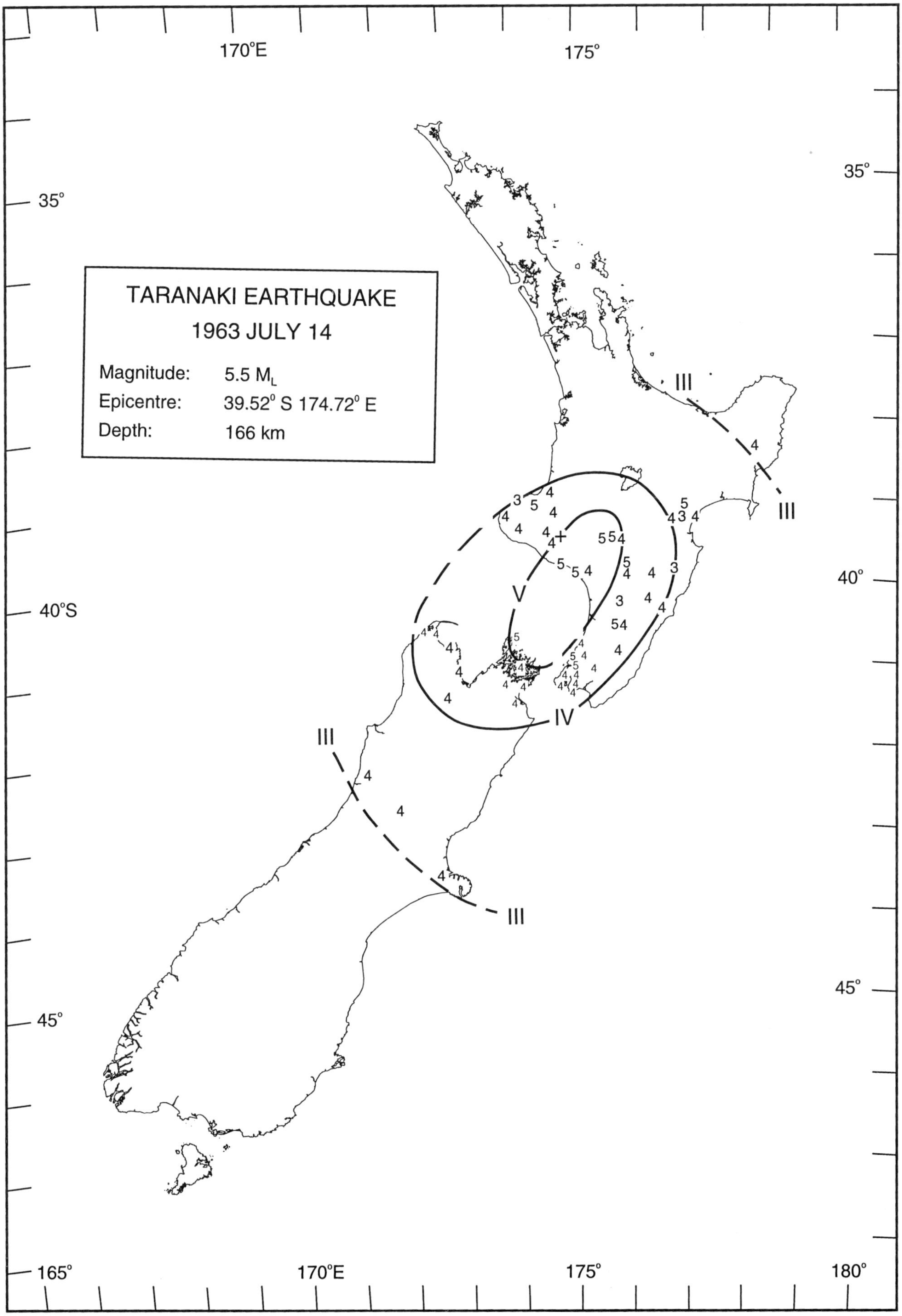

TARANAKI EARTHQUAKE
1963 JULY 14
Magnitude: 5.5 M_L
Epicentre: 39.52° S 174.72° E
Depth: 166 km
170°E
175°
35°
40°S
40°
45°
165°
180°
III
IV
V

77. ISOSEISMAL MAP OF THE MANGONUI EARTHQUAKE - 1963 NOVEMBER 16

DATE:	1963 NOVEMBER 16
TIME:	15:17:02.3 UT
MAGNITUDE:	3.5 M_L*
EPICENTRE:	35.01°S 173.54°E (1)
DEPTH:	10 km (restricted)

'The earthquakes at Mangonui on 1963 November 16 and at Peria on 1963 December 22, are the first earthquakes in Northland for which it has been possible to determine instrumental epicentres.

'About 80 questionnaires were issued to observers in the area from North Cape southwards to Mercer and Thames and including the Coromandel Peninsula. Only about 50 of the questionnaires were returned, none of them from Te Reinga, Ninety Mile Beach, and Doubtless Bay. It seems probable that the shock could have been felt in at least the southern part of these localities.

'In addition to the information gained from the questionnaires, useful accounts of the shock appeared in the Northern Advocate (Whangarei) and the Northland Age (Kaitaia) for 1963 November 18.

'The highest intensities (MM5) were experienced close to Mangonui, where the only damage reported (a cracked concrete hearth and "shattering" of the base of a chimney) took place. No intensities less than MM4 were reported. This is adequately explained by the fact that the shock occurred at 3.17 a.m. local time, and by the fact that most people in the area are unfamiliar with earthquakes. Almost all observers compare the shock to a vehicle striking the building, or to an explosion of some kind.

'The only report north of the epicentre comes from Rangiputa, about 22 km to the north-west. To the south-east, the shock was felt with sufficient intensity to awaken sleepers at a distance of 35 km. The western limit is less easy to establish. One positive report from Kaitaia indicates MM4, but other nearby reporters were unaware of the shock. The situation in the Omahuta Valley, also about 25 km from the epicentre, seems to be similar. The postmaster at Broadwood, 30 km away, reported that extensive inquiries did not disclose anyone who had felt the shock. It may safely be concluded that the radius of the MM4 isoseismal was about 25 km, and that of the MM5 isoseismal not more than 5 km. There is a slight elongation of the felt area to the south-east.' (Eiby 1964a)

The isoseismal map is reproduced from Eiby (1964a).

---

* Location and local magnitude have been revised since the publication of the New Zealand Seismological Report 1963. Refer to introduction. Recalculation of the epicentre using New Zealand Model 1 results in a change of less than 5 km.

MANGONUI EARTHQUAKE 1963 NOVEMBER 16	
Magnitude:	3.5 ML
Epicentre:	35.01° S 173.54° E
Depth:	10 km (restricted)

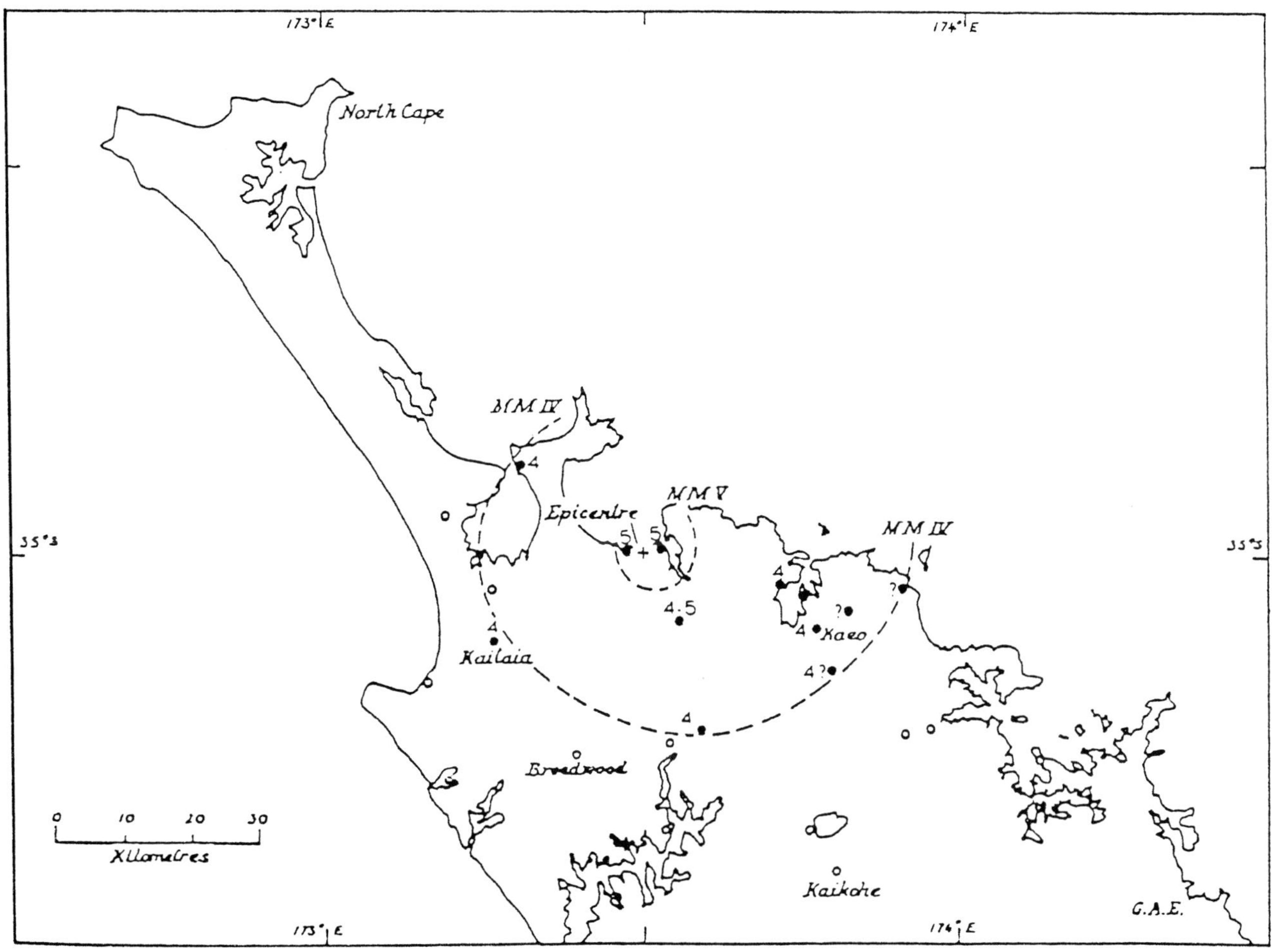

after Eiby (1964a)

78. ISOSEISMAL MAP OF THE PERIA EARTHQUAKES - 1963 DECEMBER 22

DATE:	1963 DECEMBER 22
TIME:	13:35:28 UT
MAGNITUDE:	4.5 M_L
EPICENTRE:	34.93°S 173.67 E (restricted)
DEPTH:	10 km (restricted)

DATE:	1963 DECEMBER 22
TIME:	13:35:34.4 UT
MAGNITUDE:	4.9 M_L^* ; 4.4 M_s (Dowrick & Smith 1990)
EPICENTRE:	34.93°S 173.67°E (1)
DEPTH:	10 km (restricted)

'The Peria earthquakes of 1963 December 22 were felt over a wider area than the Mangonui earthquake [1963 November 16], and were recorded at 12 New Zealand and five overseas stations. Great difficulty was experienced in reconciling the indications of the different stations, until it was realised that there had been two events, with origin times separated by only seven seconds.' (Eiby 1964a)

'As on the previous occasion [1963 November 16], a questionnaire was issued, and in view of the reports of damage to buildings it was decided to visit the district. The area visited corresponds roughly to that enclosed by the MM5 isoseismal ... some 500 miles [800 km] were covered in visiting damaged properties. In addition interviews with people from outlying districts were held in Kaeo, and telephone calls were made to places it proved impracticable to visit. This information forms the basis of the isoseismal map.

'The highest intensities experienced have been assessed at MM7, and appear to have been fairly uniform over an elliptical area whose longer axis extended from Peria, near the instrumental epicentre, to Saies and Totara North on the northern shores of Whangaroa Harbour, a distance of about 25 km. The minor axis of the ellipse measures about 15 km. Within this area, there were many examples of damage to both brick and concrete-block chimneys. A few had actually fallen, and others showed the typical fracture at or near roof level.

'In the zones of lesser intensity, the most typical effects were the movement of refrigerators, stoves, and other heavy furniture, and the fall of goods from shelves. It was also noted that material had been freshly dislodged from unstable roadside slips in cuttings on the northern side of the Mangamuka hill, and the road from Pupuke to Takakuri ...' (Eiby 1964b)

The isoseismal map is reproduced from Eiby (1964a).

---

* Location and local magnitude have been revised since the publication of the New Zealand Seismological Report 1963. Refer to introduction. Recalculation of the epicentre using New Zealand Model 1 results in a change in the location of the epicentre of about 24 km to the north-east. Given that two earthquakes occurred within seven seconds of one another the accuracy of the epicentre calculation cannot be considered high. The 1963 calculated epicentre and the recent recalculation are located at opposite extremes of the MM7 isoseismal.

PERIA EARTHQUAKE
1963 DECEMBER 22

Magnitude: 4.5 M_L
Epicentre: 34.93° S 173.67° E
Depth: 10 km (restricted)

PERIA EARTHQUAKE
1963 DECEMBER 22

Magnitude: 4.9 M_L
Epicentre: 34.93° S 173.67° E
Depth: 10 km (restricted)

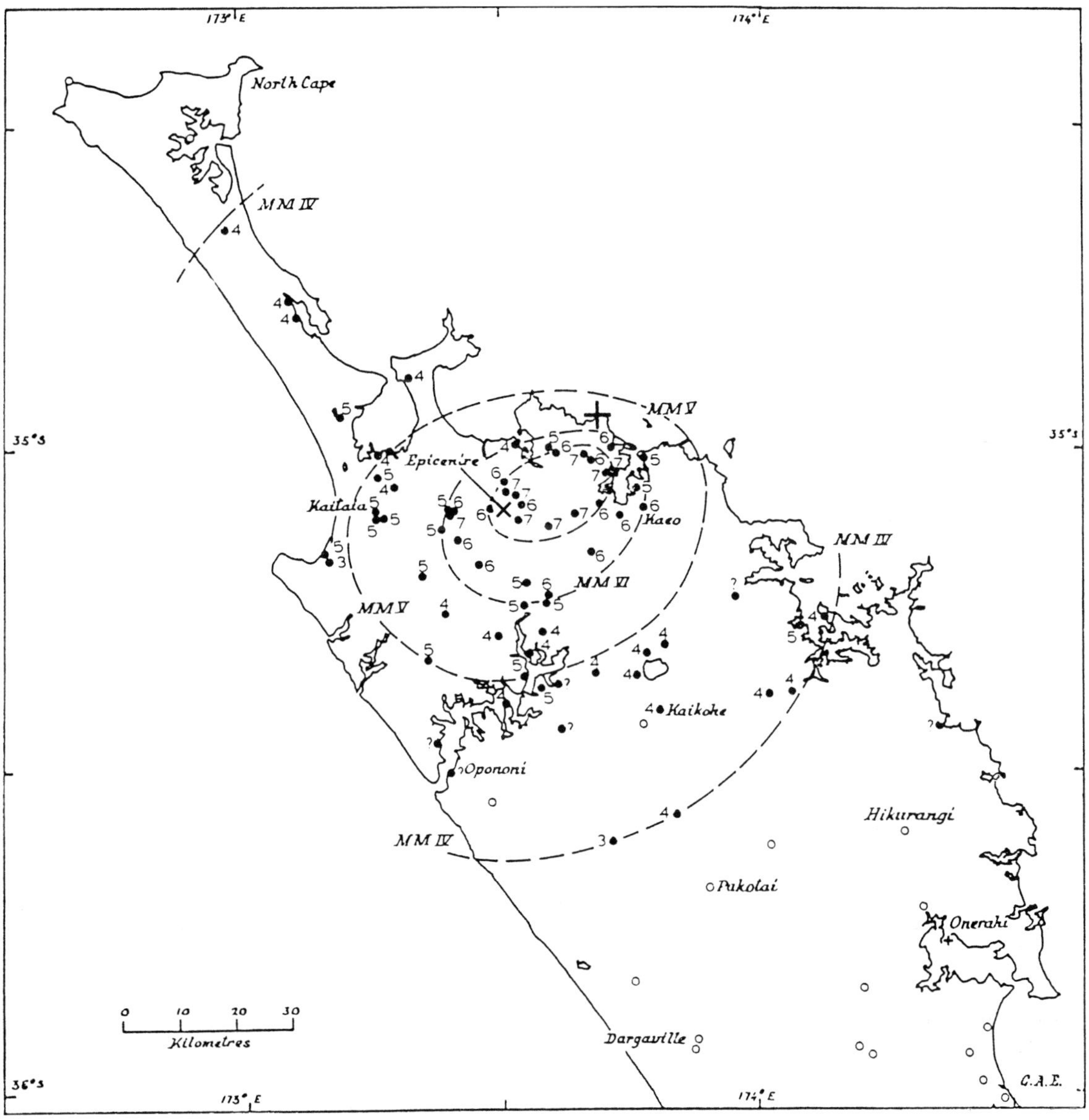

after Eiby (1964a)

79. ISOSEISMAL MAP OF THE BIG BAY EARTHQUAKE - 1964 MARCH 08

DATE:	1964 MARCH 08
TIME:	01:35:47.2 UT
MAGNITUDE:	5.8 M_L* ; 5.8 M_s (Dowrick & Smith 1990); 5.9 M_w (Anderson et al. 1993)
EPICENTRE:	44.26°S 167.72°E (1)
DEPTH:	12 km (restricted)

'The most important shallow earthquake [in 1964] in the immediate vicinity of New Zealand occurred on March 8, and was centred off the west coast of the South Island near Big Bay. This shock, of magnitude 6.3 [5.8 M_L*], was the largest since the magnitude 7.0 one in the same region in May 1960. No damage resulted, but the intensity reached MM5 or more over most of western Otago and southern Westland. The felt area extended over the whole of these provinces, and to Southland and the southern parts of Canterbury.' (Seismological Observatory bulletin E 145)

The isoseismal map has been reproduced from the Seismological Observatory bulletin E 145.

* Location and local magnitude have been revised since the publication of the Seismological Observatory bulletin E 145. Refer to introduction.

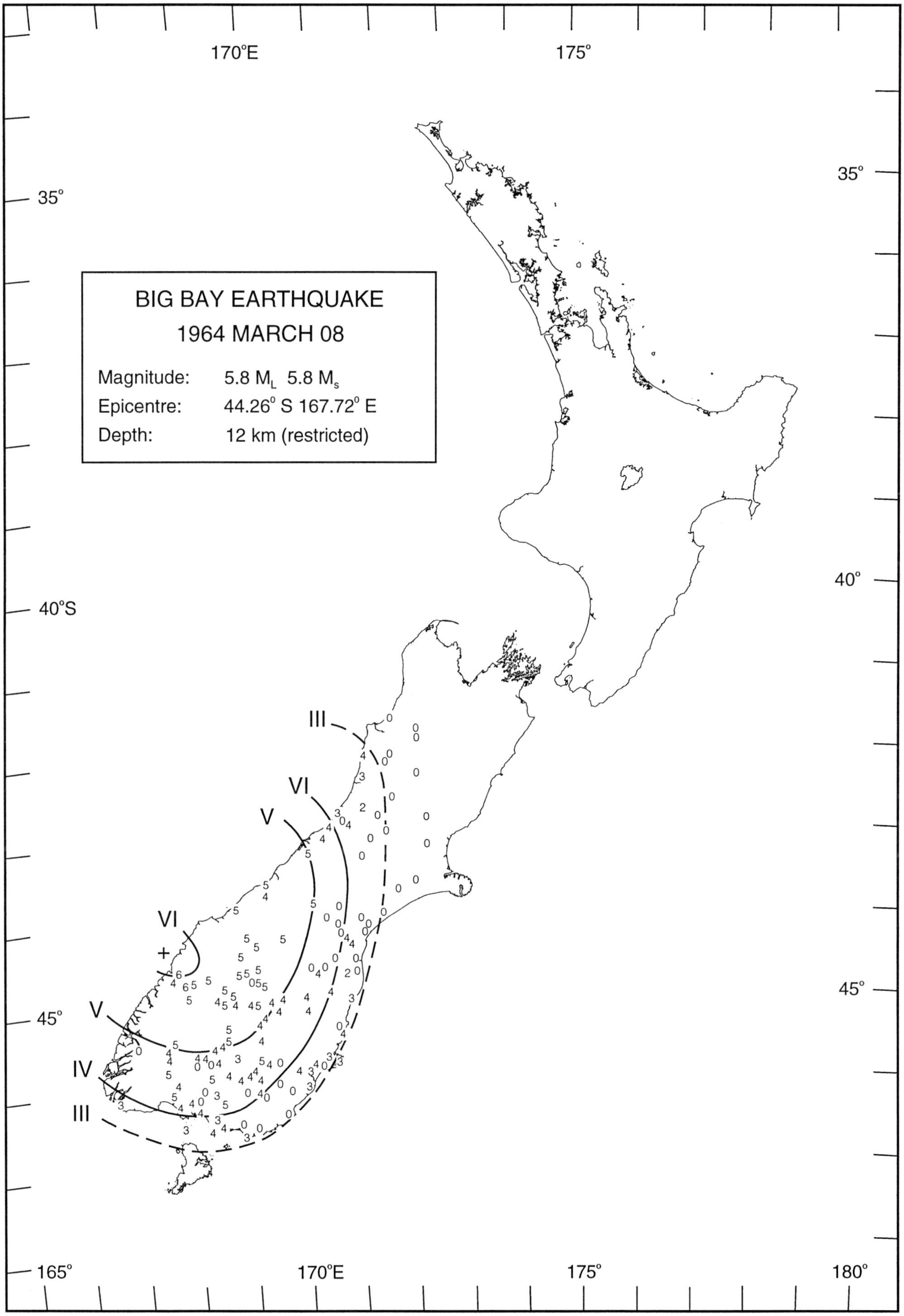
BIG BAY EARTHQUAKE
1964 MARCH 08
Magnitude: 5.8 M_L 5.8 M_s
Epicentre: 44.26° S 167.72° E
Depth: 12 km (restricted)
170°E
175°
35°
40°S
40°
45°
165°
180°
III
VI
V
IV

80. ISOSEISMAL MAP OF THE CHATHAM RISE EARTHQUAKE - 1965 APRIL 11

DATE:	1965 APRIL 11
TIME:	00:11:10.3 UT
MAGNITUDE:	6.1 M_L*; 5.8 M_s (ISC); 6.1 M_w (Anderson et al. 1993)
EPICENTRE:	42.80°S 174.31°E (1)
DEPTH:	12 km (restricted)

The earthquake of 1965 April 11 of magnitude 6.1 M_L was one of two large shallow shocks in 1965 and was centred off the coast to the south-east of Kaikoura. It was felt from Wellington to Christchurch and on the West Coast at Westport. The maximum intensity of MM5 was experienced on the coast from Kaikoura to Motunau and at Akaroa.

The isoseismal map has been reproduced from the Seismological Observatory bulletin E 146.

* Location and local magnitude have been revised since the publication of the Seismological Observatory bulletin E 146. Refer to introduction.

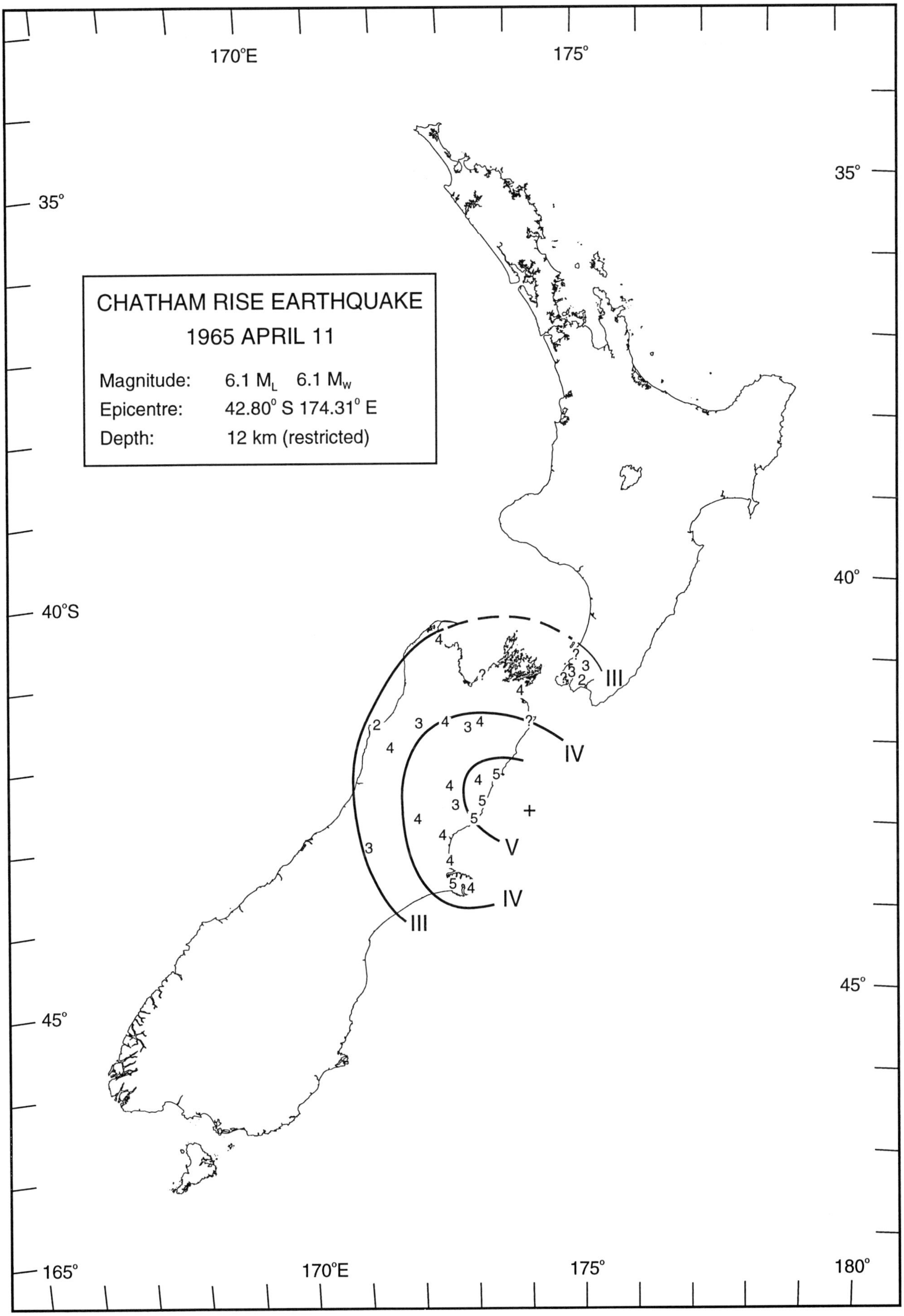

CHATHAM RISE EARTHQUAKE
1965 APRIL 11
Magnitude: 6.1 M_L 6.1 M_w
Epicentre: 42.80° S 174.31° E
Depth: 12 km (restricted)
170°E
175°
165°
180°
35°
40°S
40°
45°
III
IV
V

81. ISOSEISMAL MAP OF THE LAKE TE ANAU EARTHQUAKE - 1965 MAY 20

DATE:	1965 MAY 20
TIME:	20:37:41.0 UT
MAGNITUDE:	5.6 M_L*
EPICENTRE:	45.05°S 167.51°E (1)
DEPTH:	105 km

'The earthquake of 1965 May 20 is of particular interest because it was the first shock in the Fiordland Seismic Region that has been shown beyond dispute to have greater than normal depth. The epicentre lies in the north-western part of Lake Te Anau, and the focus is at a depth of 103 ± 5 km. The existence of shocks at depths of this order in the Fiordland Region has long been suspected, but the absence of good recording stations at short distances has hitherto prevented adequate confirmation.

'The shock was felt over the southern half of the South Island with maximum intensities of about MM5.' (Seismological Observatory bulletin E 146)

The isoseismal map has been reproduced from the Seismological Observatory bulletin E 146.

* Local magnitudes and hypocentres have been revised since the publication of the Seismological Observatory bulletin E 146. Refer to introduction.

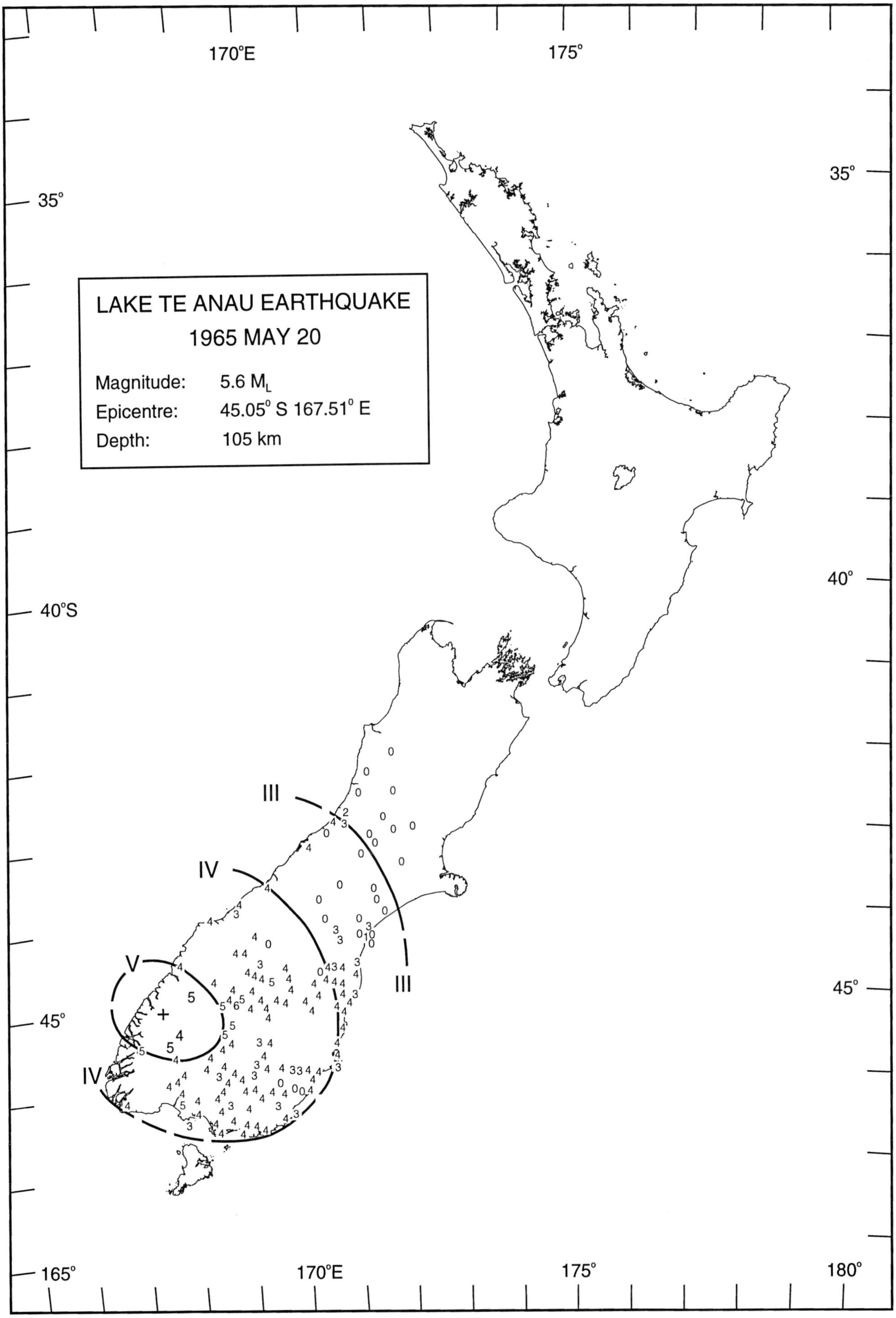
LAKE TE ANAU EARTHQUAKE
1965 MAY 20
Magnitude: 5.6 M_L
Epicentre: 45.05° S 167.51° E
Depth: 105 km
170°E
175°
35°
40°
45°
35°
40°S
45°
165°
170°E
175°
180°
III
IV
V
IV
III

82. ISOSEISMAL MAP OF THE BAY OF PLENTY EARTHQUAKE - 1965 JUNE 15

DATE:	1965 JUNE 15
TIME:	09:20:31.7 UT
MAGNITUDE:	5.8 M_L^* ; 5.3 M_s (Dowrick & Smith 1990)
EPICENTRE:	37.90°S 177.53°E (1)
DEPTH:	33 km (restricted)

'Of the large shallow shocks during the year, the most noteworthy are those of April 11 ... and of June 15 of magnitude 6.0 [5.8 M_L^*] , centred near the mouth of the Motu River, in the eastern Bay of Plenty. Neither of these shocks produced reported intensities above MM5. The earthquake was felt over most of the North Island.' (Seismological Observatory bulletin E 146)

The isoseismal map has been reproduced from the Seismological Observatory bulletin E 146.

* Local magnitudes and hypocentres have been revised since the publication of the Seismological Observatory bulletin E 146. Refer to introduction.

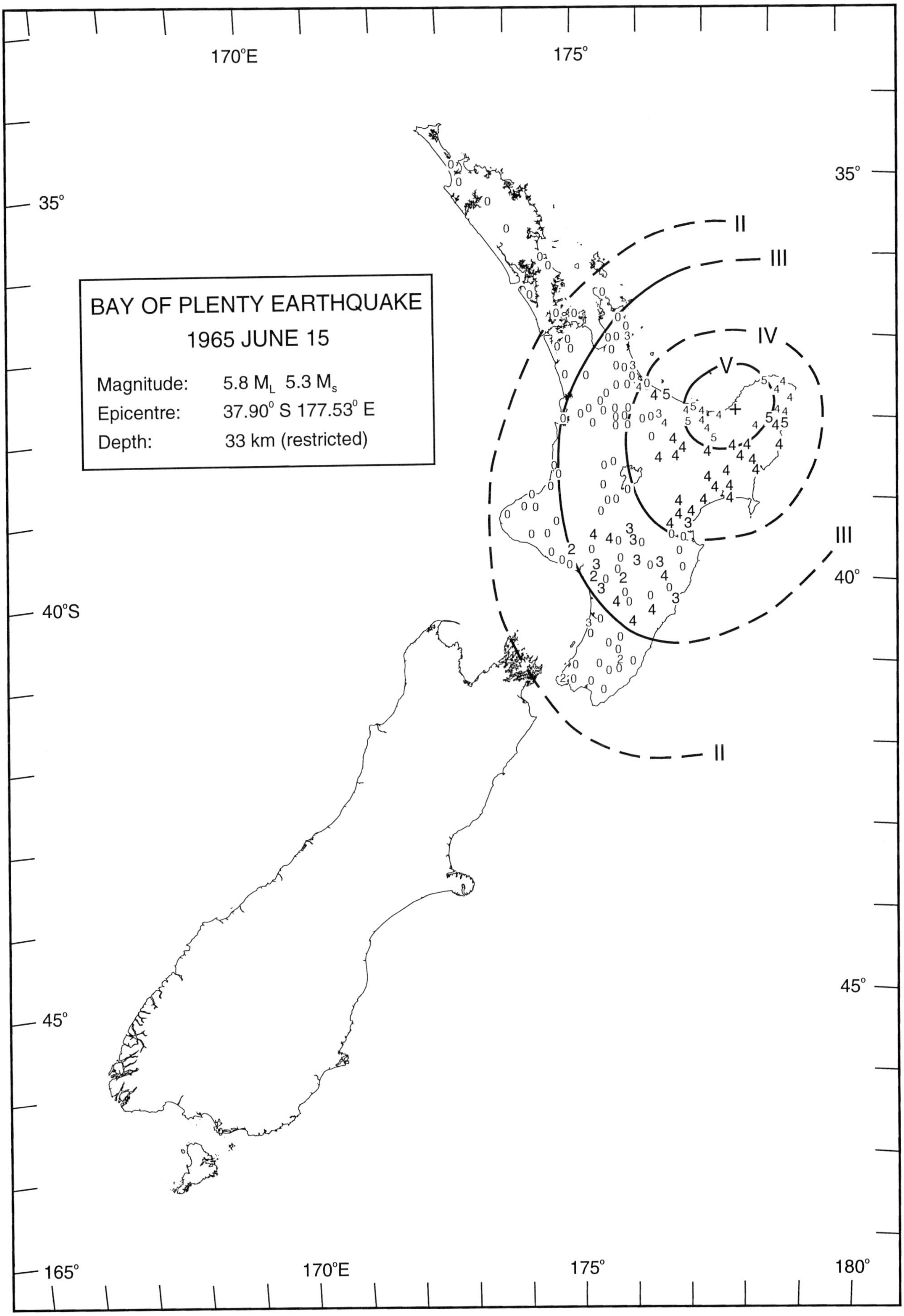

BAY OF PLENTY EARTHQUAKE
1965 JUNE 15
Magnitude: 5.8 M_L 5.3 M_s
Epicentre: 37.90° S 177.53° E
Depth: 33 km (restricted)
170°E
175°
35°
40°S
45°
165°
180°
40°
II
III
IV
V

83. ISOSEISMAL MAP OF THE BAY OF PLENTY EARTHQUAKE - 1965 DECEMBER 08

DATE:	1965 DECEMBER 08
TIME:	18:05:20.7 UT
MAGNITUDE:	6.5 M_L^*
EPICENTRE:	37.11°S 177.66°E (1)
DEPTH:	234 km

'The most widely felt of the deep earthquakes [in 1965] was on December 8, of focal depth 221 km [234 km*], magnitude 6.3 [6.5 M_L^*], and centred beneath the Bay of Plenty about 100 km north of Opotiki. This shock shows the very elongated macro-seismic region typical of the deep earthquakes of the New Zealand Sub-Crustal Rift. Intensities of MM3 or 4 were reported all over the eastern half of the North Island, and at isolated South Island localities as far south as Greymouth. No reports were received from Taranaki, or the Northland or Coromandel Peninsulas. A New Zealand Broadcasting Commission news report that the shock was felt in Auckland cannot be confirmed.' (Seismological Observatory bulletin E 146)

The isoseismal map has been reproduced from the Seismological Observatory bulletin E 146.

* Local magnitudes and hypocentres have been revised since the publication of the Seismological Observatory bulletin E 146. Refer to introduction.

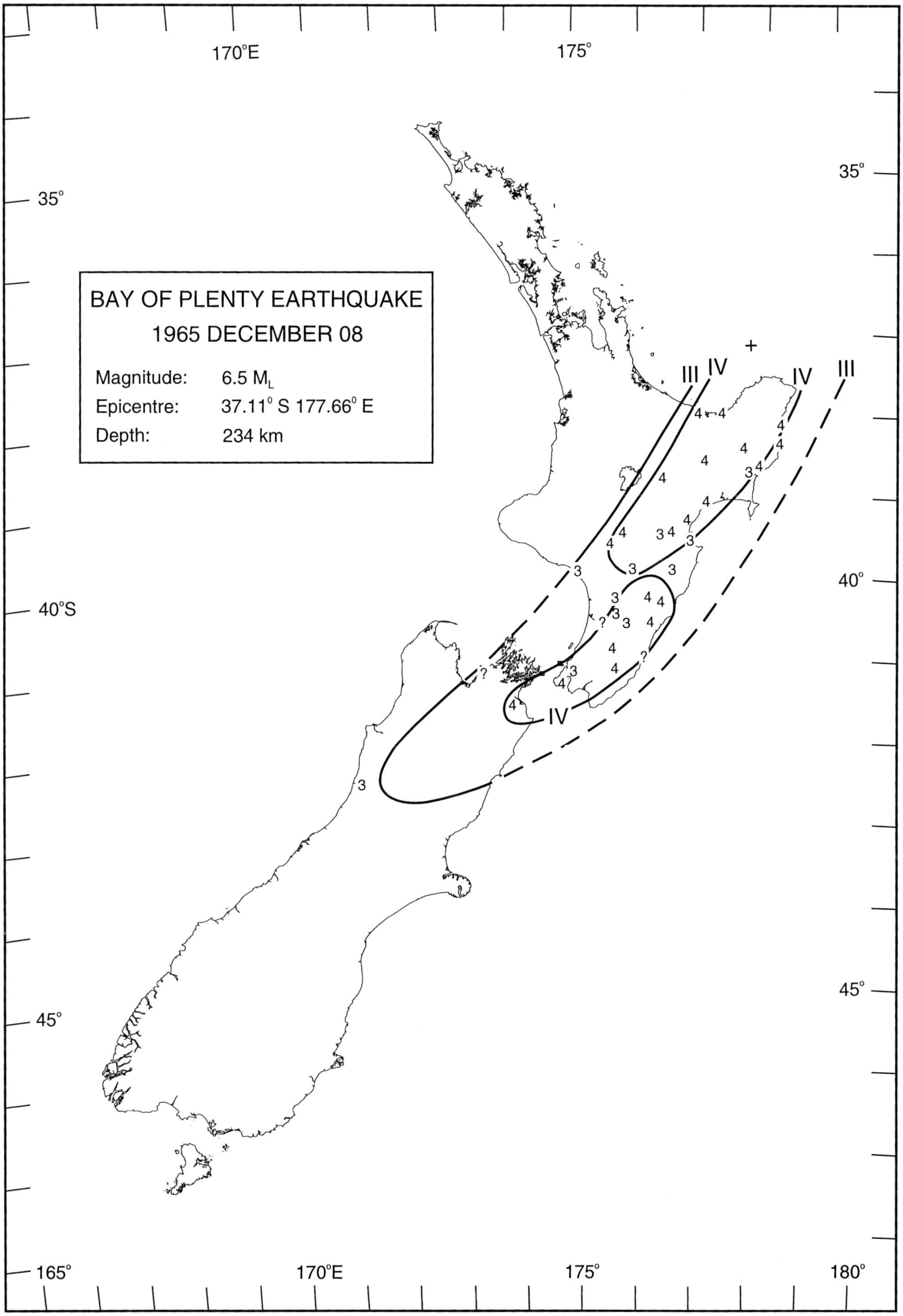

BAY OF PLENTY EARTHQUAKE
1965 DECEMBER 08
Magnitude: 6.5 M_L
Epicentre: 37.11° S 177.66° E
Depth: 234 km
170°E
175°
165°
180°
35°
40°S
40°
45°
III
IV

84a. ISOSEISMAL MAP OF THE GISBORNE EARTHQUAKE - 1966 MARCH 04

DATE:	1966 MARCH 04
TIME:	23:58:56.8 UT
MAGNITUDE:	6.0 M_L* ; 5.8 M_s (Dowrick & Smith 1990)
EPICENTRE:	38.52°S 177.85 E (Hamilton et al. 1969)
DEPTH:	25 km

'Shortly before noon on Saturday, 5 March 1966 (1966 Mar 04, UT) the city of Gisborne was shaken by one of the most damaging earthquakes in New Zealand in recent years. Since the Wairarapa earthquake of 1942, only the Westport earthquakes of 1962 had caused more damage. Although the earthquake was not an exceptionally large one, having a magnitude of 6.2 [6.0 M_L*], the resulting damage was, nevertheless, disproportionately great because the shock occurred at a shallow depth close to the city. Many chimneys were cracked, and some were broken off at the roofline, and water mains were broken in numerous places. Damage occurred in a number of buildings, but was largely confined to masonry.' (Hamilton et al. 1969)

The time, epicentre and depth and isoseismal map are taken from Hamilton et al. (1969). These differ significantly from those in the Seismological Observatory bulletin E 148. The isoseismal map from this report is reproduced on the following pages. The felt data on which the maps are based is probably identical. The epicentre recalculated using New Zealand Model 1 does not differ greatly from that calculated in 1966. Re-analysis of the seismograms would perhaps reveal which solution is preferable.

---

* Local magnitudes have been revised since the publication of the Seismological Observatory bulletin E 148. Refer to introduction.

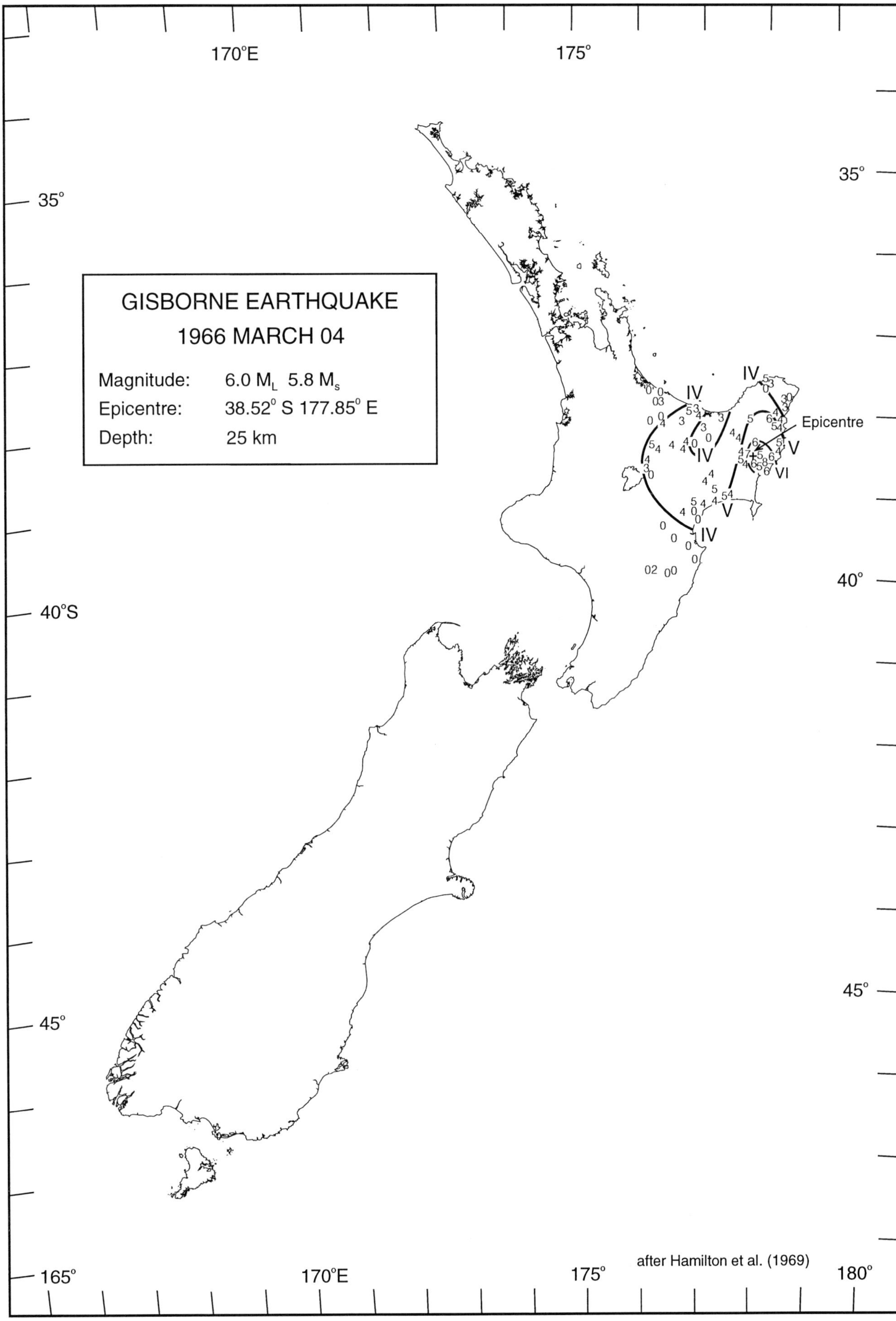
170°E
175°
35°
35°
GISBORNE EARTHQUAKE
1966 MARCH 04
Magnitude: 6.0 M_L 5.8 M_s
Epicentre: 38.52° S 177.85° E
Depth: 25 km
IV
IV
IV
IV
IV
V
V
VI
Epicentre
40°
40°S
45°
45°
after Hamilton et al. (1969)
165°
170°E
175°
180°

84b. ISOSEISMAL MAP OF THE GISBORNE EARTHQUAKE - 1966 MARCH 04

DATE:	1966 MARCH 04
TIME:	23:58:56.4 UT
MAGNITUDE:	6.0 M_L* ; 5.8 M_s (Dowrick & Smith 1990)
EPICENTRE:	38.77°S 178.14°E (1)
DEPTH:	25 km

'The radius of the felt area was about 200 km, covering the East Cape peninsula, northern Hawke's Bay and the eastern Bay of Plenty. Gisborne, with about 29,000 inhabitants, is the largest centre of population in the district, and experienced intensities of MM7, though MM8 was reached on some areas of unconsolidated sandy ground near the mouths of the Turanganui and Taruheru Rivers. Many chimneys were cracked at the roof-line, and some water and gas mains broken.

'Modern structures were not materially affected, but older brick masonry, some of it approaching 100 years old, suffered damage. In many cases damage in earlier earthquakes had been concealed rather than repaired, and again came to light. Pieces fell from the parapets of two old industrial buildings in a poor state of repair, and these and some other structures were subsequently demolished. There were no casualties or loss of life.

'The earthquake was accompanied by some 50 foreshocks and 513 aftershocks, most of them having epicentres within 10 km of the main shock.' (Seismological Observatory bulletin E 148)

The isoseismal map is taken from the Seismological Observatory bulletin E 148. This and the hypocentral parameters differ significantly from those in Hamilton et al. (1969), which are given on the previous pages. The felt data on which the maps are based are probably identical. Re-analysis of the seismograms would perhaps reveal which solution is preferable.

---

* Local magnitude and hypocentres have been revised since the publication of the Seismological Observatory bulletin E 148. Refer to introduction. The recalculated epicentre using New Zealand Model 1 does not differ greatly from that calculated in 1966.

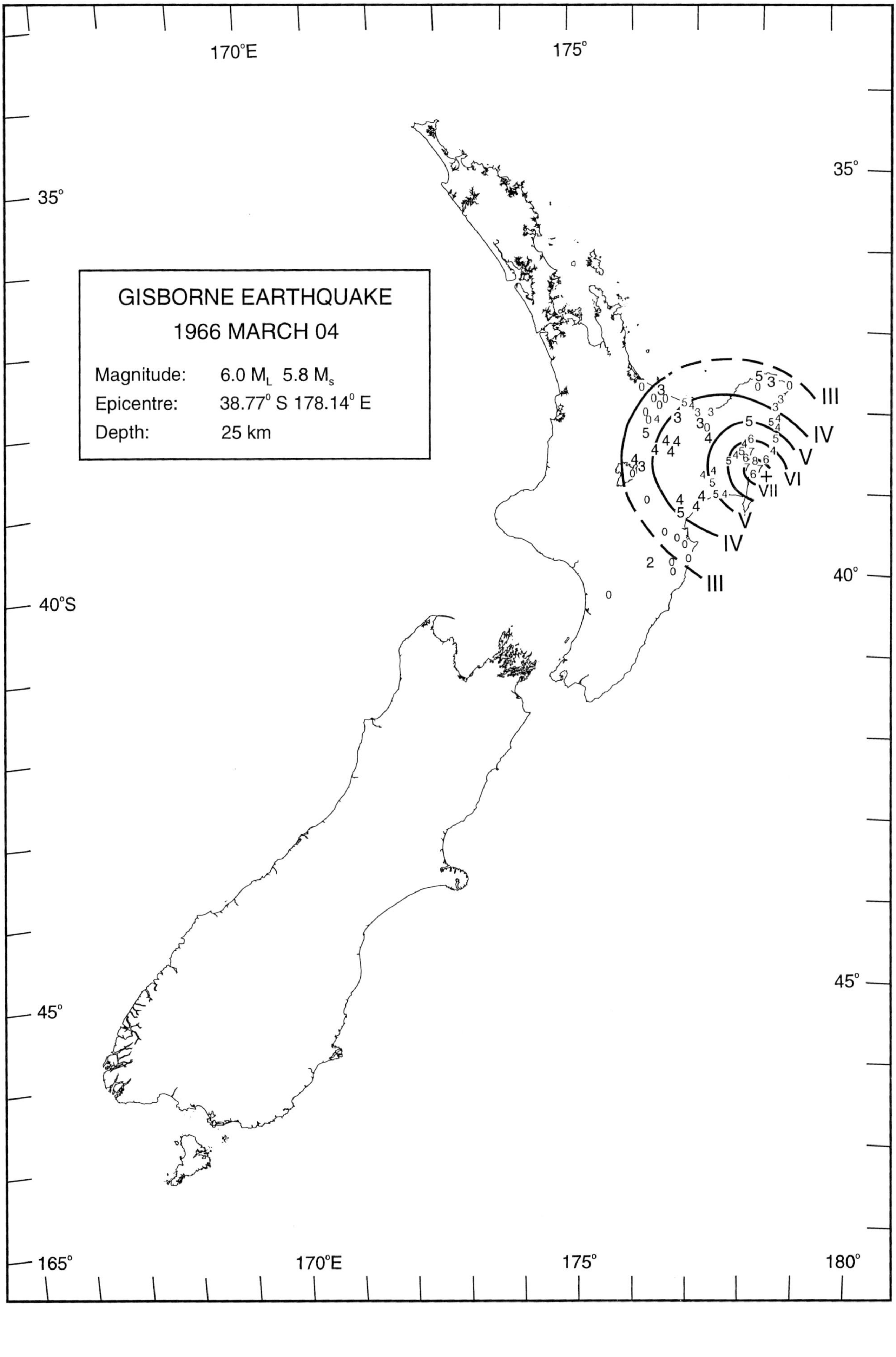
GISBORNE EARTHQUAKE
1966 MARCH 04
Magnitude: 6.0 M_L 5.8 M_s
Epicentre: 38.77° S 178.14° E
Depth: 25 km
170°E
175°
35°
40°S
40°
45°
165°
180°
III
IV
V
VI
VII

85. ISOSEISMAL MAP OF THE SEDDON EARTHQUAKE - 1966 APRIL 23

DATE:	APRIL 23 1966
TIME:	06:49:40.5 UT
MAGNITUDE:	5.8 M_L*; 5.6 M_s (Dowrick & Smith 1990); 5.8 M_w (Anderson et al. 1993)
EPICENTRE:	41.64 S 174.52 E (Adams et al. 1970)
DEPTH:	22 km

'In the early evening of 23 April 1966 an earthquake of magnitude 6.1 [5.8 M_L*], originating in Cook Strait, was felt widely in the north of the South Island and the south of the North Island. The highest reported intensities of MM VII to VIII came from the small town of Seddon, about 35 km from the epicentre, where a high proportion of chimneys was damaged.

'This earthquake was not exceptional by virtue of its magnitude or location, but, like the Gisborne shock 7 weeks earlier, attracted public attention by being close enough to centres of population to cause significant damage...

'The felt area of the Seddon earthquakes is shown in the isoseismal map ... It extends as far south as Hokitika and Banks Peninsula, and to Taranaki in the north. Isolated felt reports were received from Gisborne and Cambridge. Damage at Seddon, reported extensively in this volume [i.e. Adams et al. 1970], was mainly confined to the breaking of chimneys, and is generally consistent with an intensity of MM VII. Two isolated reports from Seddon have been assigned an intensity of MM VIII. Reports from Blenheim, where some damage was done to older chimneys and brick parapets, indicate an intensity of MM VI. At Wellington, where chimneys were reported cracked, and some windows were broken, an intensity of MM VI appears to have been reached. This intensity is the highest reported in Wellington since 1942, when MM VII was experienced during the Wairarapa earthquakes.' (Adams et al. 1970)

'Seddon is built upon silts and coarse gravels up to about 18 metres thick, overlying mudstone. This cannot be considered a good foundation from a seismic point of view. Practically every house had chimney damage and damage to household goods. Railway lines were affected, watermains were broken, and electricity and telephone services failed.

'Minor damage to old structures was also experienced in Ward, Blenheim, and Wellington. In most cases, old damage or foundation settlement contributed to the result.' (Seismological Observatory bulletin E 148)

The time, epicentre and depth are from Adams et al (1970) and are the result of a special study. The authors note that the epicentre may be in error up to 15 km and the depth by 10 km. The isoseismal map is reproduced from the Seismological Observatory bulletin E 148.

* Local magnitudes have been revised since the publication of the Seismological Observatory bulletin E 148 and the Adams et al (1970) paper. Refer to introduction.

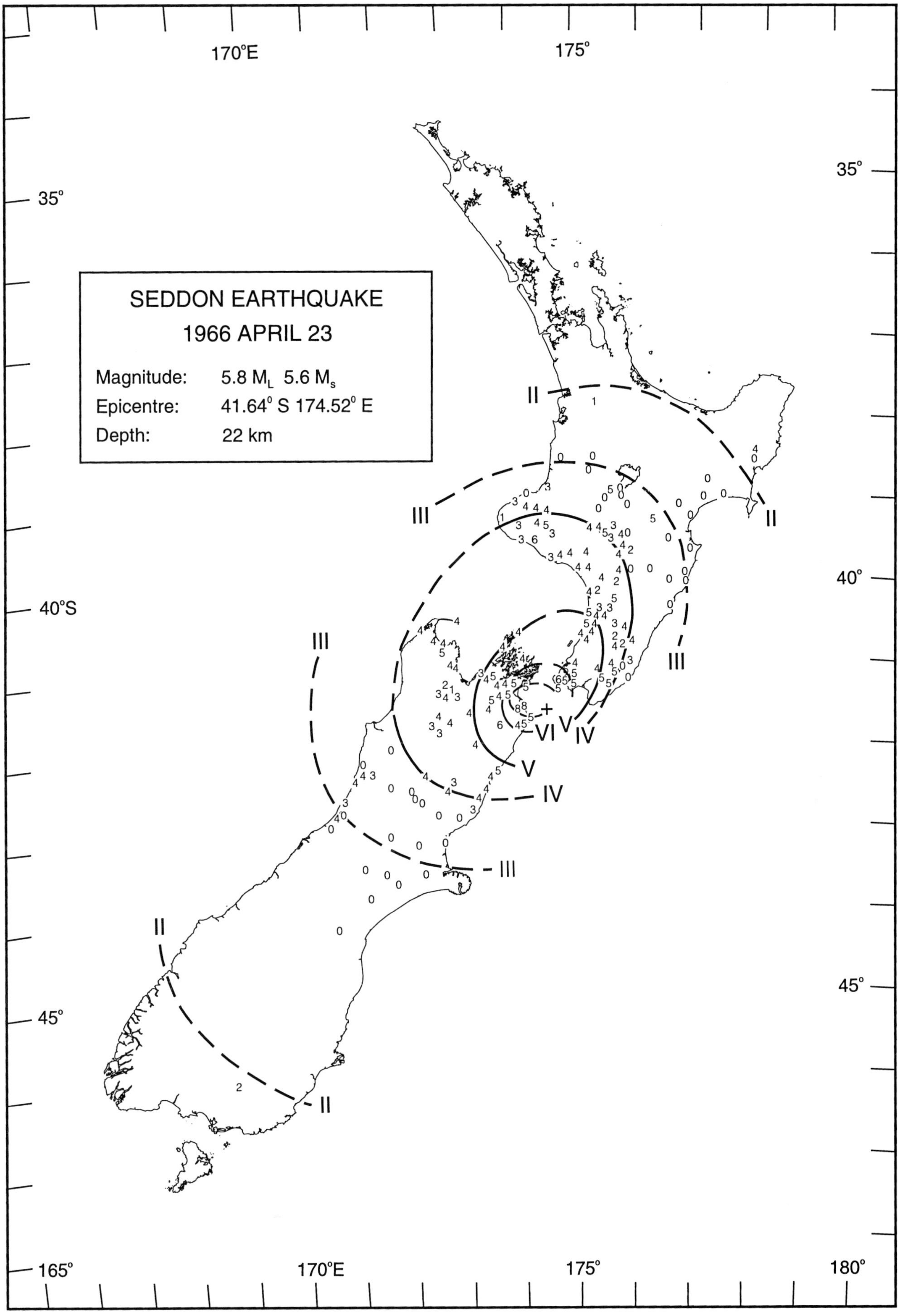

SEDDON EARTHQUAKE
1966 APRIL 23
Magnitude: 5.8 M_L 5.6 M_s
Epicentre: 41.64° S 174.52° E
Depth: 22 km
170°E
175°
180°
165°
35°
40°S
40°
45°
II
III
IV
V
VI

86. ISOSEISMAL MAP OF PALMERSTON NORTH EARTHQUAKE - 1967 JANUARY 16

DATE:	1967 JANUARY 16
TIME:	11:40:59.1 UT
MAGNITUDE:	4.9 M_L*
EPICENTRE:	40.34°S 175.52°E (1)
DEPTH:	12 km (restricted)

'The earthquake of January 16, which caused minor damage in the Foxton area, was the first of a number of moderate shallow shocks concentrated in the southern part of the North Island.' (Seismological Observatory bulletin E 149)

The isoseismal map drawn by Eiby from felt reports held at the Seismological Observatory has not been published previously.

* Local magnitudes and hypocentres have been revised since the publication of the Seismological Observatory bulletin E 149. Refer to introduction.

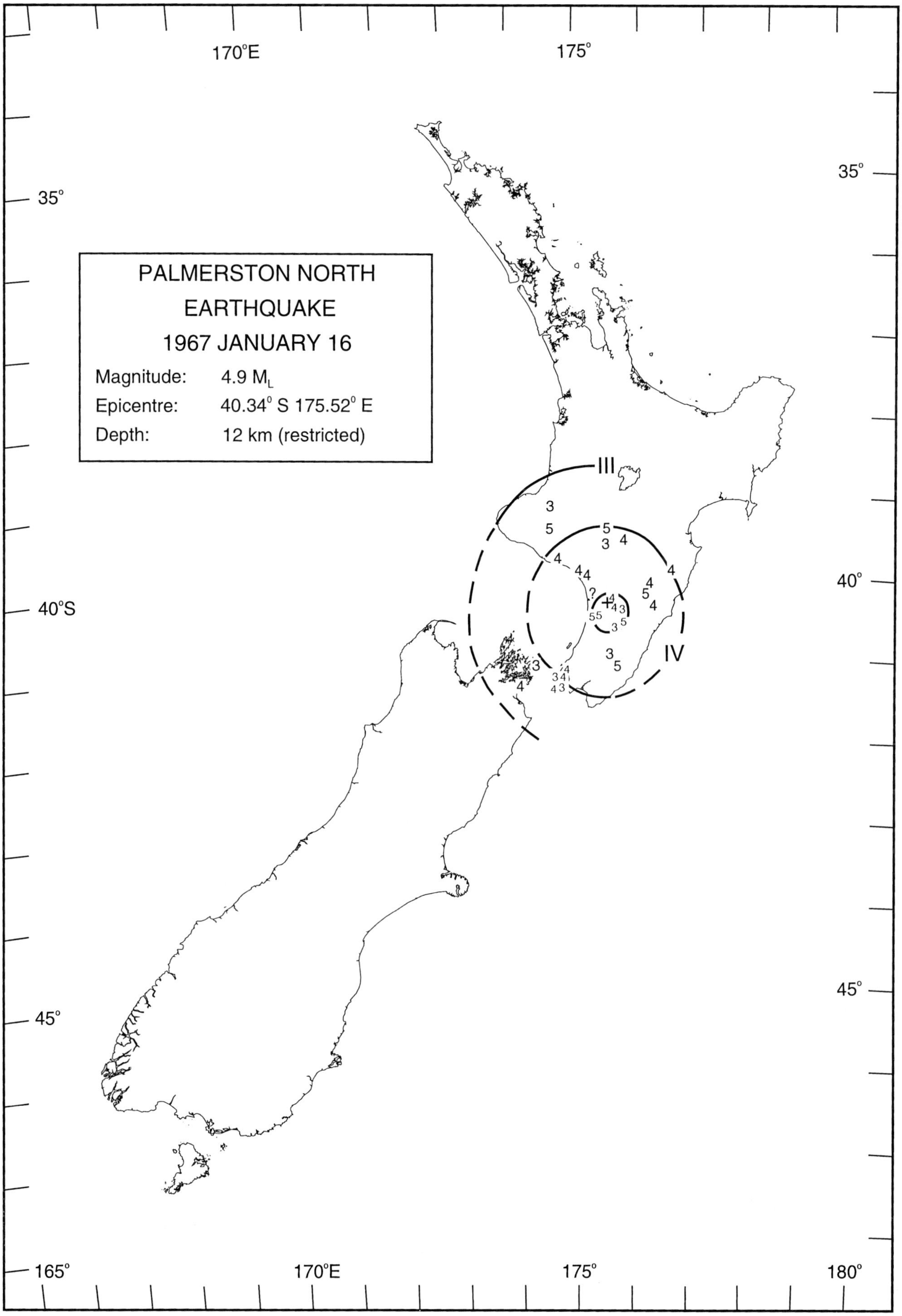

PALMERSTON NORTH
EARTHQUAKE
1967 JANUARY 16
Magnitude: 4.9 M_L
Epicentre: 40.34° S 175.52° E
Depth: 12 km (restricted)
170°E
175°
35°
40°S
40°
45°
165°
180°
III
IV

87. ISOSEISMAL MAP OF THE OWAHANGA EARTHQUAKE - 1967 MARCH 24

DATE:	1967 MARCH 24
TIME:	19:09:17.4 UT
MAGNITUDE:	5.3 M_L^*
EPICENTRE:	40.70°S 176.51°E (1)
DEPTH:	12 km (restricted)

'The largest of ... [a number of moderate shallow shocks concentrated in the southern part of the North Island] (M=5.4 [5.3 M_L^*]) occurred on March 24 and originated off the east coast, to the north of Castlepoint. Intensities on land did not exceed MM V, and there was no damage.' (Seismological Observatory bulletin E 149)

The isoseismal map previously published in the Seismological Observatory bulletin E 149 as a small scale map has been replotted and redrawn with the revised epicentre by Eiby from felt reports held at the Seismological Observatory.

* Local magnitudes and hypocentres have been revised since the publication of the Seismological Observatory bulletin E 149. Refer to introduction.

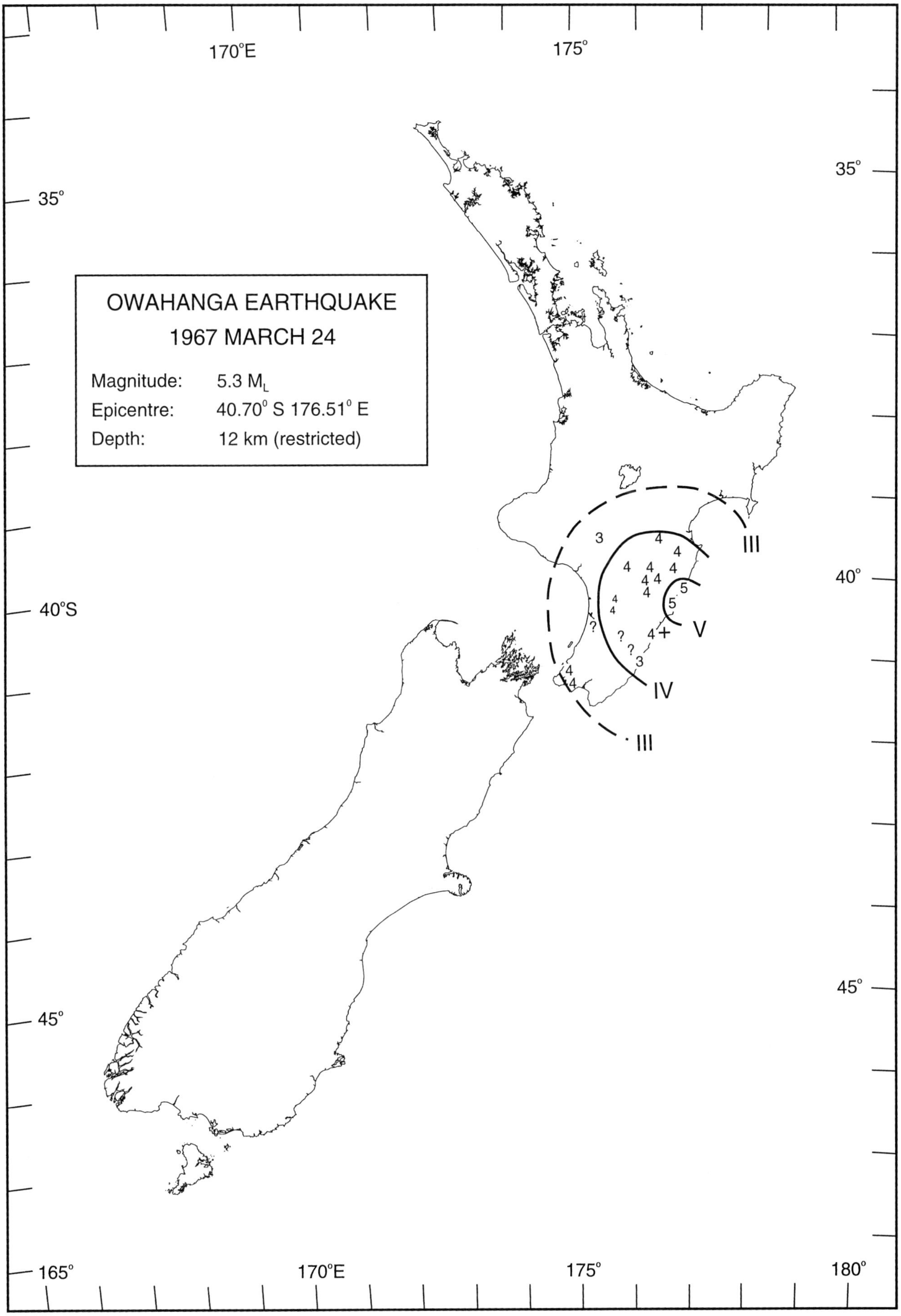
OWAHANGA EARTHQUAKE
1967 MARCH 24
Magnitude: 5.3 M_L
Epicentre: 40.70° S 176.51° E
Depth: 12 km (restricted)
170°E
175°
35°
40°S
40°
45°
165°
180°
III
IV
V

88. ISOSEISMAL MAP OF THE TE HORO EARTHQUAKE - 1967 SEPTEMBER 21

DATE:	1967 SEPTEMBER 21
TIME:	17:45:46.0 UT
MAGNITUDE:	4.7 M_L^*
EPICENTRE:	40.93°S 175.37°E (1)
DEPTH:	64 km

The earthquake, on September 21 of magnitude 4.7 M_L^* was one of a number of shallow, moderate shocks concentrated in the southern part of the North Island in 1967 and it reached a maximum intensity of MM6 at Ohau and Paraparaumu.

The isoseismal map previously published in the Seismological Observatory bulletin E 149 as a small scale map has been replotted and redrawn with the revised epicentre from felt reports held at the Seismological Observatory.

---

* Local magnitudes and hypocentres have been revised since the publication of the Seismological Observatory bulletin E 149. Refer to introduction.

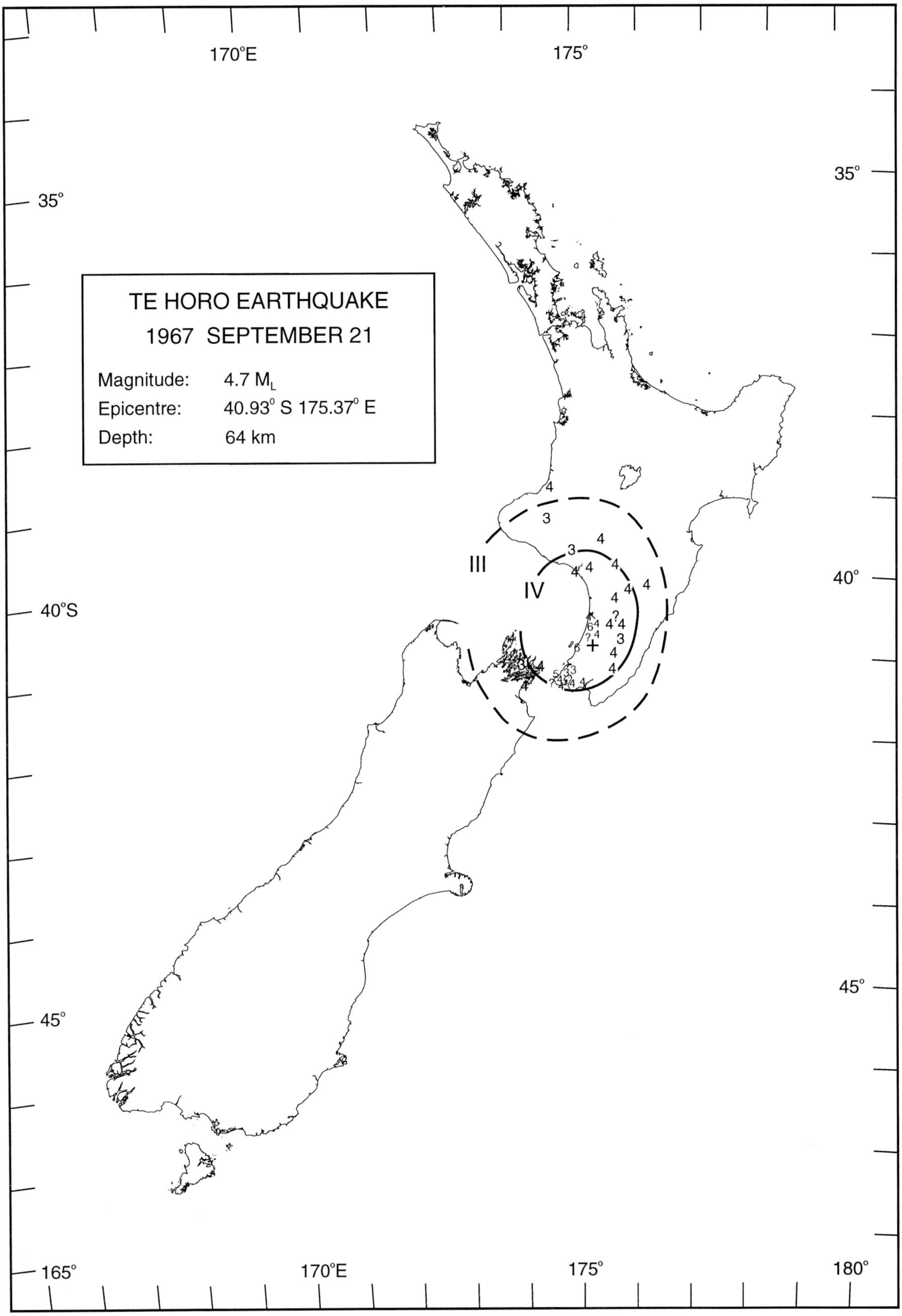
TE HORO EARTHQUAKE
1967 SEPTEMBER 21
Magnitude: 4.7 M_L
Epicentre: 40.93° S 175.37° E
Depth: 64 km
170°E
175°
35°
40°S
40°
45°
165°
180°
III
IV

89. ISOSEISMAL MAP OF THE WHANGAEHU EARTHQUAKE - 1967 DECEMBER 20

DATE:	1967 DECEMBER 20
TIME:	14:56:30.5 UT
MAGNITUDE:	4.8 M_L*
EPICENTRE:	39.98°S 175.11 E
DEPTH:	12 km (restricted)

'The last in this group [of a number of shallow moderate shocks concentrated in the southern part of the North Island], on December 20 was centred near Wanganui, and its felt area extended to Taranaki and the central North Island rather than to the south. The size of the felt area and the number of reports of MM V are a little surprising for a shock of this magnitude, and the report of MM VI at Parihauhau should probably be treated with caution.' (Seismological Observatory bulletin E 149)

The isoseismal map drawn by Eiby from felt reports held at the Seismological Observatory has not been published previously.

* Local magnitudes and hypocentres have been revised since the publication of the Seismological Observatory bulletin E 149. Refer to introduction.

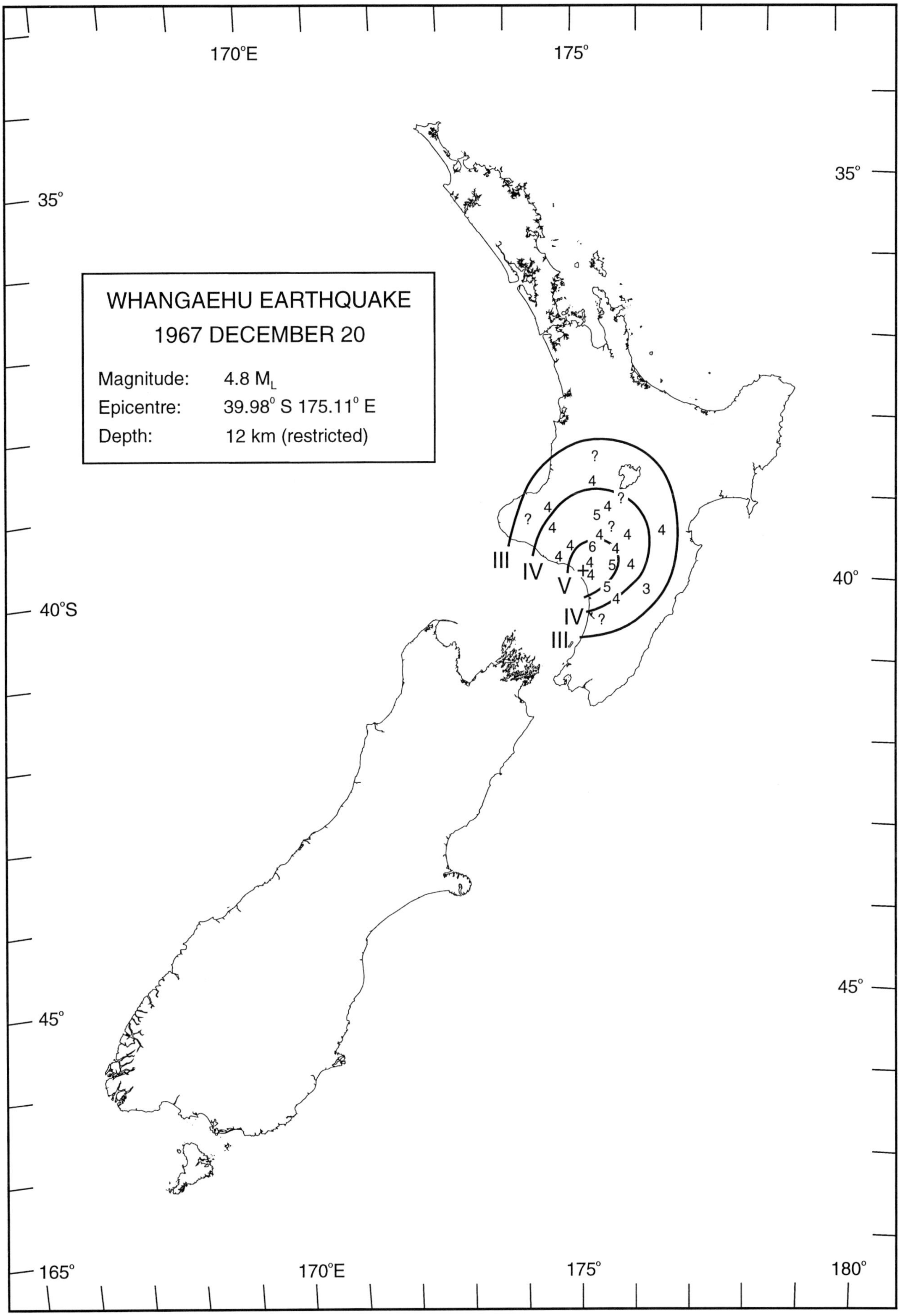
WHANGAEHU EARTHQUAKE
1967 DECEMBER 20
Magnitude: 4.8 M_L
Epicentre: 39.98° S 175.11° E
Depth: 12 km (restricted)
170°E
175°
35°
40°S
40°
45°
165°
180°
III
IV
V
IV
III

90a. ISOSEISMAL MAP OF THE INANGAHUA EARTHQUAKE - 1968 MAY 23

DATE:	1968 MAY 23
TIME:	17:24:15.6 UT
MAGNITUDE:	7.0-7.1 M_L (Adams et al. 1969);
	7.4 M_s (Dowrick & Smith 1990);
	7.1 M_w (Anderson et al. 1993)
EPICENTRE:	41.76°S 171.96 E ± 10 km (Anderson et al. 1994)
DEPTH:	15 km ± 5 km

'The Inangahua earthquake of 1968 May 23 (May 24, 5.24 a.m. local time) was the first New Zealand earthquake to reach magnitude 7 since that in Fiordland on 1960 May 24. It became the sixteenth New Zealand earthquake believed to have reached magnitude 7 or greater since 1848, since when all such earthquakes should have been observed.

'The macroseismic effects of the Inangahua earthquake were pronounced and widespread. In the immediate vicinity of the epicentre few structures were undamaged; chimneys were damaged or destroyed at distances of more than 150 km, and all of the country with the exception of northeasterly parts of the North Island and southeastern Otago experienced intensities of at least MM IV. Because the shock occurred at 5.24 a.m. local time, when most people were sleeping, lower intensities were observed in only a few cases.

'The maximum intensities reported are close to Inangahua, some 15 km south of the epicentre. Large landslides (one of them responsible for two deaths and another temporarily blocking the river), serious damage to wooden structures including houses and bridges, bending of railway lines, breaking of underground pipes, slumping and cracking of roads, and ejection of groundwater clearly establish an intensity of MM X. Some of the heaviest damage would possibly justify an intensity a rating of MM XI, but it is impossible to be certain how typical this is and over what area it extends.

'The information on which they [i.e. the isoseismal maps] are based was gathered in several ways. In the outer parts of the felt area reports by regular observers entered upon the Seismological Observatory's standard reporting form predominate. There were over 100 of these. Officers of the New Zealand Geological Survey who visited the damaged area collected another 110 reports, also on the standard forms. All of these, together with the press reports, many of which are authenticated by clear photographs, and some private correspondence have been assessed at the Observatory in a uniform manner ...' (Adams et al. 1969)

The isoseismal map is from the special issue of the New Zealand Seismological Report on the 1968 Inangahua earthquakes. A detail map of the epicentral area appears on the next pages. The epicentre is from Anderson et al. (1994)

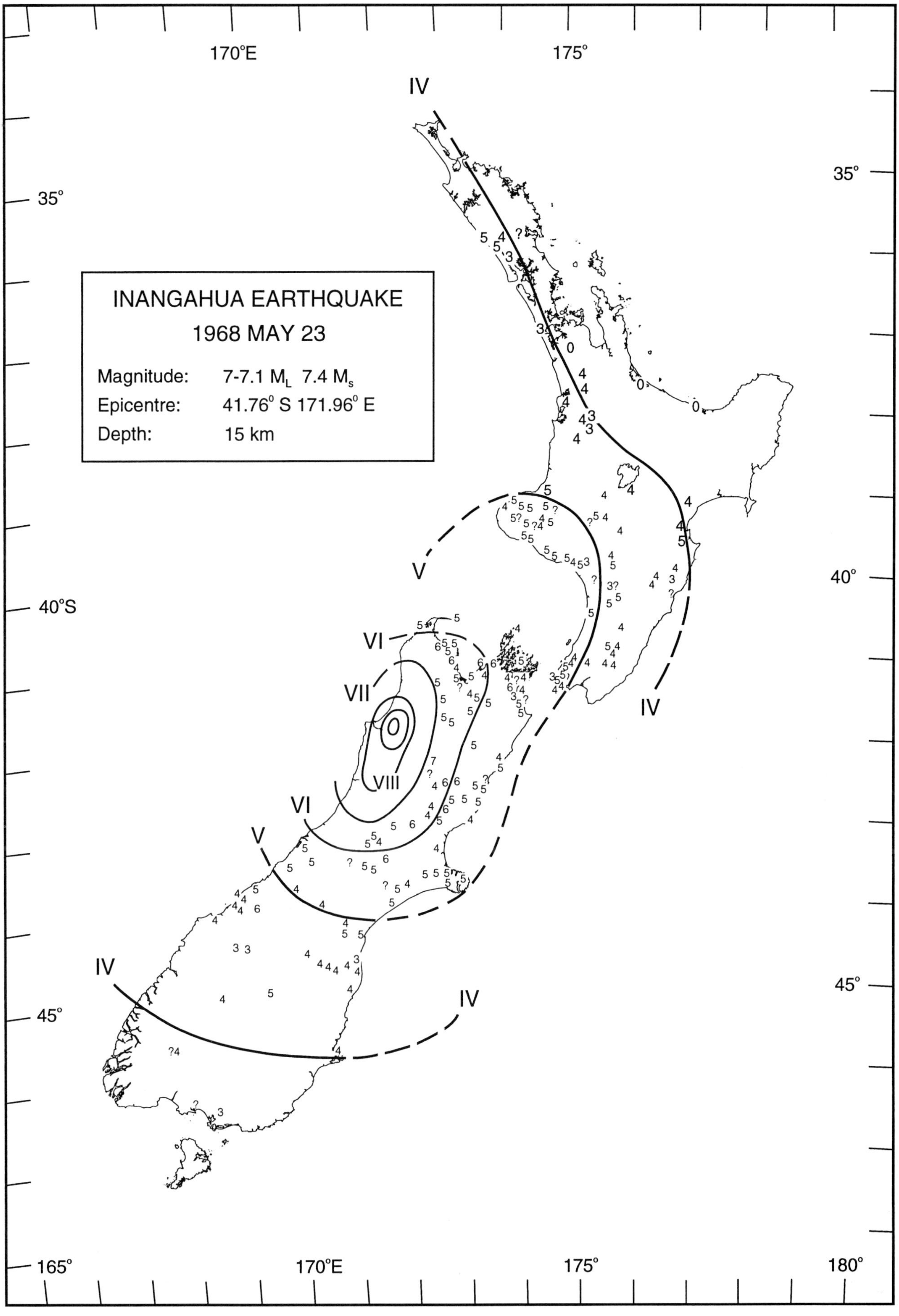
INANGAHUA EARTHQUAKE
1968 MAY 23
Magnitude: 7-7.1 M_L 7.4 M_s
Epicentre: 41.76° S 171.96° E
Depth: 15 km
170°E
175°
35°
40°S
40°
45°
165°
180°
IV
V
VI
VII
VIII

90b. ISOSEISMAL MAP OF THE INANGAHUA EARTHQUAKE - 1968 MAY 23

DATE:	1968 MAY 23
TIME:	17:24:15.6 UT
MAGNITUDE:	7.0-7.1 M_L (Adams et al. 1969); 7.4 M_s (Dowrick & Smith 1990); 7.1 M_w (Anderson et al. 1993)
EPICENTRE:	41.76°S 171.96 E ± 10 km (Anderson et al. 1994)
DEPTH:	15 km ± 5 km

'Faulting took place at two localities, near Inangahua Junction railway station and about three miles [5 km] north of Rotokohu. In both places movement comprised horizontal, vertical and thrusting components.

'The [Inangahua] trace was most clearly seen at the railway line and diminished to a small roll in the turf about ¼ mile [250 m] to the south. Northwards it was followed for ½ mile [500 m], diminishing in prominence and splaying into two smaller traces which trend between north and 020°. The maximum displacement is about 16 inches [0.4 m] both horizontally in a sinistral sense and vertically.

'The Rotokohu traces consist of several parallel but staggered traces upthrown to the south, trending at 060°-080° and extending over a length of 1 mile [1.6 km]. Individual traces show displacements of the terrace surfaces up to about 3 ft [1 m] both vertically and horizontally. Over a zone about 50 ft [17 m] wide parallel to and including the fault, the horizontal displacement amounts to about 6 ft [2 m]. ... The traces are marked by "mole track" features ...' (Lensen & Suggate 1969)

Anderson et al. (1994) reinterpret past observations and more recent data, in light of new structural and tectonic theories, producing a new source model for the earthquake and showing that "general acceptance that the source of the Inangahua earthquake was the Inangahua fault is inconsistent with the pattern of uplift which occurred during the event."

The isoseismal map is from the special issue of the New Zealand Seismological Report on the 1968 Inangahua earthquakes. The origin of the data on which it was based is given on the previous pages, where a map of the felt effects over the whole of New Zealand is given. The epicentre is from Anderson et al. (1994)

INANGAHUA EARTHQUAKE
1968 MAY 23

Magnitude:	7-7.1 M_L 7.4 M_s
Epicentre:	41.76° S 171.96° E
Depth:	15 km

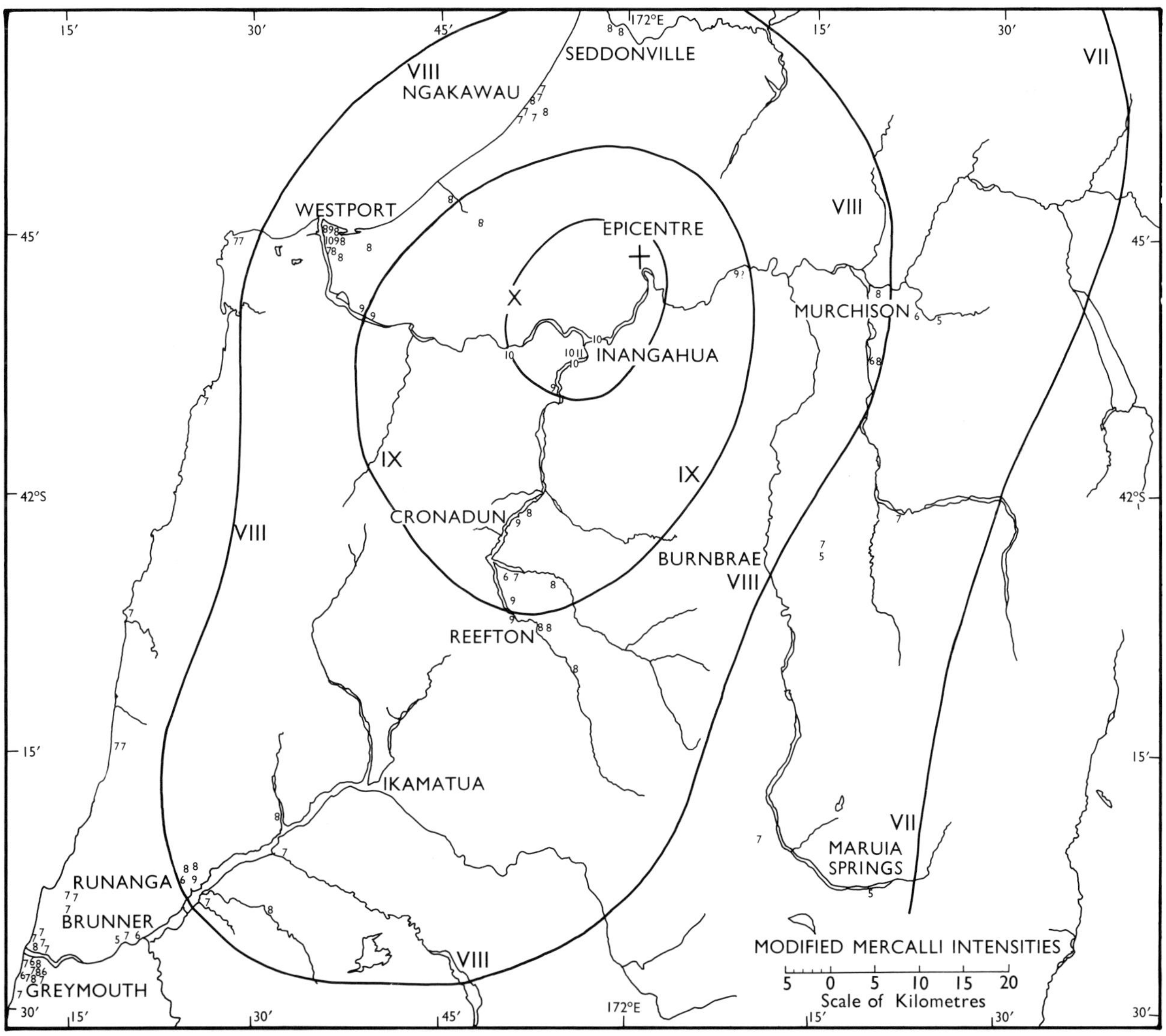

91. ISOSEISMAL MAP OF THE SOLANDER EARTHQUAKE- 1968 SEPTEMBER 25

DATE:	1968 SEPTEMBER 25
TIME:	07:02:45.5 UT
MAGNITUDE:	5.9 M_L; 6.2 M_s (Dowrick & Smith 1990); 6.3 M_w (Anderson et al. 1993)
EPICENTRE:	46.67°S 166.50°E (1)
DEPTH:	12 km (restricted)

'On September 25, minor damage occurred at Otautau in Southland. The shock responsible had a magnitude of 6.5 [5.9 M_L^*] and was centred at sea, 30 km south of Puysegur Point, where the intensity reached MM7. Intensities exceeded MM5 in many places throughout Otago and Southland. Nine of the numerous aftershocks were reported felt.' (Seismological Observatory bulletin E 151)

The isoseismal map has been reproduced from the Seismological Observatory bulletin E 151.

* Local magnitudes and hypocentres have been revised since the publication of the Seismological Observatory bulletin E 151. Refer to introduction.

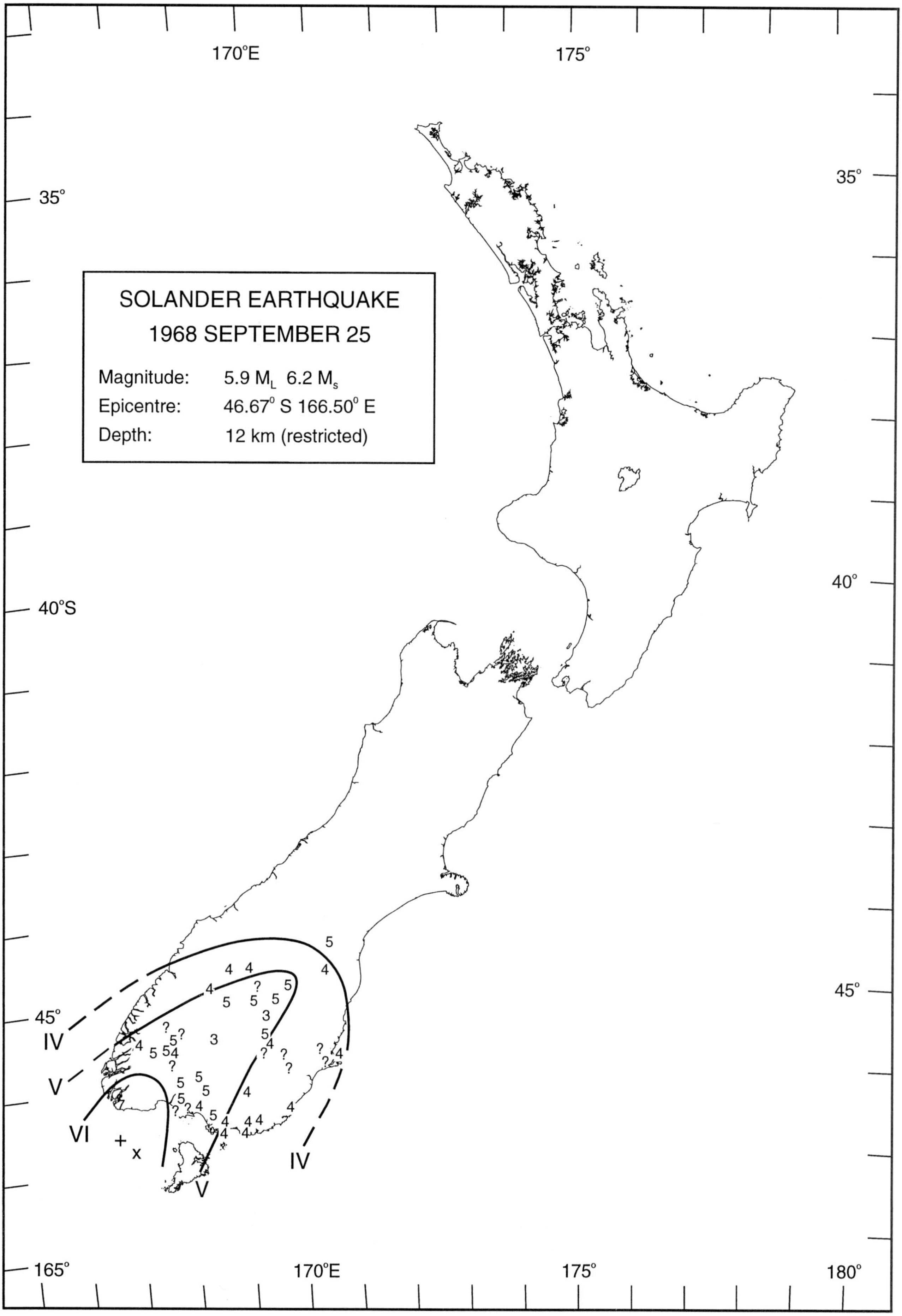
SOLANDER EARTHQUAKE
1968 SEPTEMBER 25
Magnitude: 5.9 M_L 6.2 M_s
Epicentre: 46.67° S 166.50° E
Depth: 12 km (restricted)
170°E
175°
35°
40°S
40°
45°
165°
180°
IV
V
VI
IV
V

92. ISOSEISMAL MAP OF THE PALLISER BAY EARTHQUAKE - 1968 NOVEMBER 01

DATE:	1968 NOVEMBER 01
TIME:	01:32:25.1 UT
MAGNITUDE:	5.4 M_L; 5.0 M_s (Dowrick & Smith, 1990)
EPICENTRE:	41.62°S 175.05°E (1)
DEPTH:	33 km (restricted)

'On November 1 [1968] a magnitude 5.5 [5.4 M_L*] earthquake centred in Palliser Bay gave rise to an intensity of MM6 in Wellington city, deranging goods and causing superficial damage to older structures. With the possible exception of the magnitude 6.0 Seddon earthquake in April 1966 this shock was more strongly felt in Wellington than any other since the Wairarapa earthquakes in 1942.' (Seismological Observatory bulletin E 151)

As a result of the earthquake a survey of a large number of buildings in Wellington was executed and

'Initial reports indicated that some building damage had occurred in Wellington City, and a brief survey that afternoon, ... indicated that there was some variation in the amount of damage. Three days later a rapid assessment ... confirmed this and suggested that the pattern might be zonal.

'A building by building survey was continued by the author [Grant-Taylor, T. L.], who examined most brick and concrete commercial buildings of 2 to 6 storeys in the metropolitan area.' (Grant-Taylor et al. 1974)

The results of the survey and their significance in terms of microzonation of Wellington City are discussed in Grant-Taylor et al. (1974)

The isoseismal map has been reproduced from the Seismological Observatory bulletin E 151.

* Local magnitudes and hypocentres have been revised since the publication of the Seismological Observatory bulletin E 151. Refer to introduction.

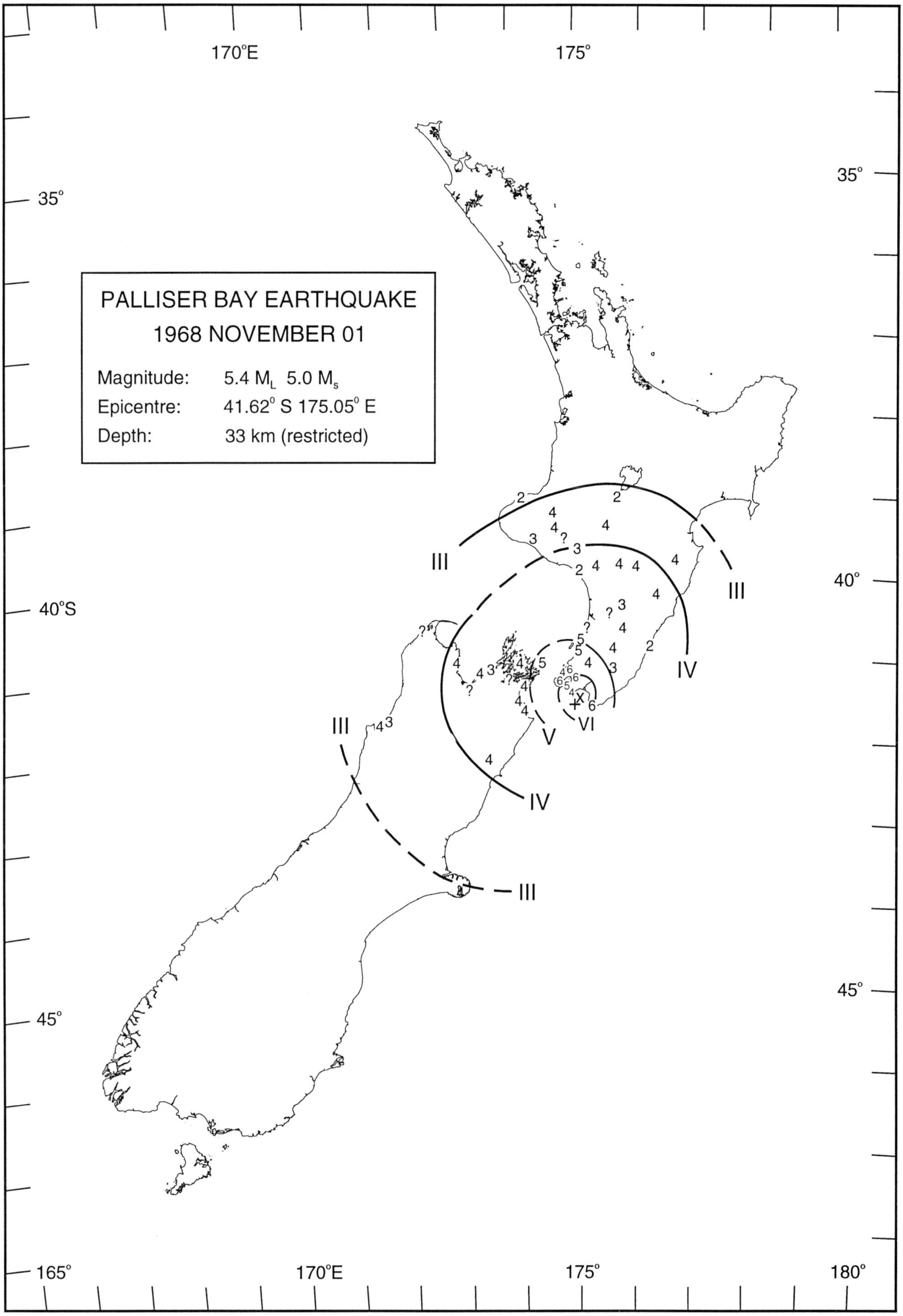

PALLISER BAY EARTHQUAKE
1968 NOVEMBER 01
Magnitude: 5.4 M$_L$ 5.0 M$_s$
Epicentre: 41.62° S 175.05° E
Depth: 33 km (restricted)
170°E
175°
35°
40°S
40°
45°
165°
180°
III
IV
V
VI

93. ISOSEISMAL MAP OF THE FIORDLAND EARTHQUAKE - 1969 JANUARY 02

DATE:	1969 JANUARY 02
TIME:	10:25:21.5 UT
MAGNITUDE:	5.8 M_L*
EPICENTRE:	45.12°S 167.66°E (1)
DEPTH:	139 km

The 139 km deep earthquake of magnitude 5.8 M_L* on 1969 January 02, at 10:25 UT, was widely felt throughout Otago and Southland, with maximum intensities of about MM5.

The isoseismal map has been reproduced from the Seismological Observatory bulletin E 153.

* Local magnitudes and hypocentres have been revised since the publication of the Seismological Observatory bulletin E 153. Refer to introduction.

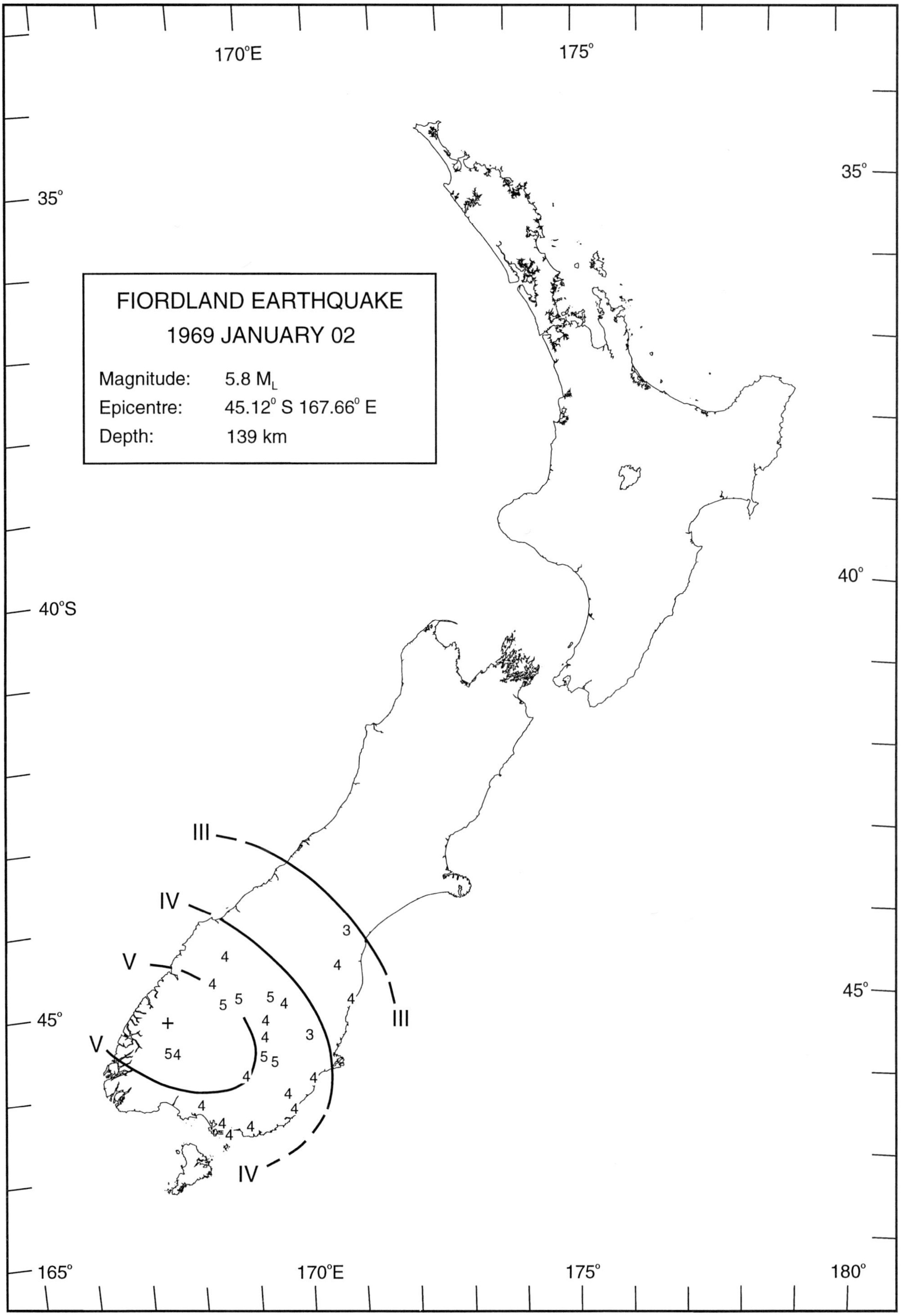
FIORDLAND EARTHQUAKE
1969 JANUARY 02
Magnitude: 5.8 M_L
Epicentre: 45.12° S 167.66° E
Depth: 139 km
170°E
175°
35°
40°
45°
40°S
165°
180°
III
IV
V

94. ISOSEISMAL MAP OF THE NORTH ISLAND EARTHQUAKE - 1969 MAY 23

DATE:	1969 MAY 23
TIME:	14:29:42.4 UT
MAGNITUDE:	5.6 M_L*
EPICENTRE:	39.70°S 175.42°E (1)
DEPTH:	110 km

'Two deep shocks in the central North Island were widely felt [in 1969], though with only moderate intensities. The first, on May 23 had a magnitude of 5.7 [5.6 M_L*] and a focal depth of 110 km, and was centred about 20 km west of Taihape. Its felt area covered the central North Island, and extended southwards to include Wellington and Collingwood in West Nelson. Intensities did not exceed MM5.' (Seismological Observatory bulletin E 153)

The isoseismal map has been reproduced from the Seismological Observatory bulletin E 153.

* Local magnitudes and hypocentres have been revised since the publication of the Seismological Observatory bulletin E 153. Refer to introduction.

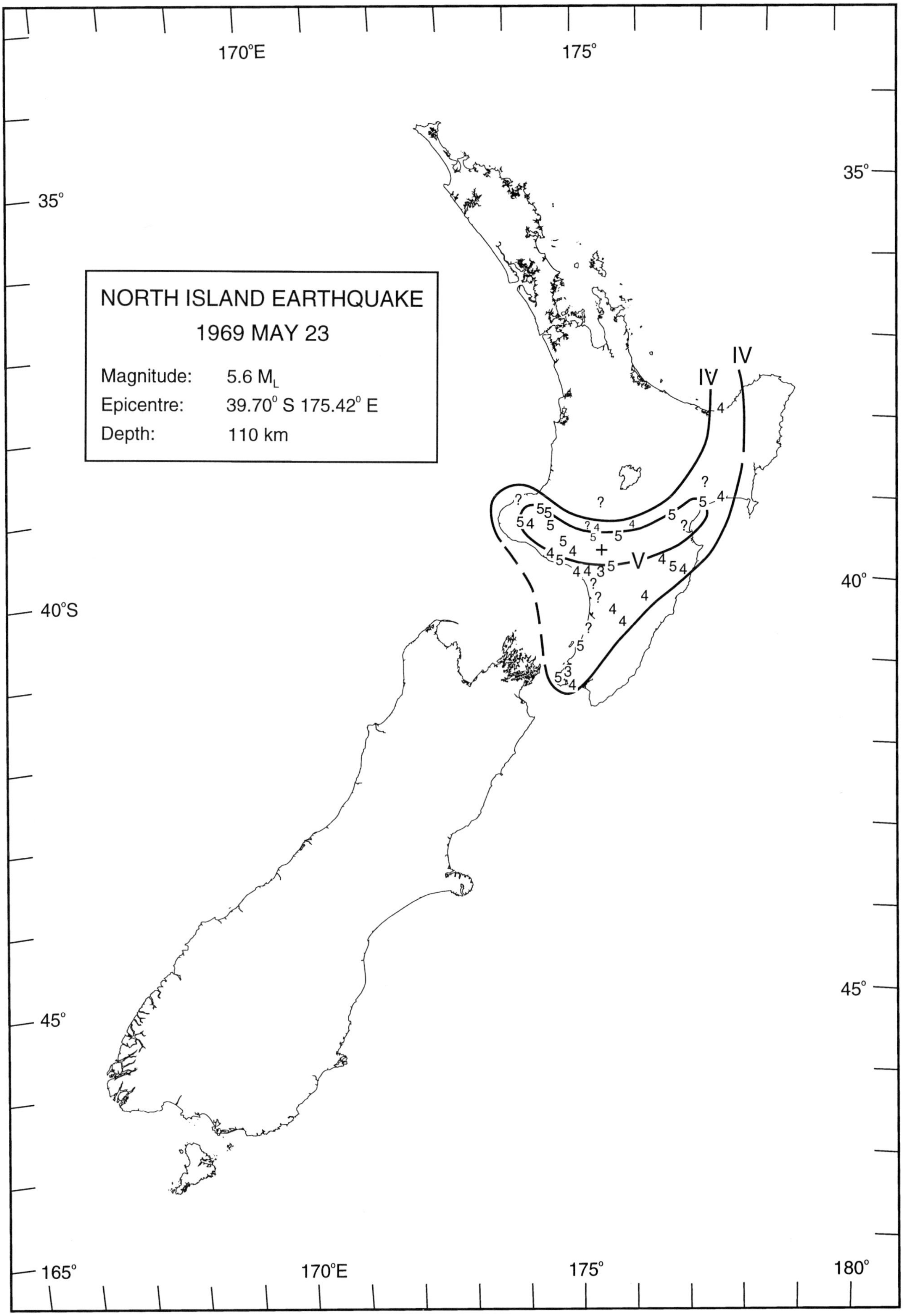

NORTH ISLAND EARTHQUAKE
1969 MAY 23
Magnitude: 5.6 M_L
Epicentre: 39.70° S 175.42° E
Depth: 110 km
170°E
175°
180°
165°
35°
40°S
40°
45°
IV
V

95. ISOSEISMAL MAP OF THE BAY OF PLENTY EARTHQUAKE - 1970 JULY 27

DATE:	1970 JULY 27
TIME:	12:31:18.9 UT
MAGNITUDE:	5.7 M_L*
EPICENTRE:	37.91°S 177.53°E (1)
DEPTH:	138 km

The largest event in 1970 within the main New Zealand seismic region was a shock of magnitude 5.7 M_L* on 1970 July 27 with a depth of 138 km, centred about 50 km to the east of Whakatane.

'The felt area was confined to the eastern Bay of Plenty and northern Hawke's Bay, except for an isolated report from Paraparaumu Beach, with no intensities above MM4.' (Seismological Observatory bulletin E 157)

The isoseismal map has been reproduced from the Seismological Observatory bulletin E 157.

* Local magnitudes and hypocentres have been revised since the publication of the Seismological Observatory bulletin E 157. Refer to introduction.

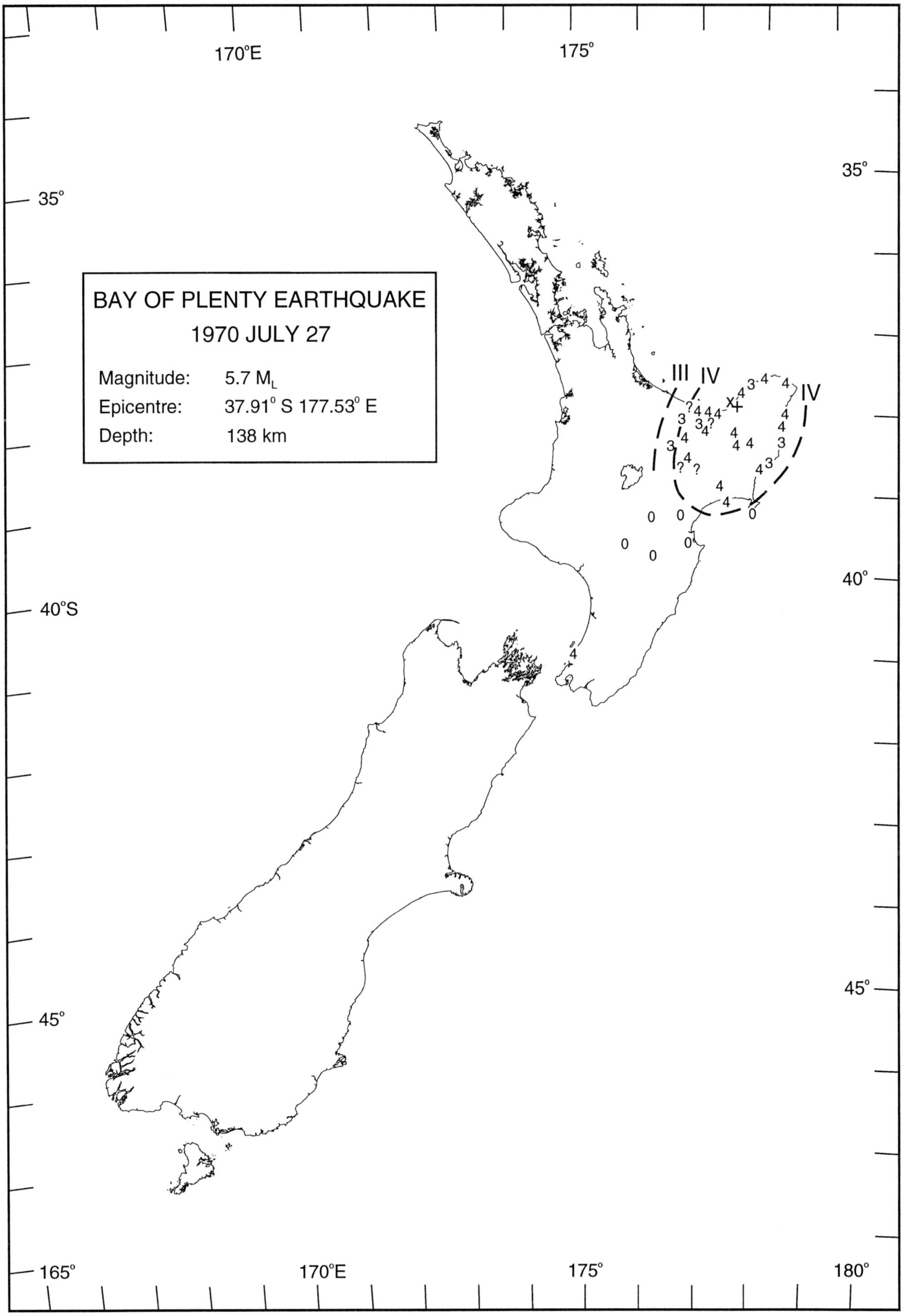
BAY OF PLENTY EARTHQUAKE
1970 JULY 27
Magnitude: 5.7 M_L
Epicentre: 37.91° S 177.53° E
Depth: 138 km
170°E
175°
35°
40°S
40°
45°
165°
180°
III
IV
IV

96. ISOSEISMAL MAP OF THE MARUIA SPRINGS EARTHQUAKE - 1971 AUGUST 13

DATE:	1971 AUGUST 13
TIME:	14:42:42.8 UT
MAGNITUDE:	5.8 M_L^*; 5.5 M_s (Dowrick and Smith 1990); 5.7 M_w (Anderson et al. 1993)
EPICENTRE:	42.18°S 172.30°E (1)
DEPTH:	12 km (restricted)

'The largest shallow shock [in 1971] occurred on August 13 near Maruia about 30 km east of Reefton. Its magnitude was 5.9 [5.8 M_L^*] and intensities up to MM6 were reported from the epicentral region, which is not heavily populated. Some minor damage occurred at Westport, Greymouth, and Reefton. The felt area covered the northern half of the South Island, with isolated reports from Wellington and Hawera in the North Island.' (Seismological Observatory bulletin E 150)

The isoseismal map has been reproduced from the Seismological Observatory bulletin E 150.

---

* Local magnitudes and hypocentres have been revised since the publication of the Seismological Observatory bulletin E 150. Refer to introduction. Observatory bulletin E-150.

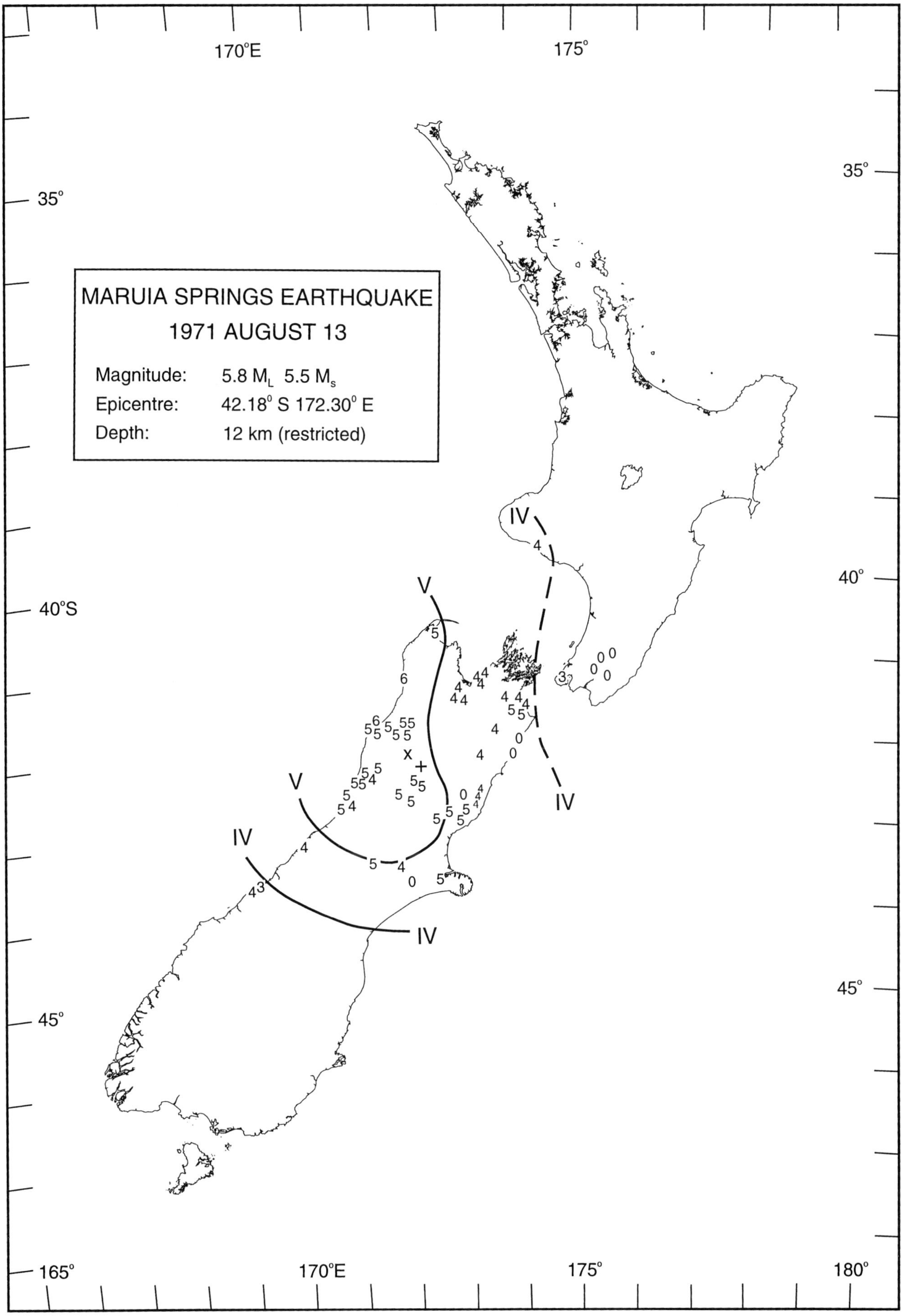

MARUIA SPRINGS EARTHQUAKE
1971 AUGUST 13
Magnitude: 5.8 M_L 5.5 M_s
Epicentre: 42.18° S 172.30° E
Depth: 12 km (restricted)
170°E
175°
35°
40°S
40°
45°
165°
180°
V
V
IV
IV
IV
IV

97. ISOSEISMAL MAP OF THE HAWKE'S BAY EARTHQUAKE - 1971 OCTOBER 31

DATE:	1971 OCTOBER 31
TIME:	12:10:18.3 UT
MAGNITUDE:	5.9 M_L*
EPICENTRE:	38.96°S 176.37°E (1)
DEPTH:	141 km

'The shock on October 31 (... magnitude 5.7 [5.9 M_L*]) was centred to the east of Lake Taupo at a depth of 140 km. Intensities as high as MM5 were reported from many places in Hawke's Bay, and the felt area extended to Farewell Spit in the South Island. As usual for deep shocks in this region, the main felt area is limited to the west by the peculiarities of the upper mantle structure beneath the North Island, but there are in this case isolated observations from the Northland Peninsula, Taranaki, and West Nelson ...' (Seismological Observatory bulletin E 150)

The isoseismal map has been reproduced from the Seismological Observatory bulletin E 150.

* Local magnitudes and hypocentres have been revised since the publication of the Seismological Observatory bulletin E 150. Refer to introduction.

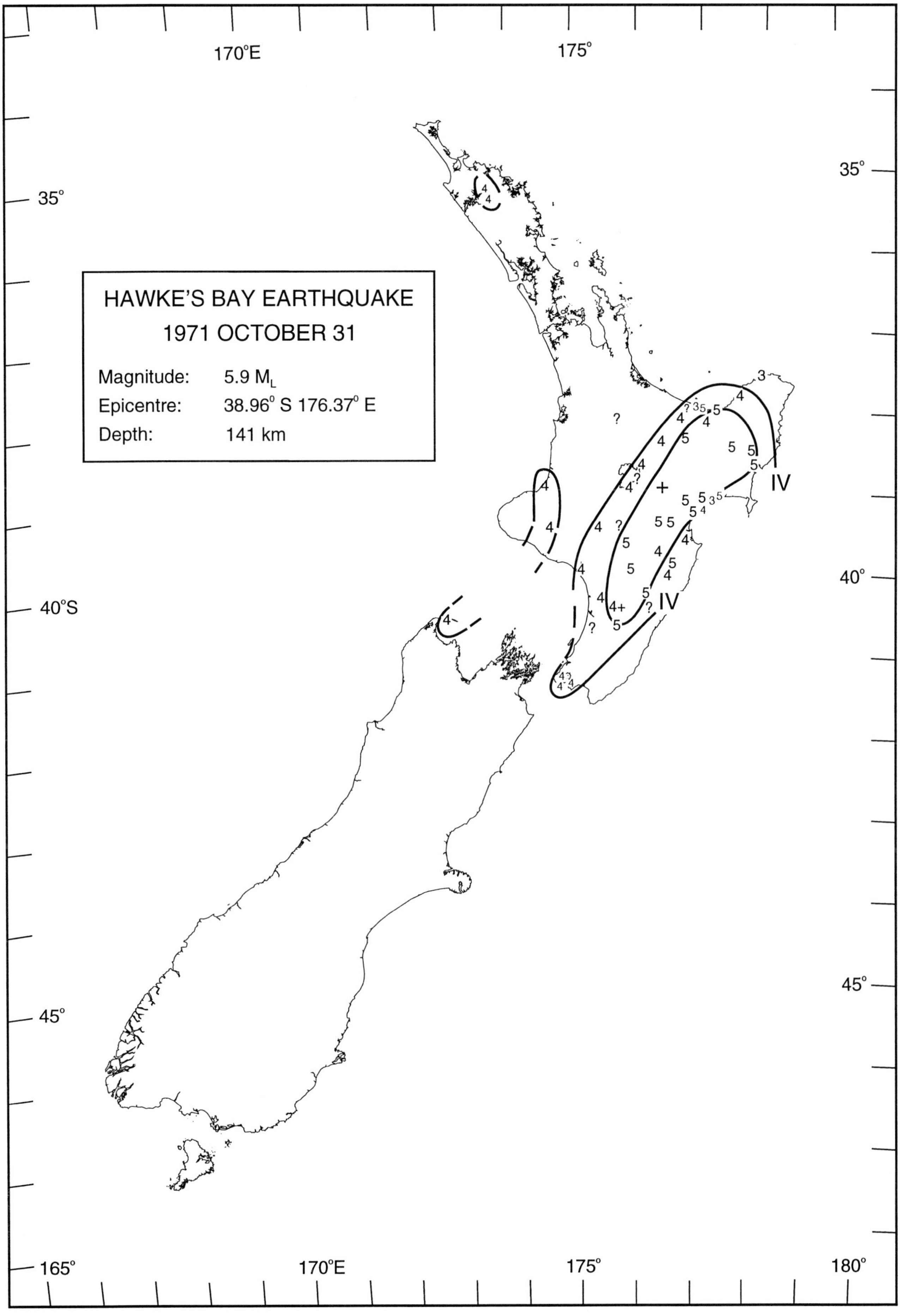
HAWKE'S BAY EARTHQUAKE
1971 OCTOBER 31
Magnitude: 5.9 M_L
Epicentre: 38.96° S 176.37° E
Depth: 141 km
170°E
175°
180°
165°
35°
40°S
40°
45°
IV
IV

98. ISOSEISMAL MAP OF THE TE AROHA EARTHQUAKE - 1972 JANUARY 08

DATE:	1972 JANUARY 08
TIME:	21:33:00.1 UT
MAGNITUDE:	5.3 M_L^* ; 5.0 M_s (Dowrick & Smith 1990)
EPICENTRE:	37.57°S 175.69°E (1)
DEPTH:	12 km (restricted)

'An earthquake was felt extensively in the Waikato and surrounding regions at 9.33 a.m. (N.Z.S.T.) on Sunday, 9th January 1972. Its centre was within a few kilometres of Te Aroha, a town of some 3,500 inhabitants, where a substantial amount of minor damage occurred.

'The main earthquake had a magnitude of 5.1 [5.3 M_L^*] and was followed three minutes later by another shock of magnitude 4.6 [4.6 M_L^*].

'.... shows an isoseismal map and a selection of the observations of intensity on which it is based. These observations are those reported by the Seismological Observatory's regular network of felt reporters, supplemented by the distribution of additional questionnaires.

'It is evident ... that the highest intensities were confined to the epicentral area. MM VII was reported from Te Aroha, Waihou (4 km to the south-west), Te Aroha West (5 km to the south-east) and Elstow (7 km to the north-west). One report from Waihou was assigned an intensity of MM VIII.

'.... The MM VI isoseismal includes Waitoa (8 km to the south-west of Te Aroha) and Waihi. The intensity had in general fallen to MM V at Morrinsville, but some minor chimney damage would indicate MM VI. The MM V isoseismal extends to Hamilton and includes Matamata, Tauranga, and the southern half of the Coromandel Peninsula. The limits of the felt area extend to Auckland, Mangakino, Rotorua and Whakatane.' (Adams et al. 1972)

The isoseismal map has been reproduced from Adams et al. (1972).

* Local magnitudes and hypocentres have been revised since the publication of the Seismological Observatory bulletin E 152. Refer to introduction.

TE AROHA EARTHQUAKE	
1972 JANUARY 08	
Magnitude:	5.3 M_L 5.0 M_s
Epicentre:	37.57° S 175.69° E
Depth:	12 km (restricted)

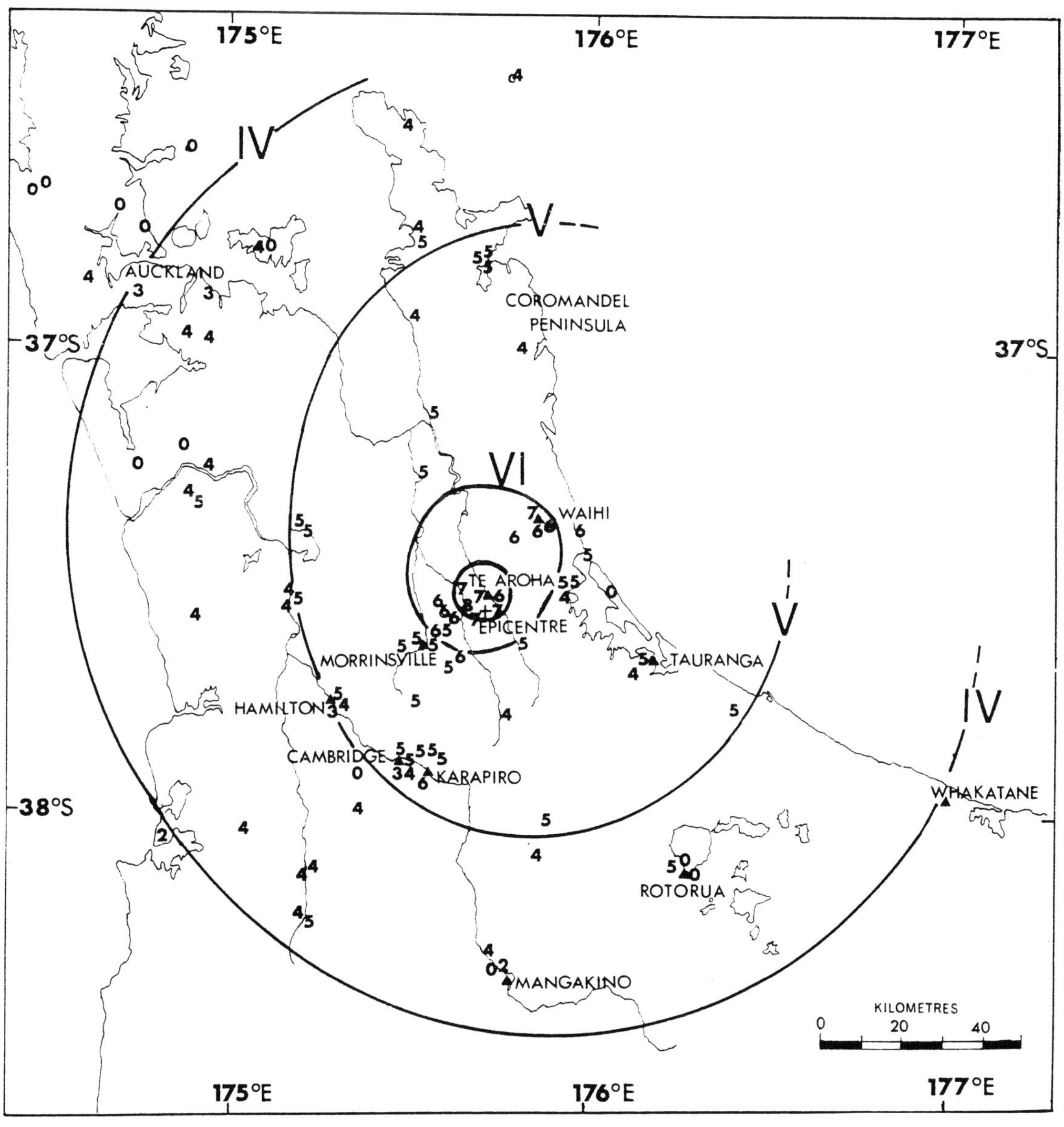

after Adams et al (1972)

99. ISOSEISMAL MAP OF THE NORTH ISLAND EARTHQUAKE - 1973 JANUARY 05

DATE:	1973 JANUARY 05
TIME:	13:54:27.6 UT
MAGNITUDE:	7.0 M_L^* ; 6.3 M_s (Dowrick & Smith 1990)
EPICENTRE:	39.13°S 175.18°E (1)
DEPTH:	173 km

'The largest earthquake during the year, on January 5 with a magnitude of 6.7 [7.0 M_L^*] and a focal depth of 173 km, was centred about 30 km west of National Park in the centre of the North Island. In the epicentral region and in parts of Hawke's Bay the intensity exceeded MM6, and chimney damage, minor cracks in foundations and plaster, and the fall of articles from shelves were fairly widespread. The Earthquake and War Damage Commission received more than 2500 insurance claims, and paid out a total of about $180 000 to claimants. This was the first occasion since the major Napier earthquake in 1931 when intensities as great as this were generally experienced in Hawke's Bay, it being noted that a grandfather clock in a homestead near Waipukurau was stopped for the first time since the earlier event. The felt area covered the North Island south of a line from Whakatane to Kawhia and extended southwards to include Hokitika and Timaru...

'An unusual feature of a shock at so great a depth is the number of apparent aftershocks. There are at least four with approximately the same epicentre, and focal depths lying in the small range 170 to 175 km. The largest of these reached magnitude 5.3 [5.6 M_L^*] and produced an isolated felt report from Eketahuna.' (Seismological Observatory bulletin E 154)

The isoseismal map has been reproduced from the Seismological Observatory bulletin E 154.

* Local magnitudes and hypocentres have been revised since the publication of the Seismological Observatory bulletin E 154. Refer to introduction.

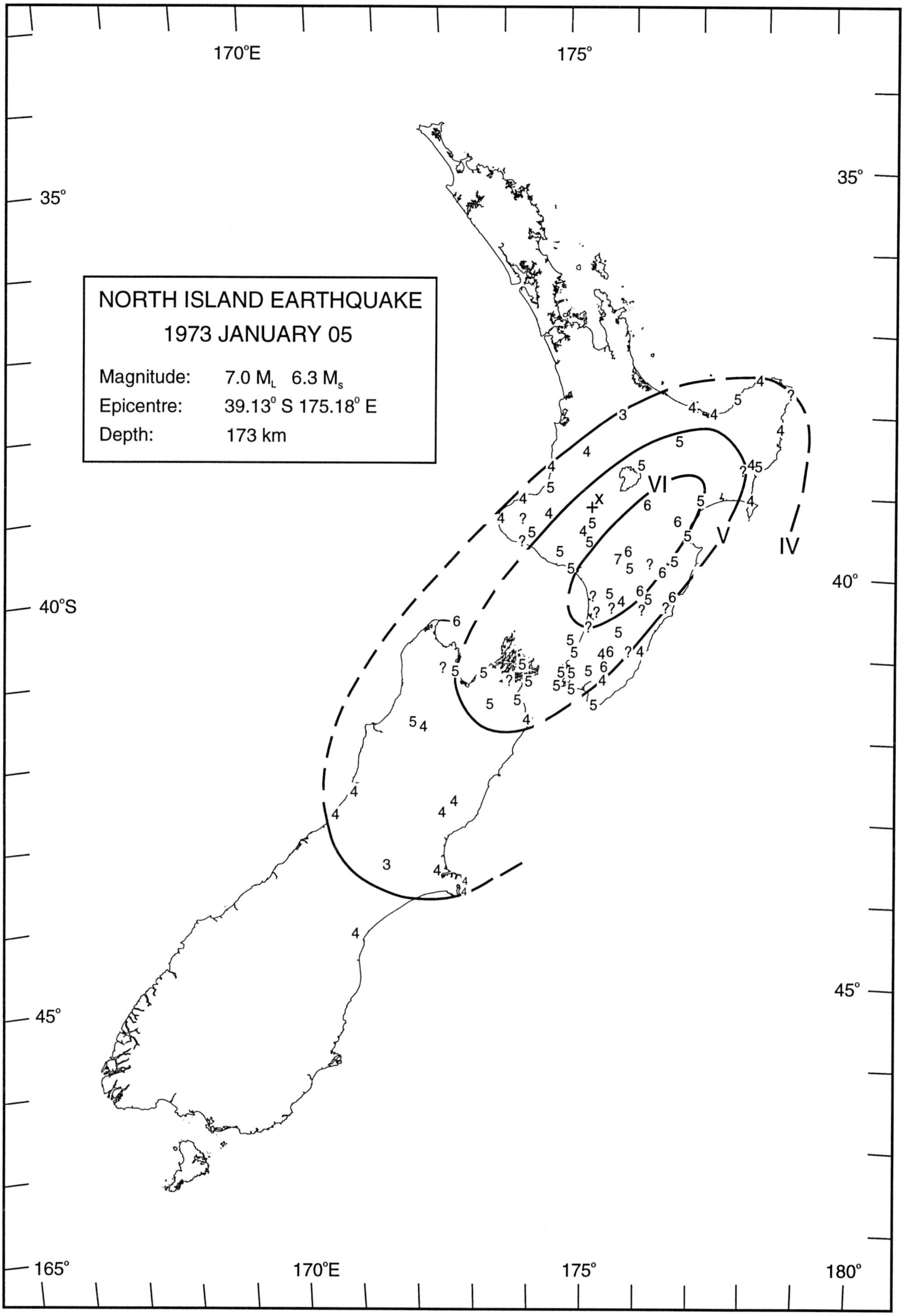

NORTH ISLAND EARTHQUAKE
1973 JANUARY 05
Magnitude: 7.0 M_L 6.3 M_s
Epicentre: 39.13° S 175.18° E
Depth: 173 km
170°E
175°
35°
40°S
45°
40°
165°
180°
VI
V
IV

100. ISOSEISMAL MAP OF THE DUNEDIN EARTHQUAKE - 1974 APRIL 09

DATE:	1974 APRIL 09
TIME:	07:49:46.1 UT
MAGNITUDE:	4.9 M_L*
EPICENTRE:	45.97°S 170.52°E (1)
DEPTH:	12 km (restricted)

'The earthquake in the early evening of Tuesday, 9 April, 1974 was the strongest known to have affected the city of Dunedin since its founding in the middle of last century. The earthquake was of moderate magnitude ... and centred close to the city. A large amount of minor damage was caused, and in particular many domestic brick chimneys were damaged.

'The main earthquake was generally felt out to Oamaru, Roxburgh and Balclutha, with isolated reports from as far afield as Waipapa Point and Gore. The isoseismals have the regular configuration usually found for shallow earthquakes, with the radii of the MM V, MM IV and MM III isoseismals being about 35, 90 and 150 km respectively.

'Damage was almost entirely confined to the Dunedin area ... From standard felt report forms submitted to the Seismological Observatory, and a visit to Dunedin two months after the earthquake, it was established that the intensity reached MM VII in the worst affected areas of Dunedin. These were mainly the southern suburbs on the alluvium between Otago Peninsula and St. Clair. Little damage beyond that to domestic chimneys occurred even in these areas, but chimney damage was consistent and widespread, and although the standard of construction was commonly poor, the proportion of chimneys in which the brick-work was broken prevents the allocation of a lower intensity. Chimney damage occurred, but less densely, over most of the rest of the city, including the hill suburbs, but was usually more superficial and restricted to fall of pots and plaster. In these areas, MM VI would be appropriate.' (Adams & Kean 1974)

The isoseismal map has been reproduced from Adams and Kean (1974).

---

* Local magnitudes and hypocentres have been revised since the publication of the Seismological Observatory bulletin E 155. Refer to introduction.

DUNEDIN EARTHQUAKE
1974 APRIL 09

Magnitude: 4.9 M_L
Epicentre: 45.97° S 170.52° E
Depth: 12 km (restricted)

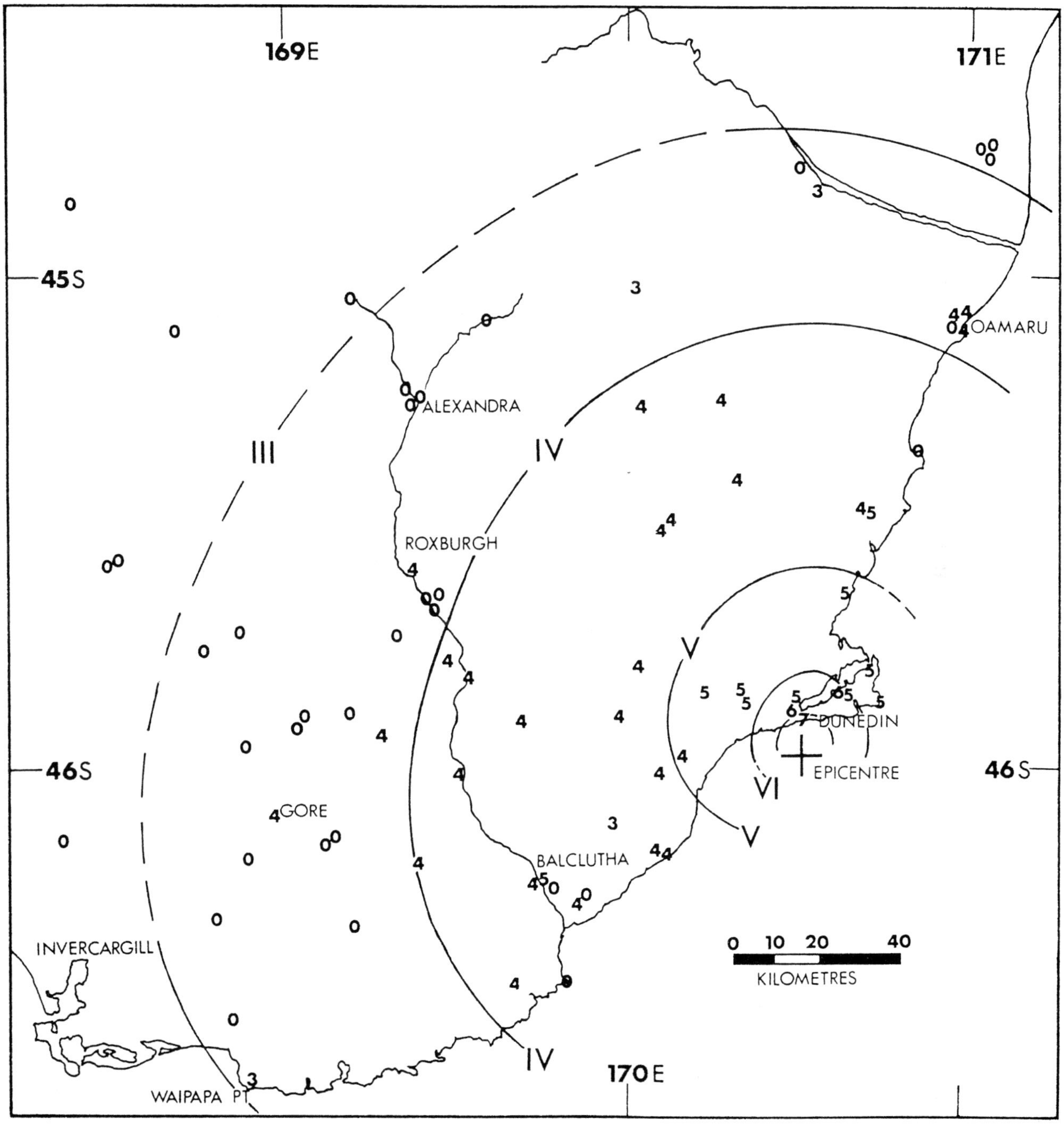

after Adams and Kean (1974)

101. ISOSEISMAL MAP OF THE MILFORD SOUND EARTHQUAKE - 1974 SEPTEMBER 20

DATE:	1974 SEPTEMBER 20
TIME:	19:48:39.5 UT
MAGNITUDE:	5.5 M_L*; 5.3 M_s (Dowrick & Smith 1990)
EPICENTRE:	44.40°S 167.99°E (1)
DEPTH:	12 km (restricted)

On 1974 September 20, at 19:48 UT (07:48 Sept 21 local time) a shallow earthquake of magnitude 5.5 M_L occurred near Milford Sound followed by another shock of the same magnitude and in a similar location 10 hours later.

'... they were felt over much of the southern part of the South Island, with reported intensities up to MM5. Aftershocks were very numerous, and although the epicentral region is very sparsely populated, at least ten of them were felt.

'A shock of magnitude 5.2 [5.1 M_L*], centred 30 km north-west of Christchurch, occurred only three minutes after the first of the Milford Sound earthquakes. It was felt in Christchurch and the Lake Coleridge district with an intensity of MM4, and caused some initial confusion by suggesting that a shock with a very large felt area had occurred. Some difficulty in relating felt reports to the correct event was experienced.' (Seismological Observatory bulletin E 155)

The isoseismal map has been reproduced from the Seismological Observatory bulletin E 155.

* Local magnitudes and hypocentres have been revised since the publication of the Seismological Observatory bulletin E 155. Refer to introduction.

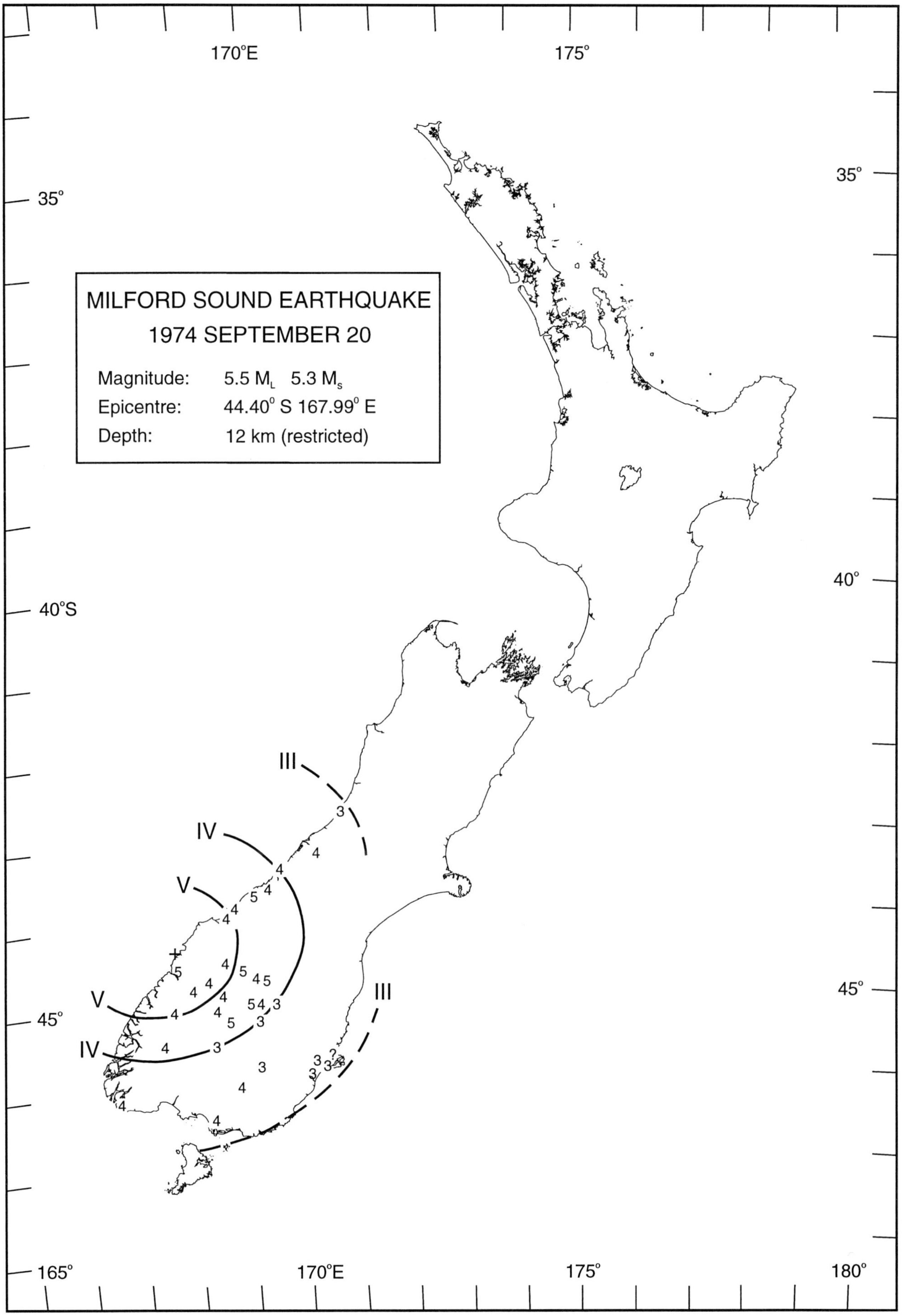

MILFORD SOUND EARTHQUAKE
1974 SEPTEMBER 20
Magnitude: 5.5 M_L 5.3 M_s
Epicentre: 44.40° S 167.99° E
Depth: 12 km (restricted)
170°E
175°
35°
40°S
40°
45°
165°
180°
III
IV
V

102. ISOSEISMAL MAP OF THE OPUNAKE EARTHQUAKE - 1974 NOVEMBER 05

DATE:	1974 NOVEMBER 05
TIME:	10:38:38.9 UT
MAGNITUDE:	6.0 M_L^* ; 5.4 M_s (Dowrick & Smith 1990)
EPICENTRE:	39.54°S 173.46°E (1)
DEPTH:	12 km (restricted)

'The earthquake in the late evening of Tuesday, 5th November 1974, was felt throughout a large part of central New Zealand. Although this event was of moderate magnitude (M_L=6.1 [6.0 M_L^*]), little serious damage was done because it was centred off the SW Taranaki coast some 35 km west of Opunake. Shallow earthquakes of similar or larger magnitude occur in New Zealand about once a year, on a long-term average, but this was the first in the Taranaki region since 1928, and the first in the whole of New Zealand since 1968.

'The earthquake, hence termed the Opunake earthquake, is also of interest because of its unusual source characteristics and its proximity to the Maui gas field ...

'... The configuration of isoseismals as derived from reports on standard Seismological Observatory questionnaires is shown [on the isoseismal map]. For events occurring late at night the outer edge of the region of felt reports corresponds to the boundary between MM III and MM IV, because by definition, MM III is not sufficient to awaken sleepers. The maximum reported intensity was MM VII.

'The damage done by the earthquake was confined to the Taranaki region. The most serious damage appears to have been a few cracked chimneys and cracked walls or ceilings in the region from Opunake to New Plymouth. Many shops west of New Plymouth and Hawera reported that goods were thrown from shelves. The Earthquake and War Damage Commission received claims for damage amounting to $12 000.

'Aftershocks of the Opunake earthquake were recorded at least to the end of November 1974, and some were felt in Taranaki for up to a week after the main event. The two largest aftershocks were of magnitude 5.2 [5.2 M_L^*]. There were no foreshocks recorded. In all some 128 aftershocks with magnitudes greater than 3.5 could be identified on records from standard New Zealand seismograph stations. (Robinson et al. 1976)

The isoseismal map has been reproduced from the Seismological Observatory bulletin E 155.

---

* Local magnitudes and hypocentres have been revised since the publication of the Seismological Observatory bulletin E 155. Refer to introduction.

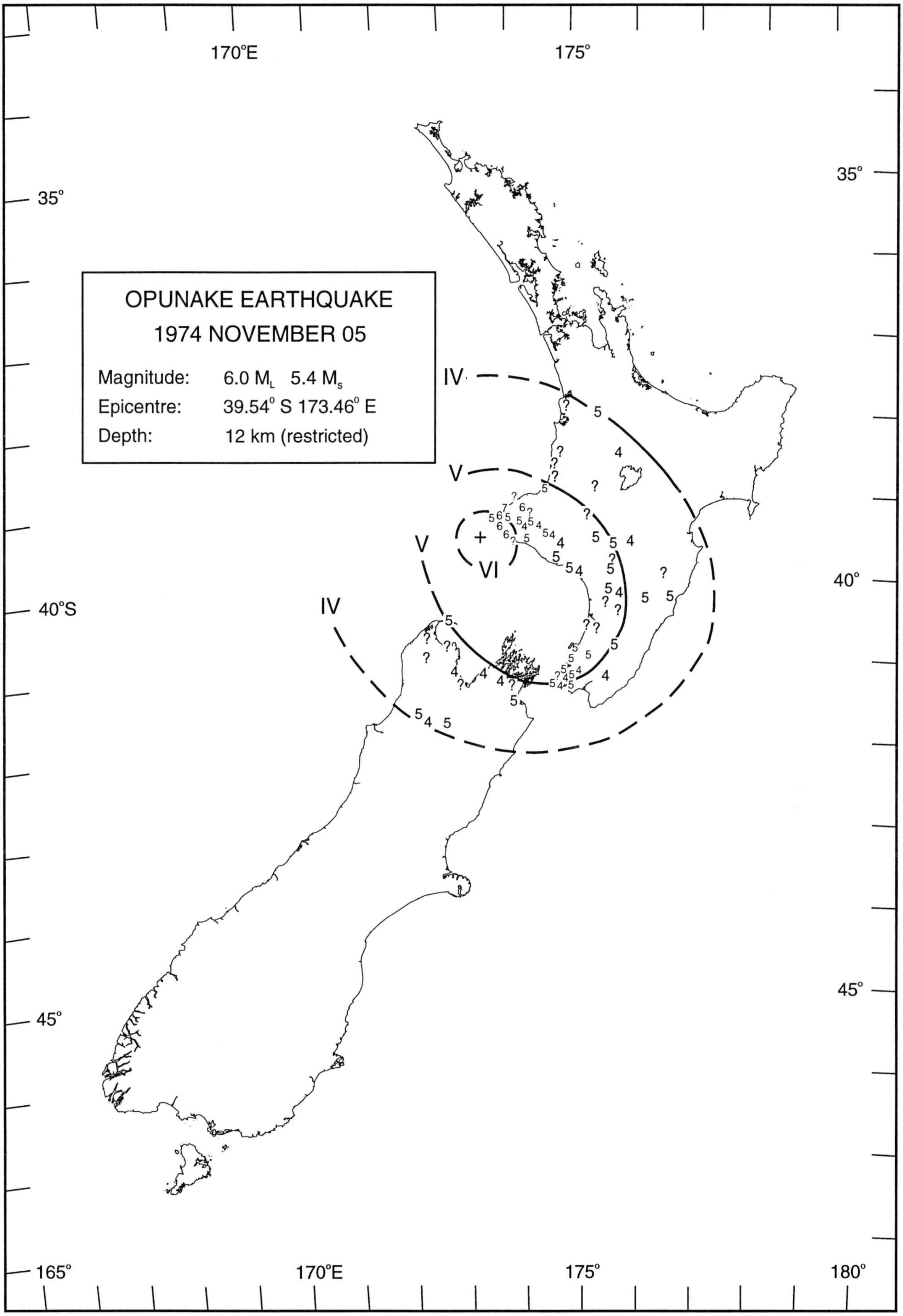

OPUNAKE EARTHQUAKE
1974 NOVEMBER 05
Magnitude: 6.0 M_L 5.4 M_s
Epicentre: 39.54° S 173.46° E
Depth: 12 km (restricted)
170°E
175°
35°
40°S
40°
45°
165°
180°
IV
V
VI

103. ISOSEISMAL MAP OF THE KAPITI ISLAND EARTHQUAKE - 1975 JANUARY 04

DATE:	1975 JANUARY 04
TIME:	20:37:17.5 UT
MAGNITUDE:	5.9 M_L^* ; 4.6 M_s (Dowrick & Smith 1990)
EPICENTRE:	40.77°S 174.67°E (Adams & Ware 1977)
DEPTH:	72 km

'A moderately deep shock of magnitude 5.9 [5.9 M_L^*], occurred on January 4 and brought down goods and crockery from shelves in coastal districts to the north of Wellington, indicating an intensity of MM6. It was felt over an area that included most of Taranaki, Hawke's Bay, and parts of Nelson and Marlborough. The standard origin solution, which uses a symmetrical Earth-model, places its centre near Upper Hutt, at a depth of 128 km. Adams and Ware [1977] ... have obtained an alternative solution using a model that allows for lateral inhomogeneities in the upper mantle. This is given with the station readings, and places the origin to the west of Kapiti Island, at the smaller depth of 72 km. As this is in better agreement with the observed surface-effects and with teleseismic observations it should be preferred to the standard solution, except for studies involving other shocks and in which it is important to use a consistent model throughout.' (Seismological Observatory bulletin E 156)

The isoseismal map is from the Seismological Observatory bulletin E 156. The hypocentral parameters are taken from Adams & Ware (1977). They differ significantly from those published in the Seismological Observatory bulletin E 156 and the comments by Adams & Ware (1977) on their use should be noted.

* Local magnitudes have been revised since the publication of the Seismological Observatory bulletin E 156. Refer to introduction.

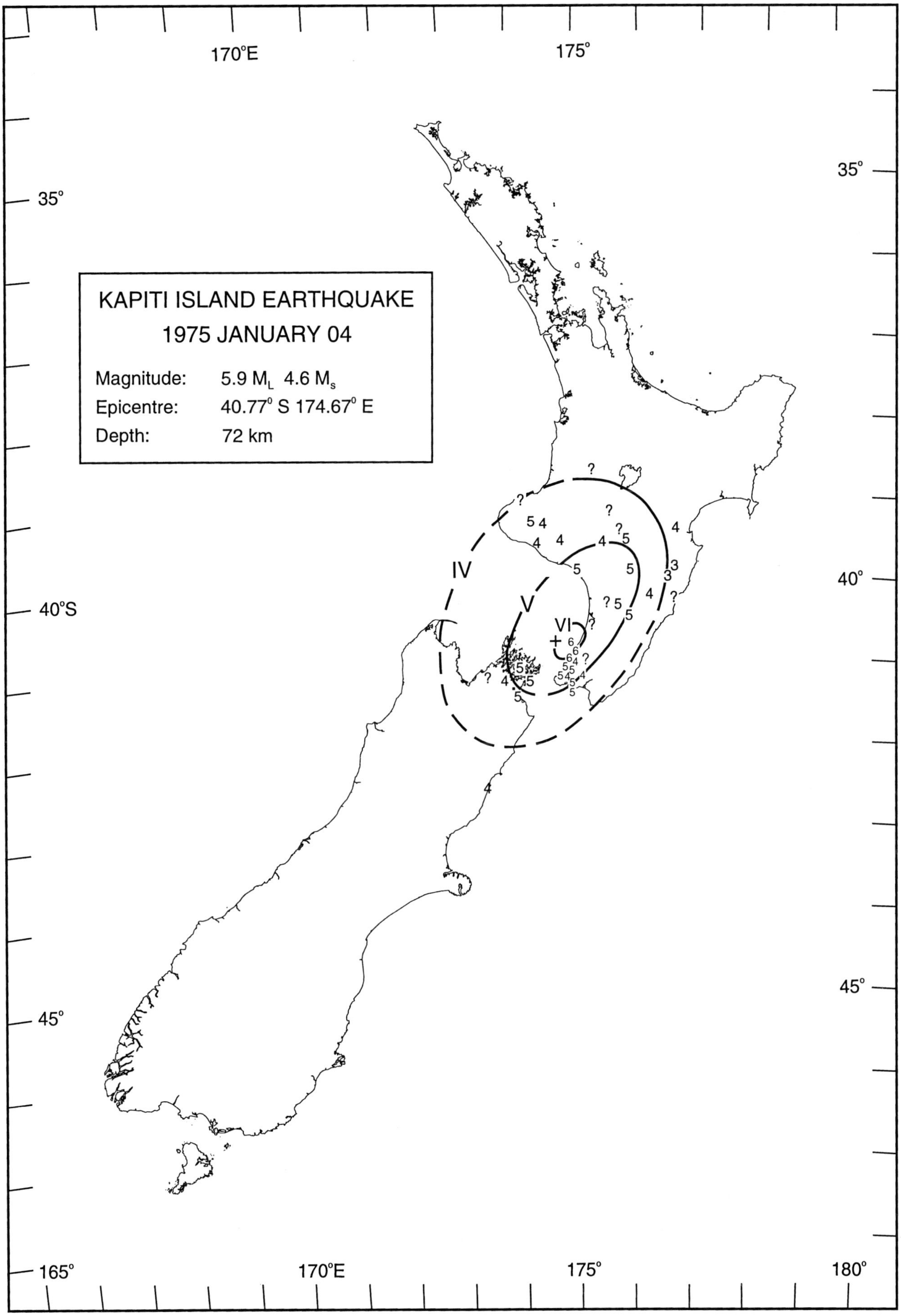

KAPITI ISLAND EARTHQUAKE
1975 JANUARY 04
Magnitude: 5.9 M$_L$ 4.6 M$_s$
Epicentre: 40.77° S 174.67° E
Depth: 72 km
170°E
175°
35°
40°S
40°
45°
165°
180°
IV
V
VI

104. ISOSEISMAL MAP OF THE HEN AND CHICKENS ISLANDS EARTHQUAKE - 1975 FEBRUARY 11

DATE:	1975 FEBRUARY 11
TIME:	16:45:20.4 UT
MAGNITUDE:	4.4 M_L*
EPICENTRE:	35.93°S 174.76°E (1)
DEPTH:	12 km (restricted)

'Several earthquakes in 1975 occurred in areas of less frequent activity. On February 11 a shock of magnitude 4.4 [4.4 M_L*] centred near the Hen and Chickens Islands, about 50 km south-east of Whangarei, was felt over a large area of Northland but although a postal questionnaire was issued, the number of observations collected remained small. A second shock in this area occurred on August 16.' (Seismological Observatory bulletin E 156)

The isoseismal map has been reproduced from the Seismological Observatory bulletin E 156.

* Local magnitudes and hypocentres have been revised since the publication of the Seismological Observatory bulletin E 156. Refer to introduction.

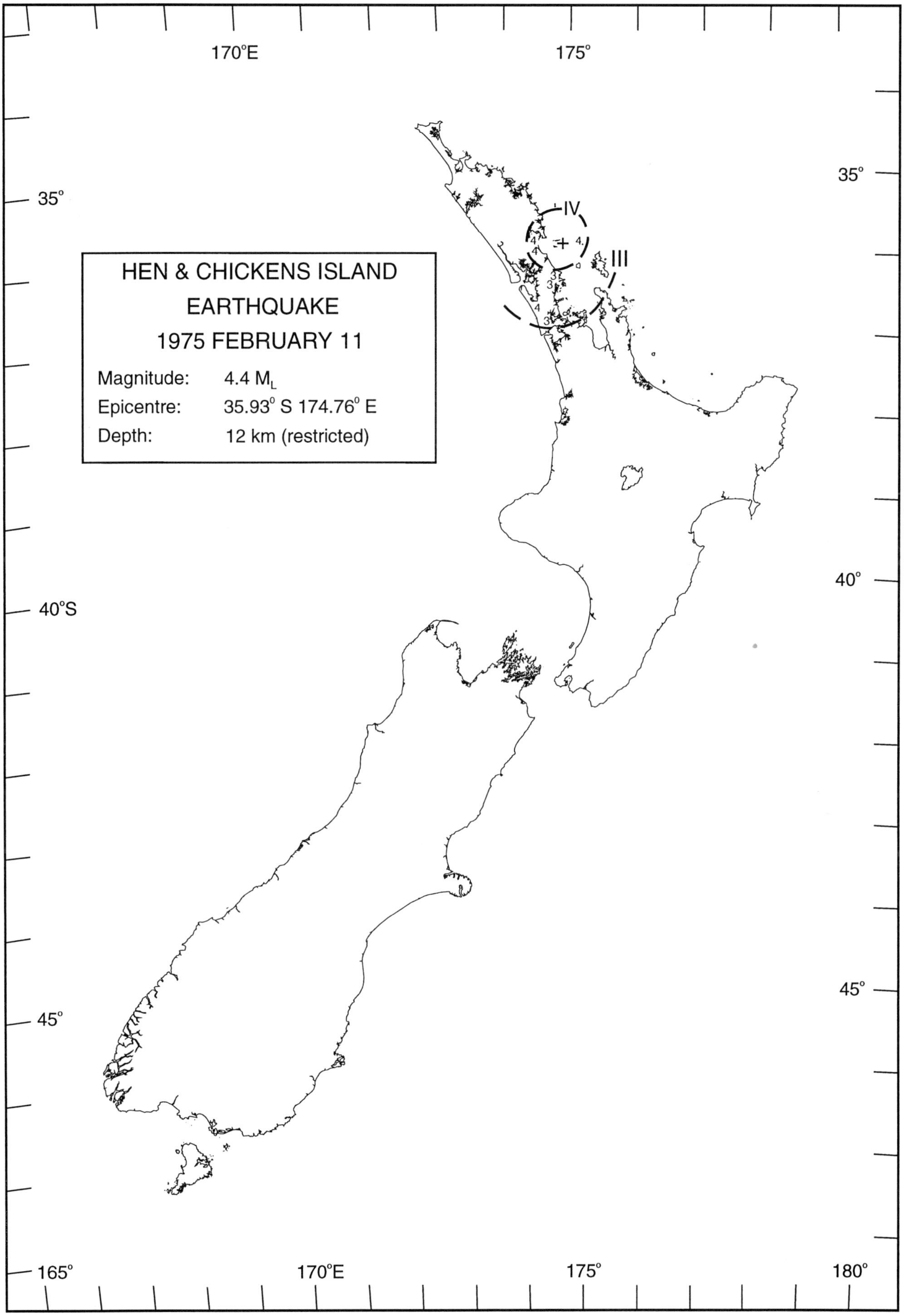

170°E
175°
35°
35°
40°
40°S
45°
45°
165°
170°E
175°
180°
IV
III
HEN & CHICKENS ISLAND
EARTHQUAKE
1975 FEBRUARY 11
Magnitude: 4.4 M_L
Epicentre: 35.93° S 174.76° E
Depth: 12 km (restricted)

105. ISOSEISMAL MAP OF THE ORINGI EARTHQUAKE - 1975 JUNE 10

DATE:	1975 JUNE 10
TIME:	10:11:20.5 UT
MAGNITUDE:	5.8 M_L* ; 5.3 M_s (Dowrick & Smith 1990)
EPICENTRE:	40.31°S 176.07°E (1)
DEPTH:	33 km (restricted)

'The largest shallow earthquake [in 1975], on the evening of June 10 had an epicentre about 15 km south of Dannevirke, where windows were broken and masonry dislodged, indicating a maximum intensity of about MM7, but MM6 appears to be a more representative value for the township. The magnitude of the shock was 5.9 [5.8 M_L*], and its felt area extended from the Bay of Plenty to Banks Peninsula.'(Seismological Observatory bulletin E 156).

The isoseismal map has been reproduced from the Seismological Observatory bulletin E 156.

* Local magnitudes and hypocentres have been revised since the publication of the Seismological Observatory bulletin E 156. Refer to introduction.

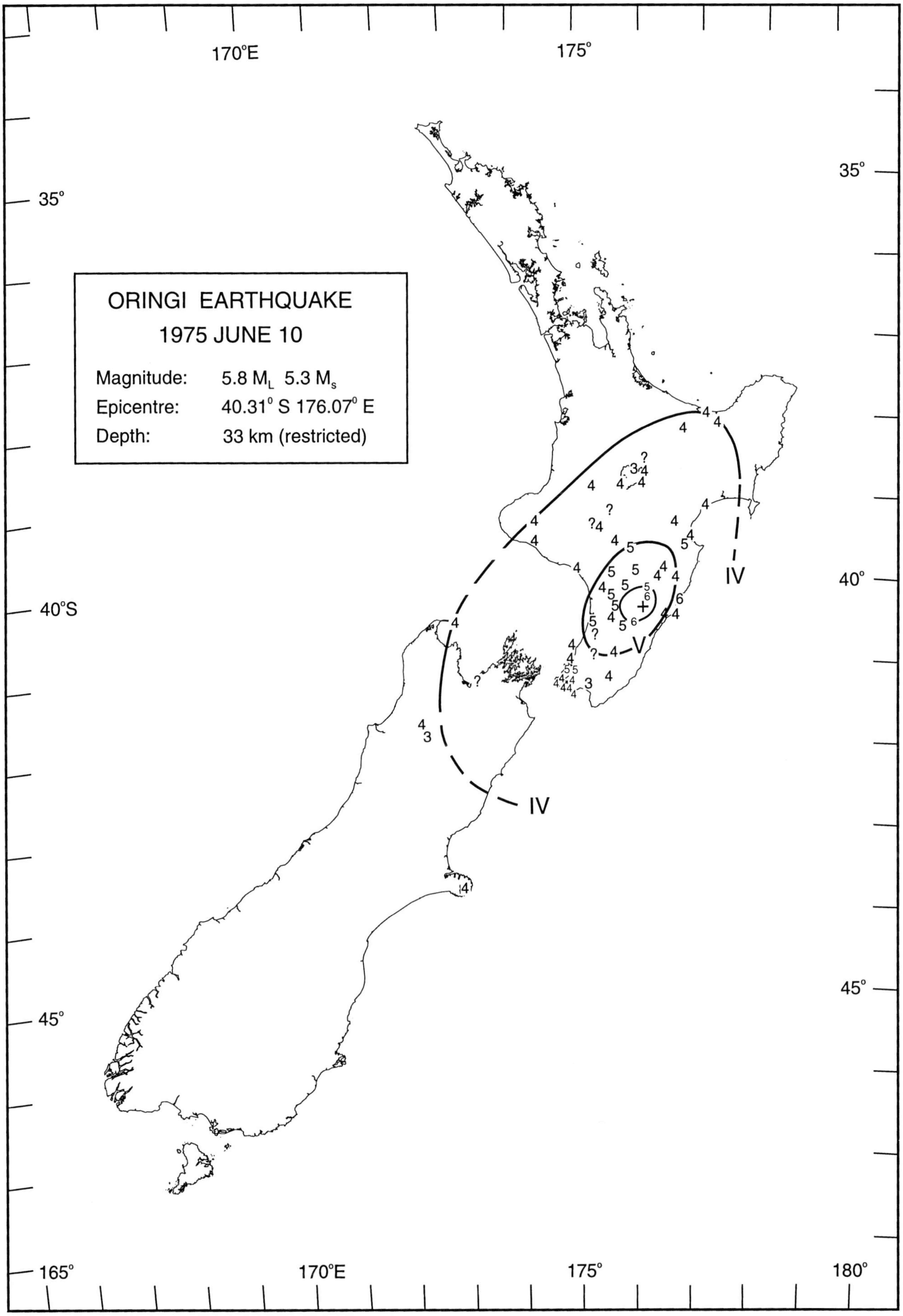

ORINGI EARTHQUAKE
1975 JUNE 10
Magnitude: 5.8 M_L 5.3 M_s
Epicentre: 40.31° S 176.07° E
Depth: 33 km (restricted)
170°E
175°
165°
180°
35°
40°S
40°
45°
IV
V

106. ISOSEISMAL MAP OF THE MILFORD SOUND EARTHQUAKE - 1976 MAY 04

DATE:	1976 MAY 04
TIME:	13:56:29.2 UT
MAGNITUDE:	6.5 M_L*; 6.4 M_s (Dowrick & Smith 1990); 6.5 M_w (Anderson et al. 1993)
EPICENTRE:	44.67°S 167.38°E (1)
DEPTH:	12 km (restricted)

'The Milford Sound earthquake of 1976 May 4 was the largest shallow earthquake in New Zealand since the Inangahua earthquake of 1968 May 23....

'The epicentre lies approximately 30 km west of the nearest recording station, at Milford Sound (MSZ). This distance and the fact that it lies outside the perimeter of the recording network limits the accuracy with which any depth less than 30 km can be determined, and the computer was therefore constrained to produce a solution at the standard shallow depth of 12 km. The appearance of crustal phases in the records, the pattern of felt intensities, and the number of aftershocks makes an appreciably greater depth unlikely....

'[The isoseismal map] shows Modified Mercalli intensities based on replies to standard questionnaires. The extent of the felt area, which covered the whole of the South Island south of a line from Westport to Banks Peninsula, is clearly limited by the fact that intensities below MM IV do not wake soundly-sleeping people. The shock was also felt on Stewart Island. The intensity of MM VI in the epicentral region should perhaps be regarded as a minimum, there being few structures or other objects that could serve as indicators of higher intensity to be found in the area. Landslides were observed on the steeper slopes, but these alone were not considered sufficient to justify a higher rating.

'Aftershocks continued until at least the end of June, about 60 shocks above magnitude 4 occurring within the first month ...

'In the absence of damage, the shock attracted little attention from the press, and the general public has remained unaware that a major earthquake occurred.' (Eiby 1978).

The isoseismal map has been reproduced from the Seismological Observatory bulletin E 158.

---

* Local magnitudes and hypocentres have been revised since the publication of the Seismological Observatory bulletin E 158. Previously this event had a magnitude 7.0 M_L. Refer to introduction.

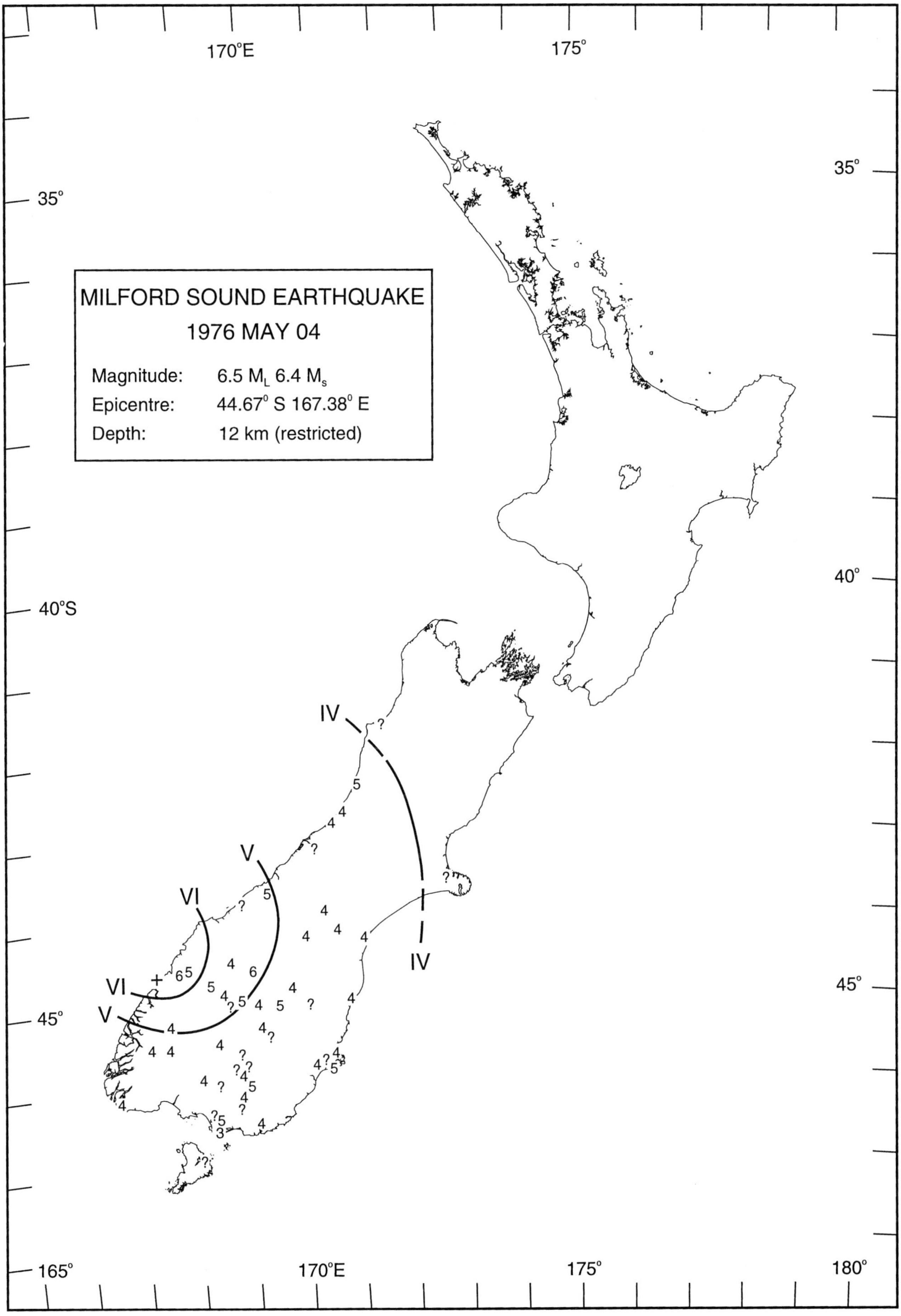

MILFORD SOUND EARTHQUAKE
1976 MAY 04
Magnitude: 6.5 M_L 6.4 M_s
Epicentre: 44.67° S 167.38° E
Depth: 12 km (restricted)
170°E
175°
35°
40°S
45°
165°
180°
40°
VI
V
IV

107. ISOSEISMAL MAP OF THE TE PUKE EARTHQUAKE - 1976 OCTOBER 27

DATE:	1976 OCTOBER 27
TIME:	20:57:09.5 UT
MAGNITUDE:	5.0 M_L*
EPICENTRE:	37.83°S 176.34°E (1)
DEPTH:	12 km (restricted)

'A swarm of smaller but more persistent shocks troubled the district around Te Puke from late August until early December, though the shocks became markedly less frequent after the occurrence of the largest earthquake of the sequence, a magnitude 4.9 [5.0 M_L*] on October 27. This threw goods from shelves in the business area of Te Puke and caused substantial structural damage to houses in a small area near Royden Downs, some 10 km to the south-east, where the intensity reached MM8. An earlier shock on September 9 also brought goods from shelves and caused other minor damage at Royden Downs. As its magnitude was only 3.8, it must be assumed to have had an unusually shallow focus.' (Seismological Observatory bulletin E 158)

The isoseismal map is assumed to have been drawn some time ago by Eiby as it was found after his death among his papers with an accompanying (unfinished) manuscript.

* Local magnitudes and hypocentres have been revised since the publication of the Seismological Observatory bulletin E 158. Refer to introduction.

TE PUKE EARTHQUAKE
1976 OCTOBER 27

Magnitude:	5.0 M_L
Epicentre:	37.83° S 176.34° E
Depth:	12 km (restricted)

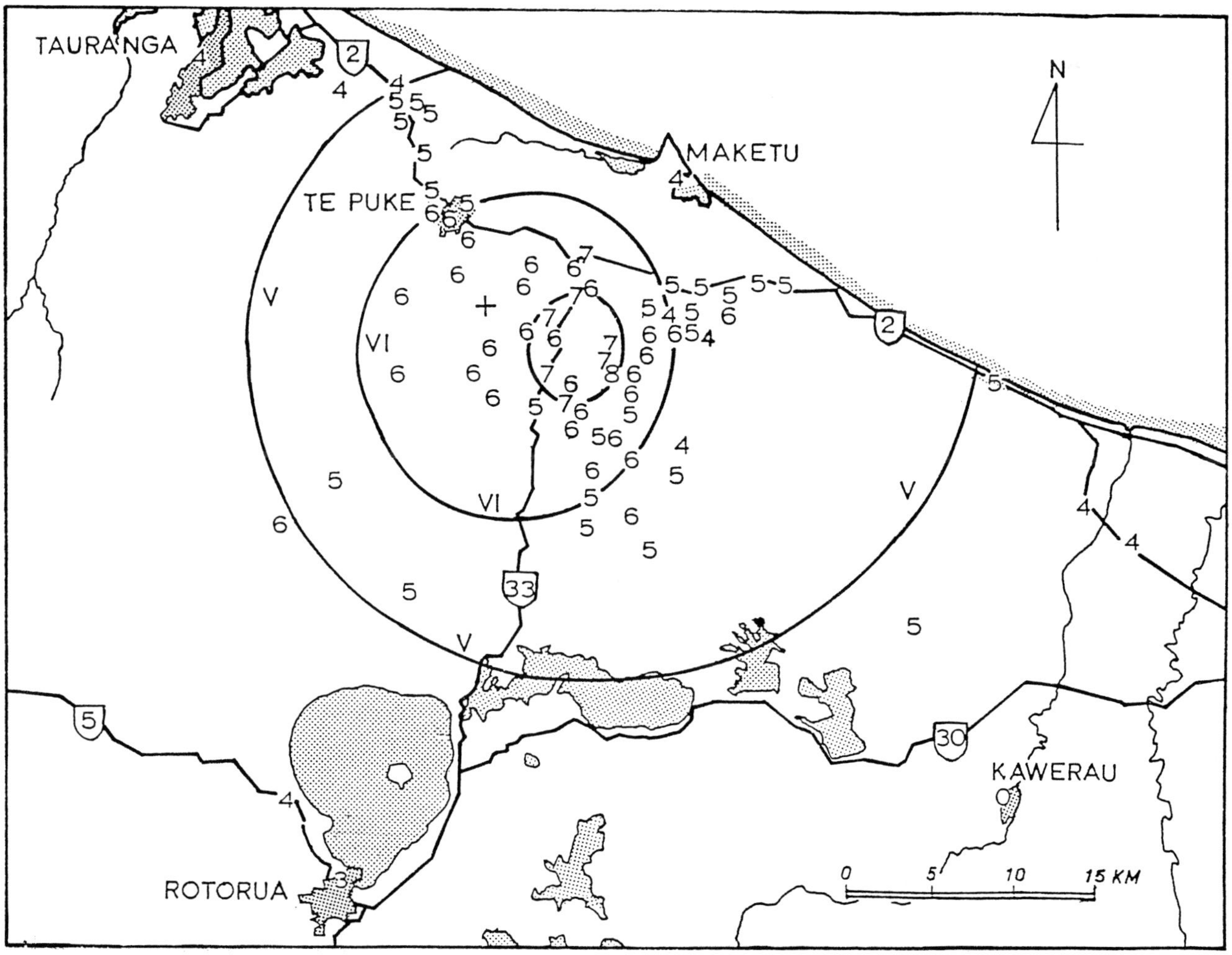

after Eiby (unpublished)

108. ISOSEISMAL MAP OF THE GOLDEN BAY EARTHQUAKE - 1976 DECEMBER 03

DATE:	1976 DECEMBER 03
TIME:	13:25:52.4 UT
MAGNITUDE:	5.6 M_L*
EPICENTRE:	40.33°S 173.47°E (1)
DEPTH:	213 km

'A shock centred to the north of Golden Bay, with a magnitude of 5.4 [5.6 M_L*] and a depth of 213 km occurred on December 3. Shocks at depths as great as this are known even further to the south, but they are not common, and this is one of the larger members of its class. It was reported felt at many places in the Wellington area, at Blenheim, Murchison and as far to the north as Wanganui.' (Seismological Observatory bulletin E 158).

The isoseismal map has been reproduced from the Seismological Observatory bulletin E 158.

* Local magnitudes and hypocentres have been revised since the publication of the Seismological Observatory bulletin E 158. Refer to introduction.

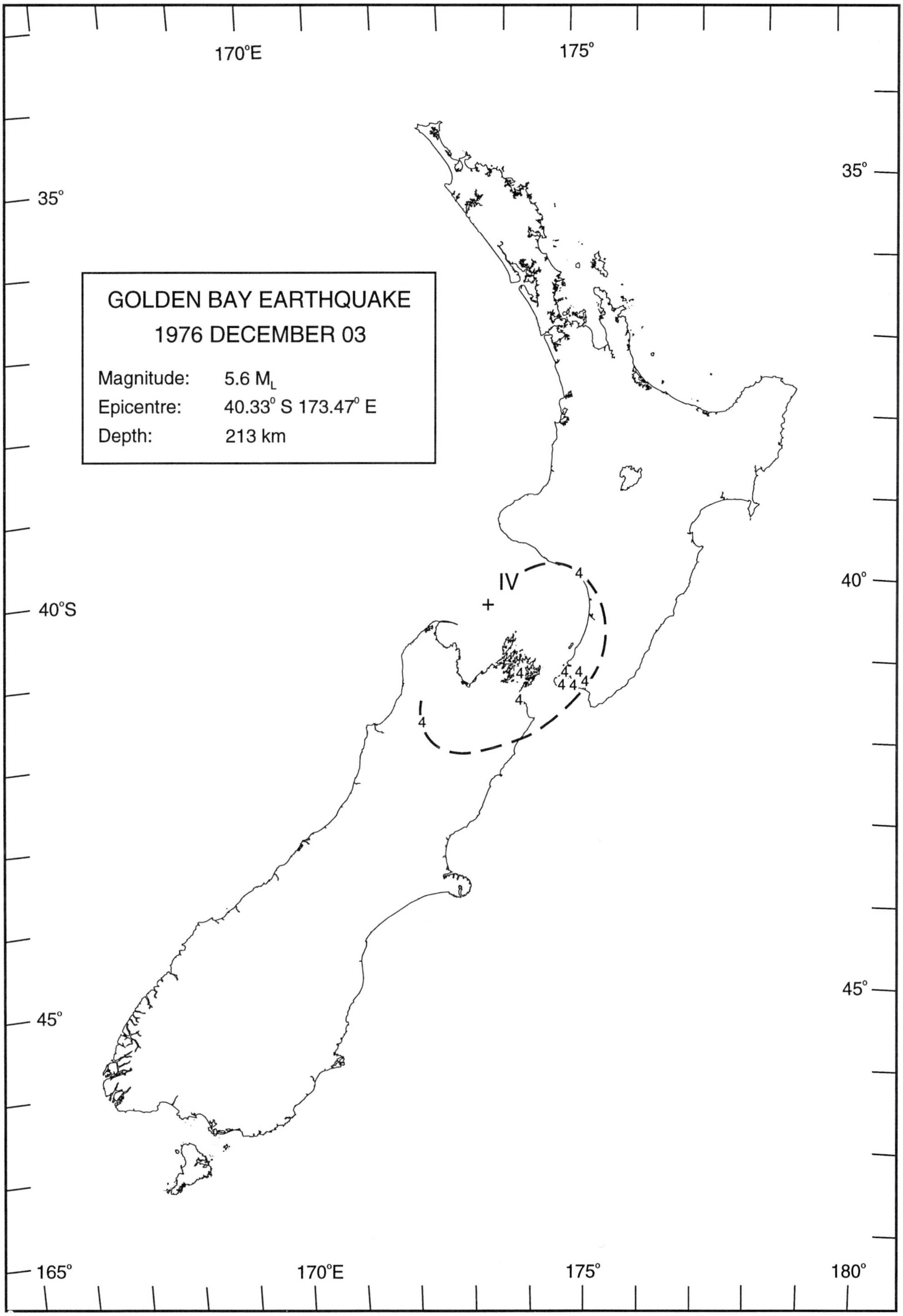

170°E
175°
35°
35°
GOLDEN BAY EARTHQUAKE
1976 DECEMBER 03
Magnitude: 5.6 M_L
Epicentre: 40.33° S 173.47° E
Depth: 213 km
40°
40°S
IV
4
4
4
4
4
4
4
4
45°
45°
165°
170°E
175°
180°

109. ISOSEISMAL MAP OF THE KORAKONUI EARTHQUAKE - 1976 DECEMBER 05

DATE:	1976 DECEMBER 05
TIME:	04:57:16.7 UT
MAGNITUDE:	5.1 M_L*
EPICENTRE:	38.17°S 175.51°E (Eiby 1977)
DEPTH:	1 km ± 2 km

'The Korakonui earthquake of 5 December 1976 occurred at 5-57 p.m. N. Z. Daylight Time (04h 57m U.T.) and was felt in most places lying west of the Waikato River, causing some public alarm. Scattered minor damage extended from Hamilton to Te Kuiti, with isolated instances at greater distances from the epicentre. This lies in a rural area, the nearest townships being Te Awamutu, Otorohanga, Tokoroa and Putaruru, all about 25 km away.

'The New Zealand Seismological Report assigns the shock a magnitude of 5.0 [5.1 M_L*] and an epicentre at 38.10°S, 175.54°E, but the preferred solution places it about six kilometres farther south west, near the settlement of Korakonui, where the greatest damage occurred. Here several houses within a radius of a few kilometres lost chimneys and suffered other damage indicating an intensity of MM VIII. Unlike similarly damaged property at Cambridge, these houses stand on firm ground, and the maximum intensity can be considered adequately established. In all some 60 insurance claims were received by the Earthquake and War Damage Commission.' (Eiby 1977)

Dowrick & Smith (1990) note no M_s could be assessed for this earthquake due to lack of observed surface waves.

The isoseismal map is taken from Eiby (1977) and is based on reports of the insurance assessors and upon 98 replies to standard questionnaires issued by the Seismological Observatory.

* Local magnitudes have been revised since the publication of the Seismological Observatory bulletin E 158. Refer to introduction.

KORAKONUI EARTHQUAKE 1976 DECEMBER 05	
Magnitude:	5.1 M_L
Epicentre:	38.17° S 175.51° E
Depth:	1 km

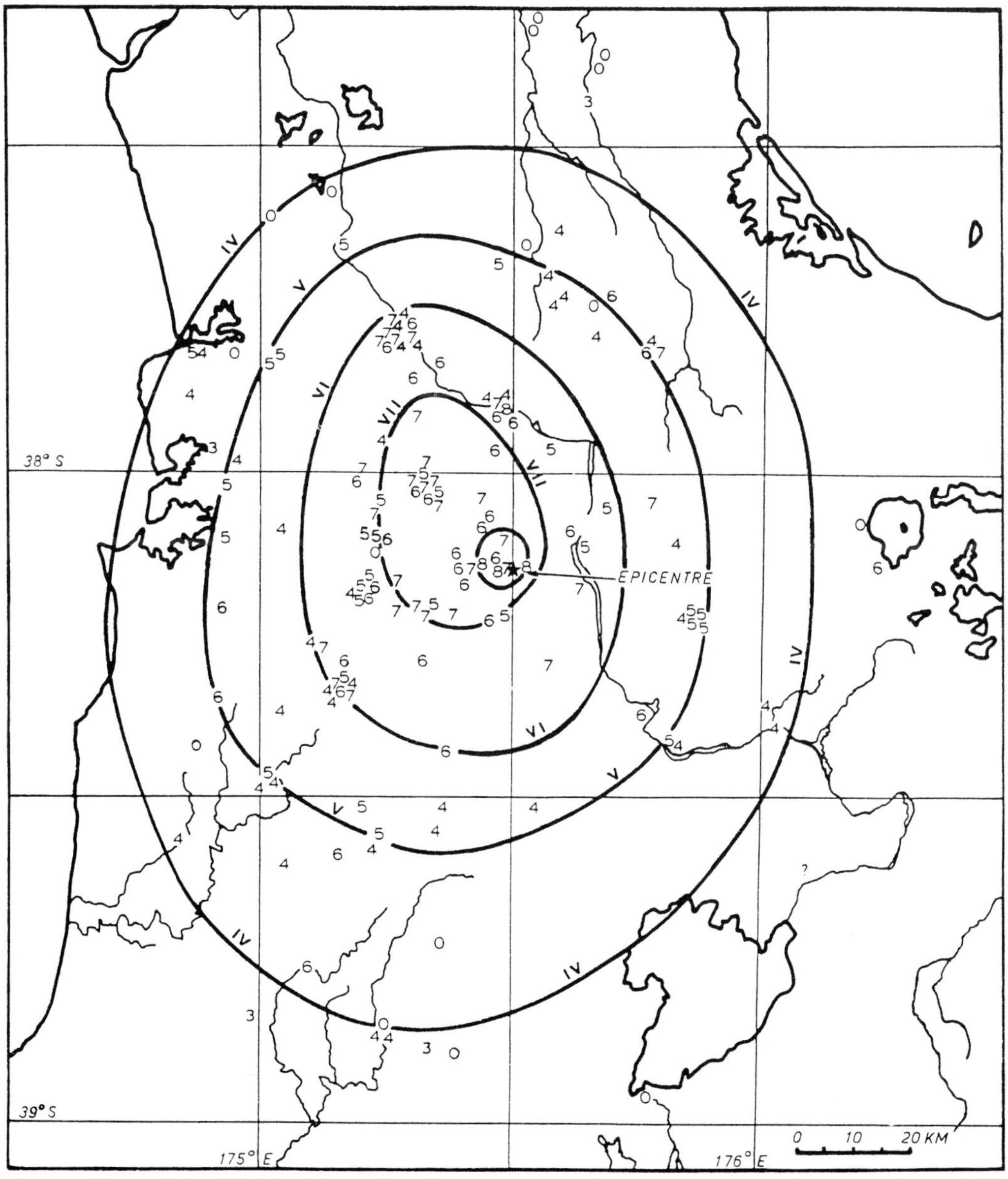

after Eiby (1977)

110. ISOSEISMAL MAP OF THE CAPE CAMPBELL EARTHQUAKE - 1977 JANUARY 18

DATE:	1977 JANUARY 18
TIME:	05:41:48.9 UT
MAGNITUDE:	6.0 M_L^*; 5.9 M_s (Dowrick & Smith 1990)
EPICENTRE:	41.84°S 174.58°E (1)
DEPTH:	33 km (restricted)

'The largest shallow event [in 1977] was centred in the eastern approaches to Cook Strait, some 25 km east of Cape Campbell. It occurred early in the year, on the evening of January 18, and had a magnitude of 6.0, and was felt from Banks Peninsula to Taranaki and parts of Hawke's Bay. There were 960 insurance claims, none of a serious nature, and mostly relating to damaged chimneys and overthrown goods. They came mainly from Wellington city and northern Marlborough, where the intensity reached MM VII in some places...' (Seismological Observatory bulletin E 159)

Two aftershocks were felt, one being of magnitude 4.7.

The isoseismal map is reproduced from the Seismological Observatory bulletin E 159.

---

* Local magnitudes have been revised since the publication of the Seismological Observatory bulletin E 159. Refer to introduction.

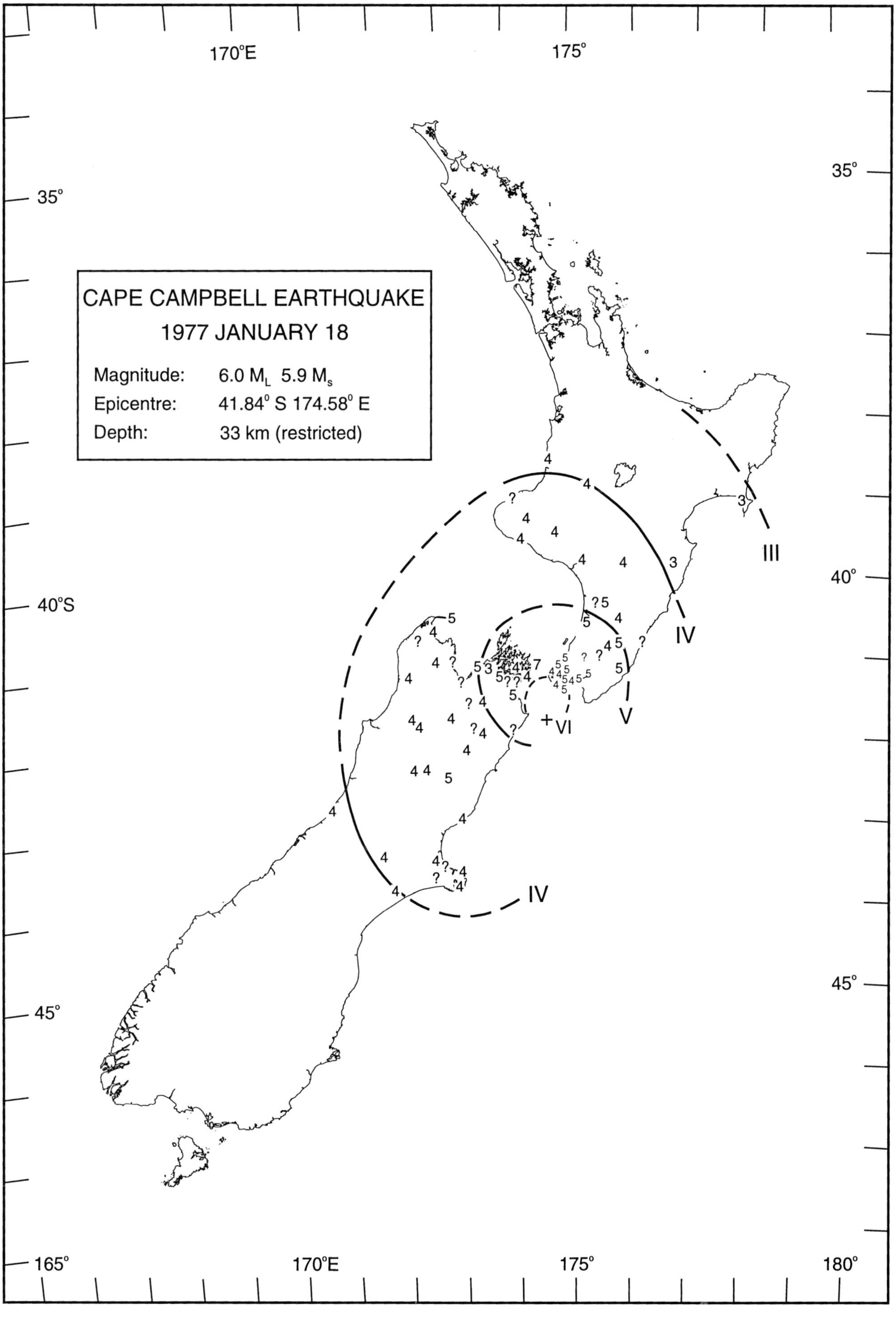
CAPE CAMPBELL EARTHQUAKE
1977 JANUARY 18
Magnitude: 6.0 M_L 5.9 M_s
Epicentre: 41.84° S 174.58° E
Depth: 33 km (restricted)
170°E
175°
35°
40°S
40°
45°
165°
180°
III
IV
V
VI

111. ISOSEISMAL MAP OF THE MATATA EARTHQUAKE - 1977 MAY 31

DATE:	1977 MAY 31
TIME:	18:50:55.3 UT
MAGNITUDE:	5.4 M_L
EPICENTRE:	37.83°S 176.83 E (Richardson 1989)
DEPTH:	9 km ± 2 km

'A magnitude M_L 5.4 earthquake and accompanying succession of aftershocks occurred in 1977 adjacent to a temporary network of 10 closely spaced, high-gain portable seismographs recording microearthquakes in the Bay of Plenty region of the North Island, New Zealand.

'The main shock, felt strongly in the Bay of Plenty region at dawn on 1977 June 1 (local time) ... occurred near Edgecumbe, produced a maximum intensity of MM VIII, and led to 360 insurance claims.' (Richardson 1989)

The isoseismal map (details of stations and other earthquakes have been omitted from the map) and the accompanying parameters of the earthquake are from Richardson (1989) and are preferred, for the purposes of an isoseismal map, to those in the Seismological Observatory bulletin E 159.

MATATA EARTHQUAKE
1977 MAY 31

Magnitude:	5.4 M_L
Epicentre:	37.83° S 176.83° E
Depth:	9 km

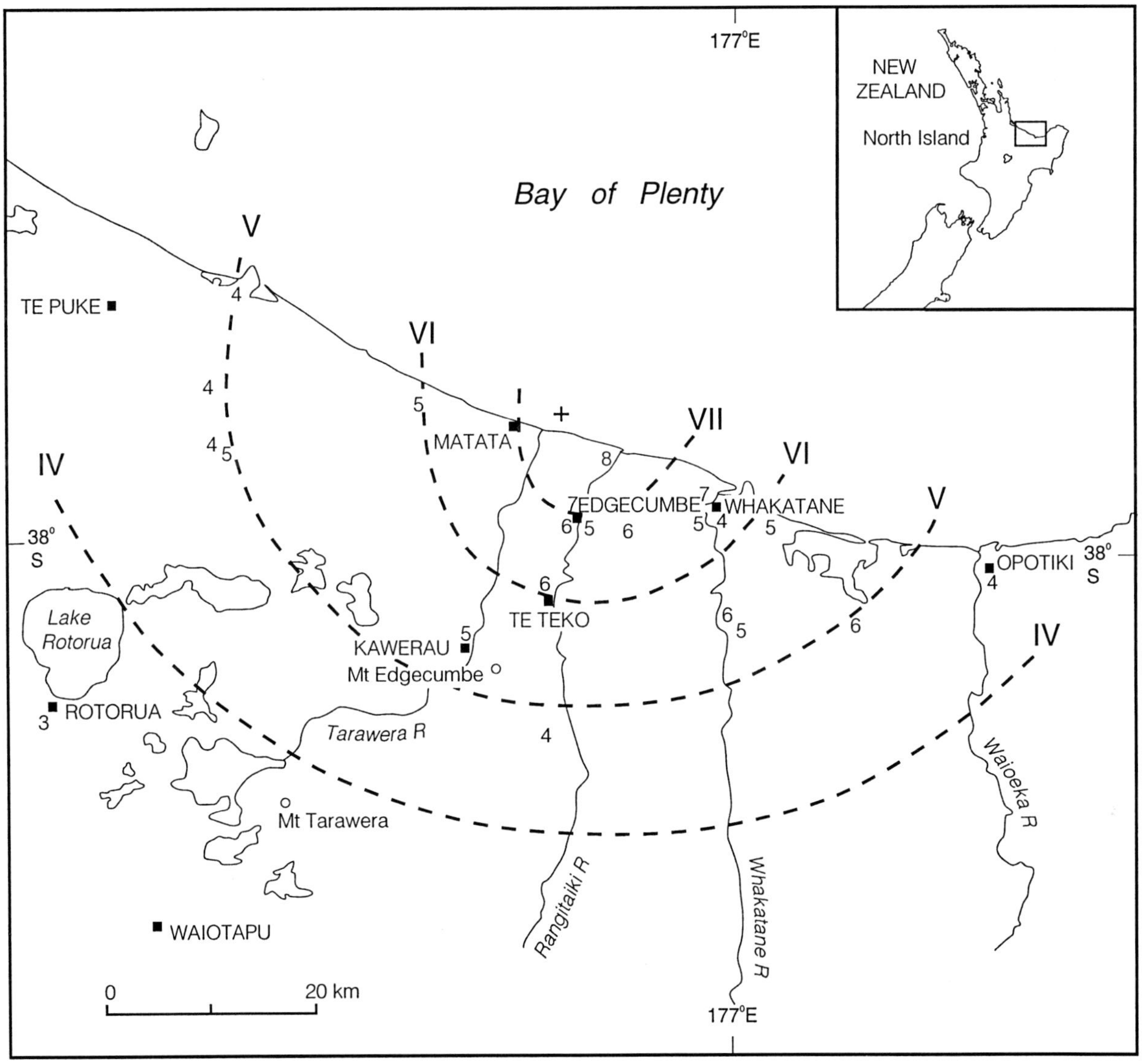

after Richardson (1989a)

112. ISOSEISMAL MAP OF THE PUYSEGUR BANK EARTHQUAKE - 1979 OCTOBER 12

DATE:	1979 OCTOBER 12
TIME:	10:25:19.5 UT
MAGNITUDE:	6.5 M_L; 7.2 M_s (Dowrick & Smith 1990); 7.3 M_w (Anderson et al. 1993)
EPICENTRE:	46.67°S 165.73°E (Webb & Lowry 1982)
DEPTH:	12 km (restricted)

'On 1979 Oct 12d 10h 25m U.T. an earthquake of local magnitude 6.4 occurred in the Puysegur Bank region, about 80 km off the southwest coast of New Zealand. Although widely felt in the South Island, intensities near the epicentre reached only MM V...

'The felt intensities (according to Modified Mercalli scale) based on information received by the Observatory from felt reporters, letters from members of the public, and damage claims are shown [on the isoseismal map] ... Circular lines have been drawn to fit the observations. Slightly elliptical lines would have fitted equally well ...

'The type of motion described in the felt reports for the event under study here is typically that of swaying or a circular or rotating motion. Several observers contrasted these felt effects with other shocks that they had felt which were described as "jolts". Some felt an initial jolt, probably due to a P phase, but the majority only noticed the low frequency motions....

'Damage claims received by the New Zealand Earthquake and War Damage Commission numbered 198 ... Of the claims for private dwellings, only six were for contents, the remainder being for structural damage. This is unusual for damage claims in New Zealand - normally about 65% of the claims would be for contents instead of the 3% in this case. Most of the structural damage was to outside walls and foundations and, in 33 cases, to chimneys....

'The long-period effects and predominance of structural damage indicate the seismic waves that radiated from this earthquake had a large low-frequency content.' (Webb & Lowry 1982)

The isoseismal map is from Webb & Lowry (1982). The hypocentral parameters of the earthquake are also taken from this special study and are preferred, for the purposes of an isoseismal map, to those in the Seismological Observatory bulletin E 161.

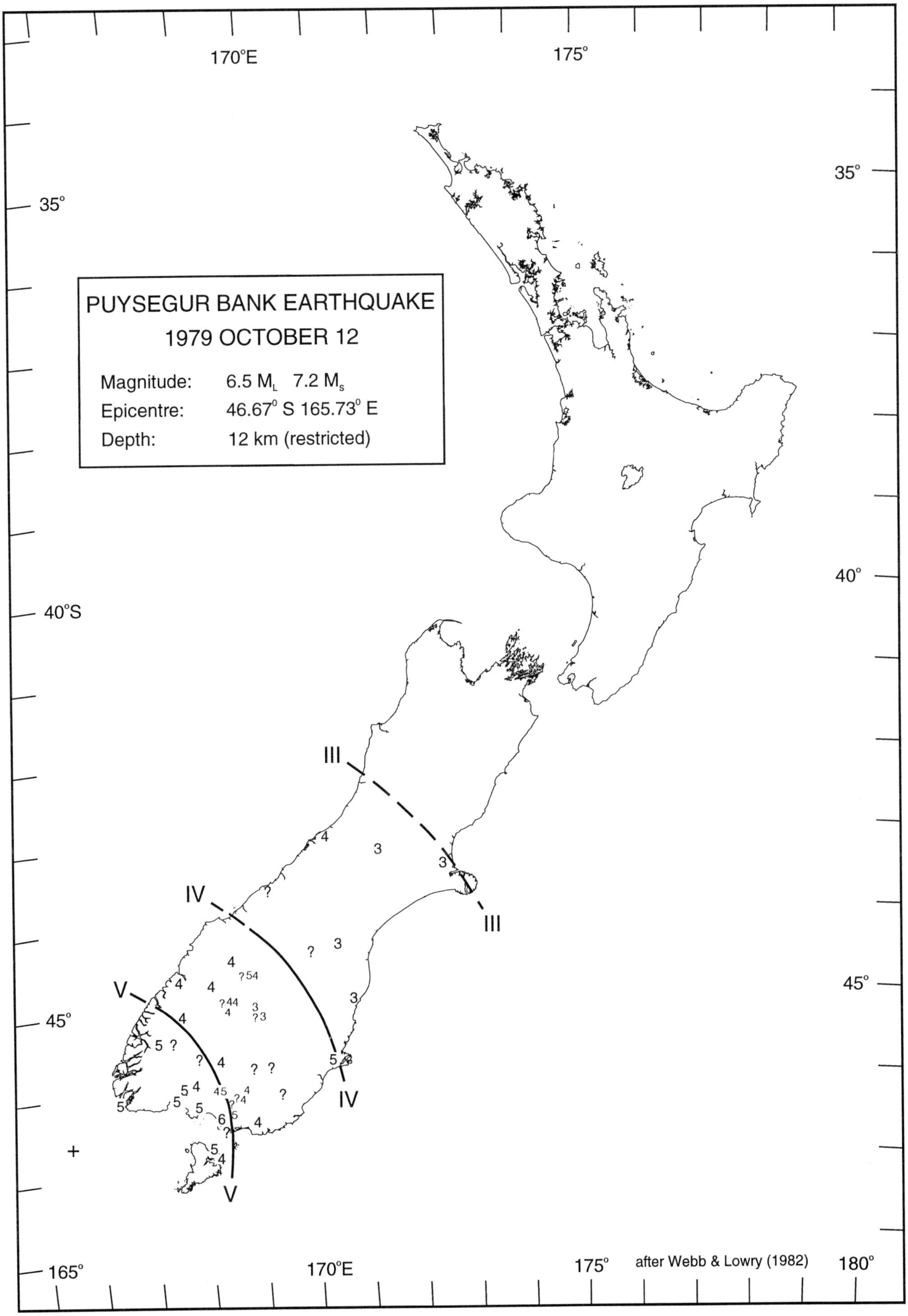

170°E
175°
35°
40°S
45°
165°
170°E
175°
180°
40°
45°
PUYSEGUR BANK EARTHQUAKE
1979 OCTOBER 12
Magnitude: 6.5 M_L 7.2 M_s
Epicentre: 46.67° S 165.73° E
Depth: 12 km (restricted)
III
IV
V
after Webb & Lowry (1982)

113. ISOSEISMAL MAP OF THE HAWKE'S BAY EARTHQUAKE - 1982 SEPTEMBER 02

DATE:	1982 SEPTEMBER 02
TIME:	15:58:54.4 UT
MAGNITUDE:	5.4 M_L; 5.6 M_b
EPICENTRE:	39.74°S 176.94°E (1)
DEPTH:	41 km

The Hawke's Bay Province earthquake of 1982 September 02 was the most damaging earthquake to occur in New Zealand in 1982. It was widely felt in the southern North Island from Omori, near Lake Taupo, to Wellington, with intensities reaching MM6 in the Hastings area. Part of the affected area experienced a power cut of about half an hour's duration. Damage to an estimated value of $100 000 was caused.

'Although routine methods of origin determination for this earthquake yield a focal depth of 41 km, observation of the aftershocks using a number of locally-deployed portable seismographs indicated that the activity was concentrated at about 48 km below the surface.' (Seismological Observatory bulletin E 165).

The isoseismal map is taken from the Seismological Observatory bulletin E 165.

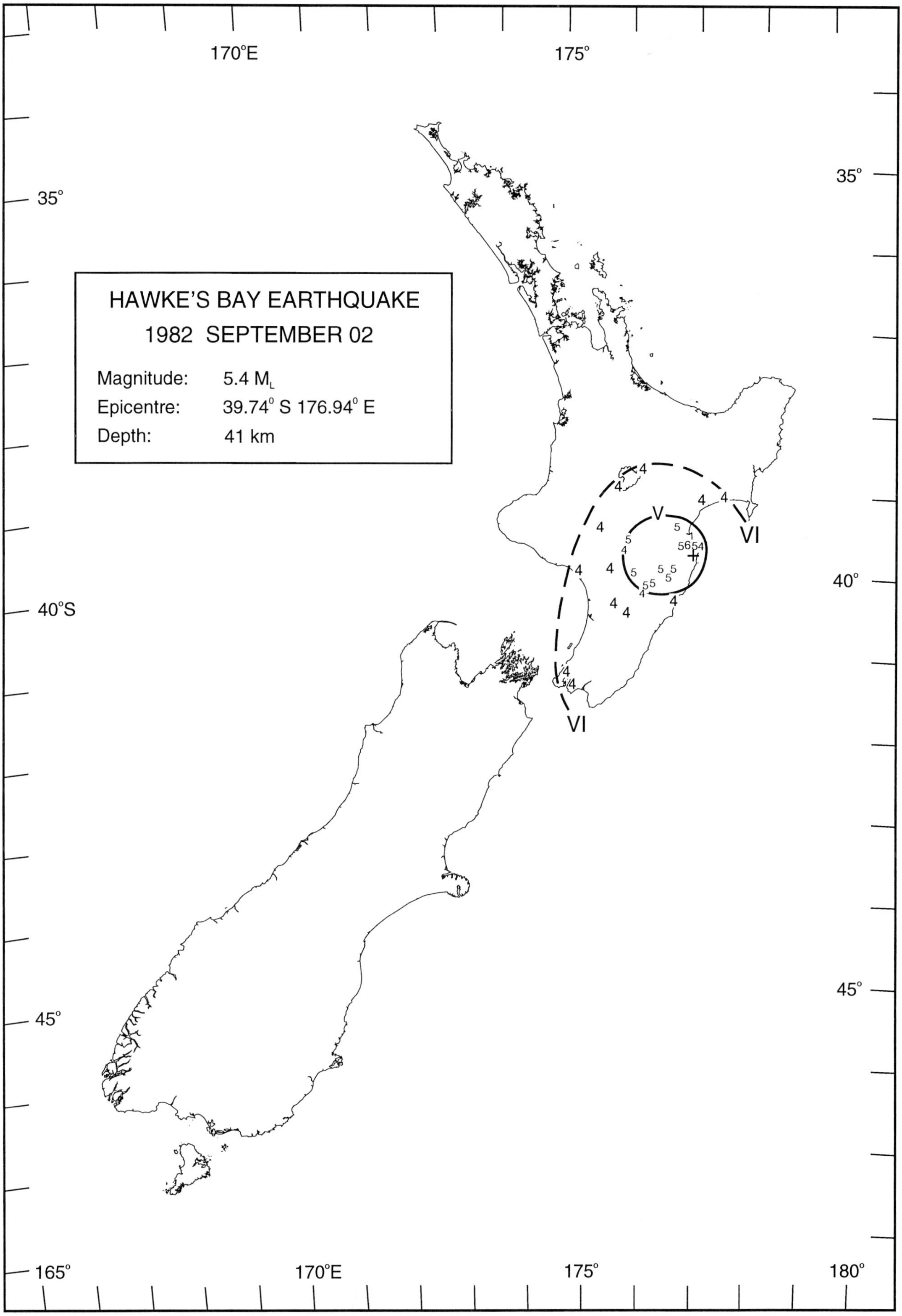

HAWKE'S BAY EARTHQUAKE
1982 SEPTEMBER 02
Magnitude: 5.4 M_L
Epicentre: 39.74° S 176.94° E
Depth: 41 km
170°E
175°
35°
40°S
40°
45°
165°
180°
V
VI
VI

114. ISOSEISMAL MAP OF THE WAIOTAPU EARTHQUAKE - 1983 DECEMBER 14

DATE:	1983 DECEMBER 14
TIME:	20:56:29.4 UT
MAGNITUDE:	5.1 M_L; 4.6 M_s (Dowrick & Smith 1990)
EPICENTRE:	38.36°S 176.33°E (1)
DEPTH:	5 km (restricted)

'In the 20 month period preceding the December shock, the epicentral region was subject to repeated swarms....

'The MM IV area extended from Wairakei to Rotorua and intensities reached MM VII and possibly MM VIII in the epicentral region....

'Examination of the area around Waiotapu, and interviews with local residents provided reasonably detailed information about felt intensities. Two observations suggested MM VIII in the epicentral region. In addition to damage to roads in areas of fill and unconsolidated material, there were large boulders up to 2 metres in diameter displaced from road cuttings on State Highway 5 up to 1.5 kilometres south of Waiotapu, and there was a landslide on the steep western face of Mount Paeroa, just south of the village of Waikite Valley. Doubt about the prior stability of the displaced material leaves the assignment of MM VIII uncertain. Undoubted instances of MM VII occurred at Waiotapu and along the Waikite Valley Road ... There were numerous instances of MM VI. It was also observed that the earthquake was accompanied by short-lived changes to the thermal activity at Waiotapu. Claims for damage totalled $29,000, including one for $3,000 for damage to a silo at the dairy factory at Reporoa.' (Smith et al. 1984)

The isoseismal map is taken from the Seismological Observatory bulletin E 167.

WAIOTAPU EARTHQUAKE
1983 DECEMBER 14

Magnitude: 5.1 M_L 4.6 M_s
Epicentre: 38.36° S 177.33° E
Depth: 5 km (restricted)

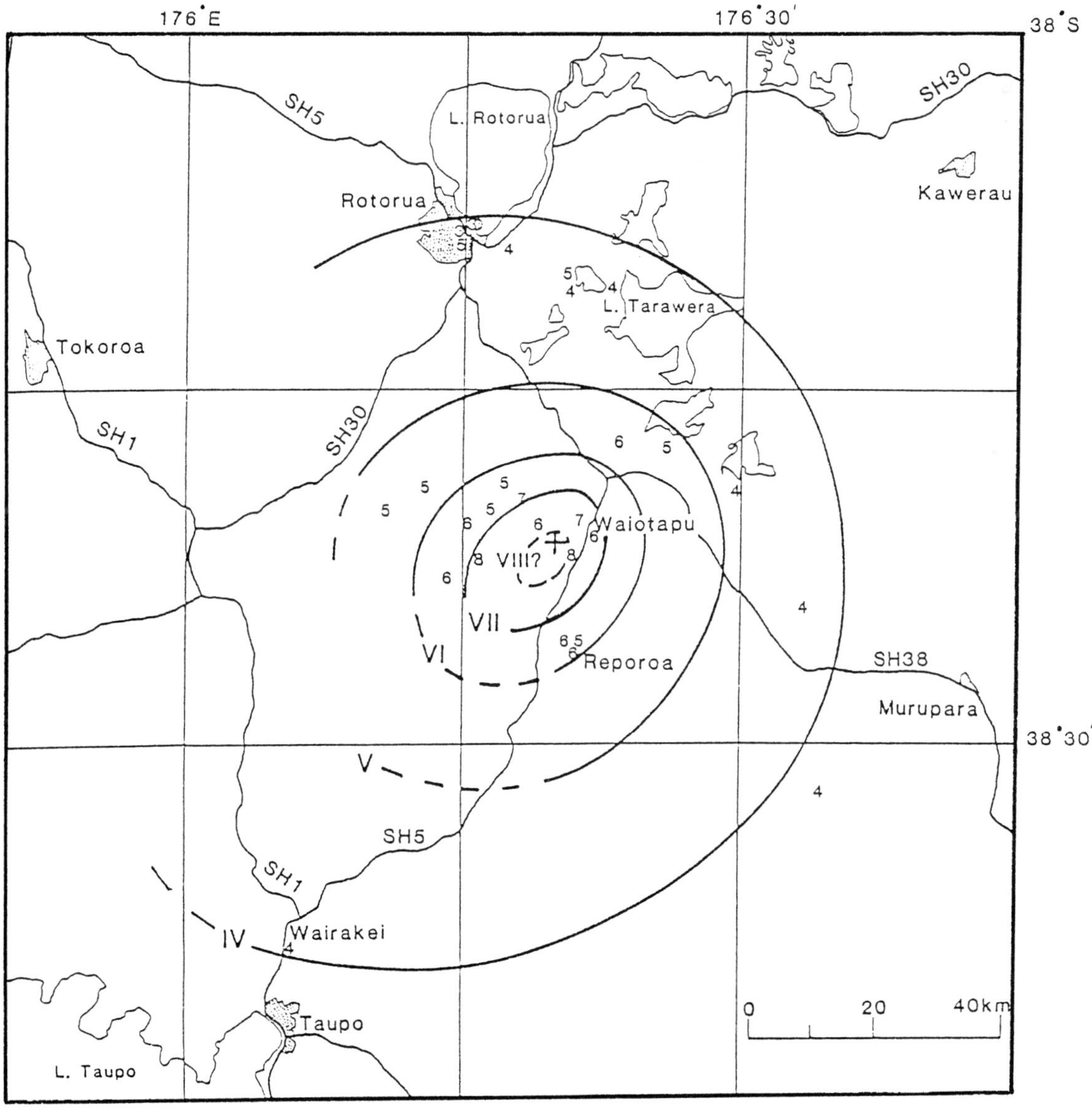

115. ISOSEISMAL MAP OF THE MOTU RIVER EARTHQUAKE - 1984 MARCH 08

DATE:	1984 MARCH 08
TIME:	00:40:52.6 UT
MAGNITUDE:	6.4 M_L
EPICENTRE:	38.22°S 177.48°E (Reyners & Hodder 1985)
DEPTH:	73 km

'The M = 6.4 earthquake which shook the upper reaches of the Motu River on 8 March 1984 was the largest event to occur at the Hikurangi Margin of New Zealand (that is, the North Island [excluding Northland] and the northern South Island) since an M = 7.0 earthquake in 1973. Like the 1973 event, the Motu River earthquake occurred at intermediate depth, and as a consequence it caused only minor damage...

'... intensities reached MM IV as far away as Greymouth and Akaroa, some 700 kilometres from the epicentre. The felt data for this earthquake fill a gap in our knowledge of the spatial distribution of felt intensities for New Zealand earthquakes [of intermediate depth].' (Reyners & Hodder 1985)

The parameters of the earthquake are from Reyners & Hodder (1985) and are preferred, for the purposes of an isoseismal map, to those in the Seismological Observatory bulletin E 168. The isoseismal map has been reproduced from the Seismological Observatory bulletin E 168.

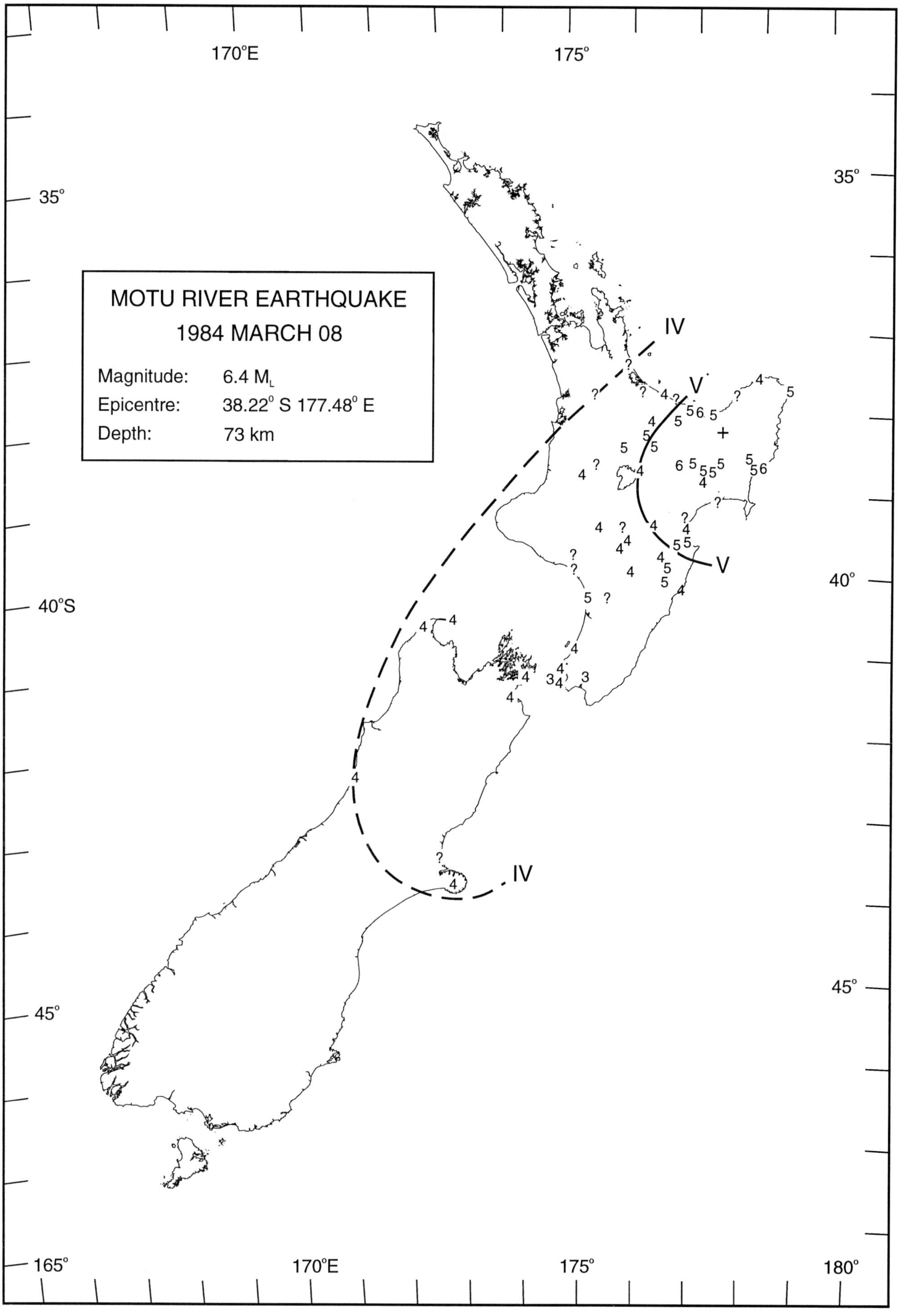

MOTU RIVER EARTHQUAKE
1984 MARCH 08
Magnitude: 6.4 M_L
Epicentre: 38.22° S 177.48° E
Depth: 73 km
170°E
175°
180°
165°
35°
40°S
40°
45°
IV
V

116. ISOSEISMAL MAP OF THE MACAULEY RIVER EARTHQUAKE - 1984 JUNE 24
(elsewhere called the Godley River earthquake)

DATE:	1984 JUNE 24
TIME:	13:29:39.9 UT
MAGNITUDE:	5.9 M_L; 6.1 M_s (Dowrick & Smith 1990); 6.1 M_w (Anderson et al. 1993)
EPICENTRE:	43.60°S 170.64°E (1)
DEPTH:	5 km (restricted)

'Intensity MM VI was reported from several places in south Westland as a result of a magnitude 5.9 earthquake in the Macauley River valley on June 24th. Single reports of the same intensity came from the thinly populated Mount Cook and Mount Somers localities, nearer the epicentre on the eastward side of the Southern Alps. The area over which the shock was felt extended from Nelson in the north to Otago in the south. In the epicentral area, car-sized boulders had been dislodged from hillsides and strewn over the valley floor. Although some of the rocks may have been monuments to earlier shocks, others had come to rest in the middle of farm tracks, and could only have arrived there recently. An aftershock sequence ensued, but because of the remoteness of the affected area, few people were inconvenienced.' (Seismological Observatory bulletin E 168)

The isoseismal map from Reyners (pers. comm.) has not been published previously.

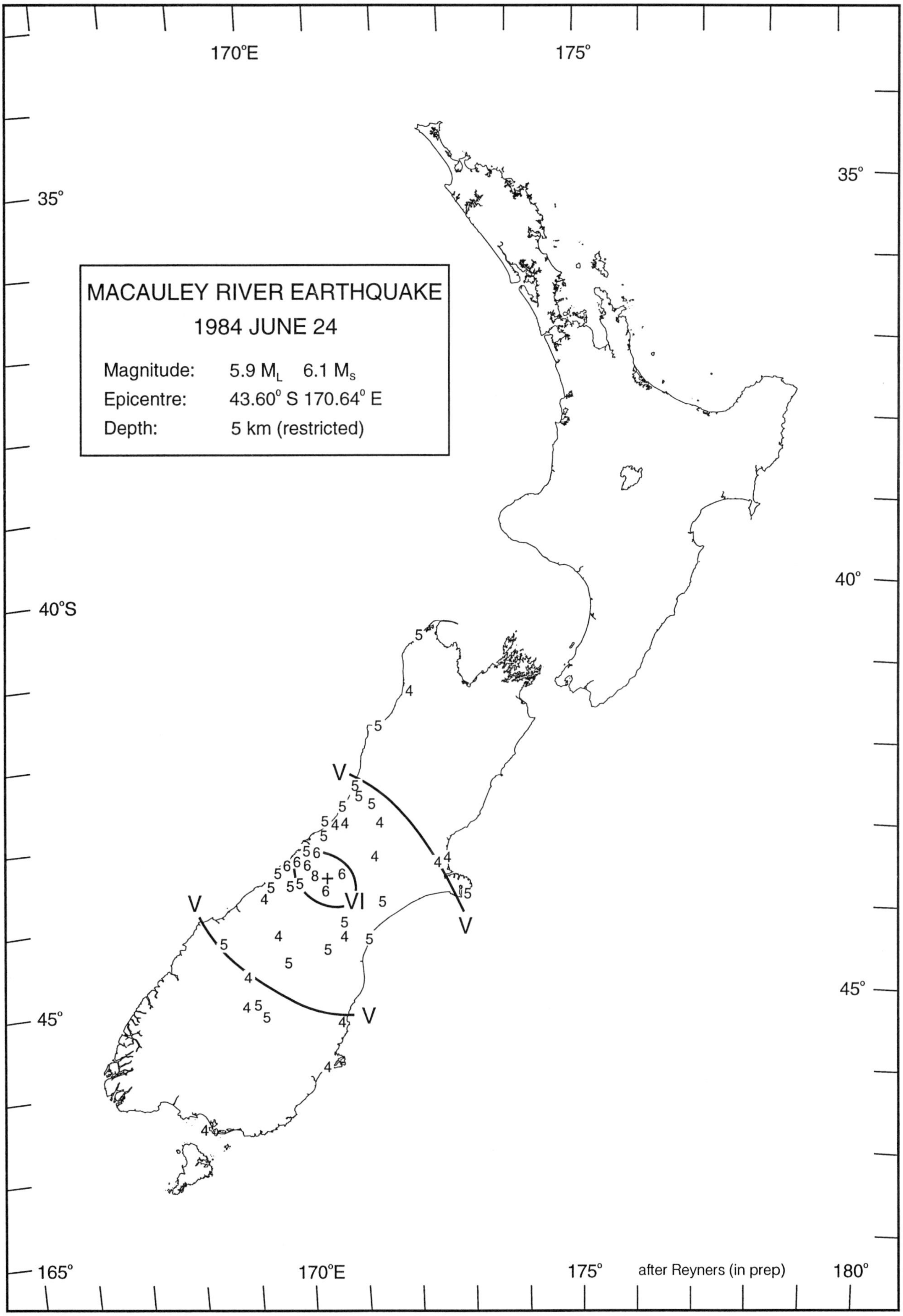

MACAULEY RIVER EARTHQUAKE
1984 JUNE 24
Magnitude: 5.9 M_L 6.1 M_S
Epicentre: 43.60° S 170.64° E
Depth: 5 km (restricted)
170°E
175°
35°
40°S
40°
45°
165°
180°
VI
V
after Reyners (in prep)

117a. ISOSEISMAL MAP OF THE EDGECUMBE EARTHQUAKE - 1987 MARCH 02

DATE:	1987 MARCH 02
TIME:	01:42:35.0 UT
MAGNITUDE:	6.1 M_L; 6.6 M_s (Dowrick & Smith 1990)
EPICENTRE:	37.89°S 176.80°E (2)
DEPTH:	10 km (restricted)

The 1987 March 2 earthquake (01:42 UT), M_L6.1, near Edgecumbe in the North Island occurred at the southeastern margin of the Central Volcanic Zone.

'With a magnitude (M_L) of 6.1, it caused extensive damage in the towns of Edgecumbe, Te Teko and Kawerau in the Bay of Plenty, and was felt as far afield as Hamilton, Taupo, Napier and Gisborne. There were also isolated reports of its being felt at greater distances. Intensities reached MM9 near the epicentre, making it the most severe earthquake in New Zealand since that at Inangahua in 1968. Apart from Inangahua, it was the most damaging earthquake since the 1942 Wairarapa shock which severely affected Masterton and Wellington.

'... The earthquake opened up a rift one metre wide and seven kilometres long, snaking across the Rangitaiki Plains. There was slumping of up to two metres on the downthrown side.

'Strong motion instruments at the Matahina Dam, some 23 km from the epicentre but rather closer to the nearest point of the main rift, recorded a peak acceleration comparable with the design level for the dam. This is the first time such records have been obtained in New Zealand.

'The earthquake was preceded by a series of foreshocks in two locations: at the coast near Thornton and northwest of there, near Matata. A long series of aftershocks followed: more than 200 were felt, and three reached M_L 5, which would be expected after a mainshock exceeding 6.0. One foreshock was of M_L 4.9. This occurred only seven minutes before the main shock and was largely responsible for the evacuation of buildings which collapsed or were severely damaged in the main shock. There were in fact no fatalities.' (Seismological Observatory bulletin E 171)

The isoseismal maps were drawn by Lowry et al. (1989) of the Seismological Observatory, based on information contained in questionnaires distributed widely in the immediate vicinity of the earthquake and also less densely over the North Island extending from Whangarei to Masterton. Press and other media reports were also used when sufficiently specific.

EDGECUMBE EARTHQUAKE
1987 MARCH 02

Magnitude:	6.1 M_L 6.6 M_s
Epicentre:	37.89° S 176.80° E
Depth:	10 km (restricted)

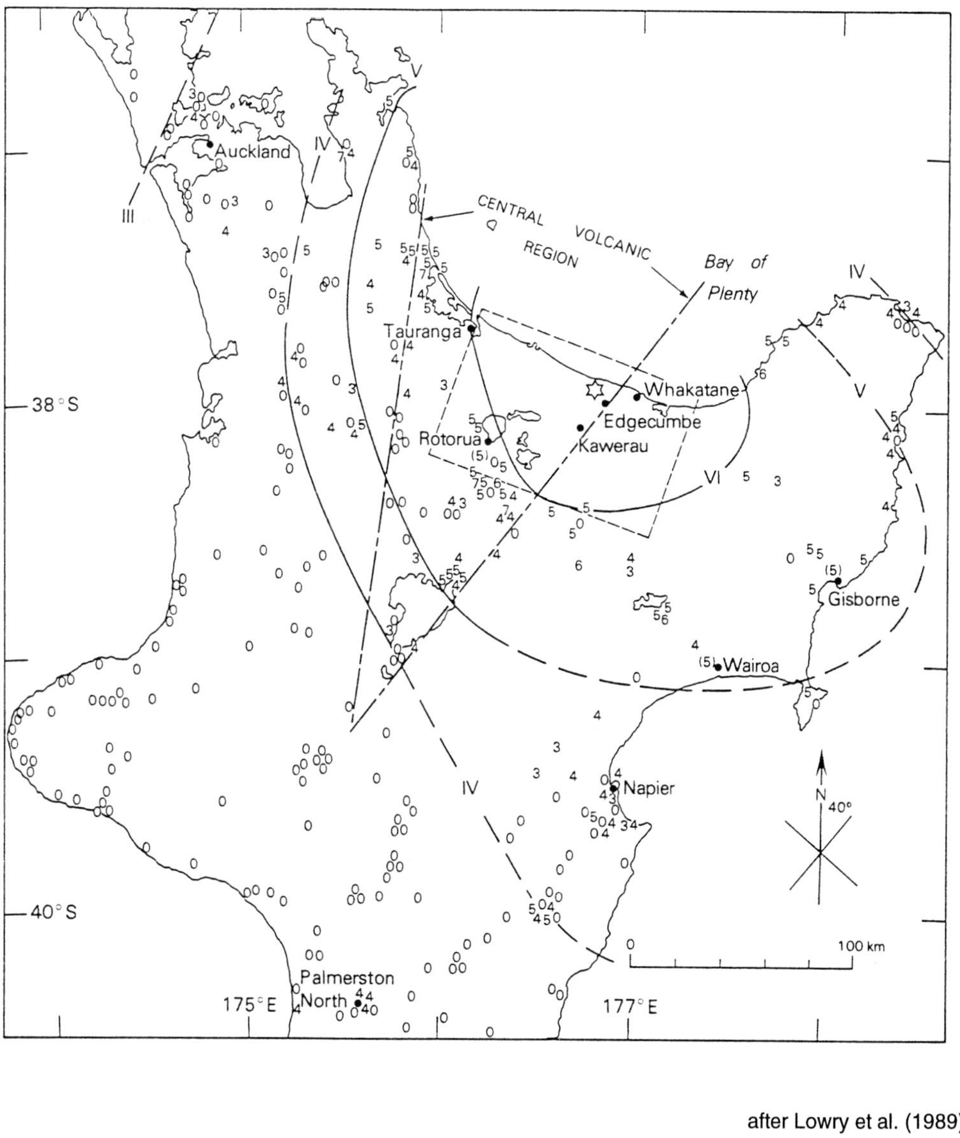

after Lowry et al. (1989)

117b. ISOSEISMAL MAP OF THE EDGECUMBE EARTHQUAKE - 1987 MARCH 02

DATE:	1987 MARCH 02
TIME:	01:42:35.0 UT
MAGNITUDE:	6.1 M_L; 6.6 M_s (Dowrick & Smith 1990)
EPICENTRE:	37.89°S 176.80°E (2)
DEPTH:	10 km (restricted)

'The highest intensity experienced was in the town of Edgecumbe, some 8 or 9 km from the instrumental epicentre, where there were a few quite well established instances of MM X. The observations justifying this high assessment include damage to river embankments (caused both by shaking and by waves formed in the river) and severe cracking in cement and asphalt roads and footpaths. Other spectacular damage in the town included a toppled locomotive and a row of large transformers, each weighing several tons, torn from their mountings ...

'The damage to railway lines, which was a widespread and notable feature of this earthquake, seemed to us to be disproportionate to other effects expected to occur at the same intensity. Although there were places where fishplates joining lengths of rail had been torn apart, the most typical damage was bending of rails which appeared to indicate compression of the ground beneath them, an indication consistent with the ramming together of underground pipes ...

'The area in which MM IX was experienced extends from Mount Edgecumbe, near Kawerau, to Matata and Thornton on the coast, and includes the town of Te Teko and parts of Kawerau as well as Edgecumbe. Within this area, damage to chimneys was widespread and structural weaknesses were exposed in houses built to superseded building codes.' (Lowry et al. 1989)

The isoseismal maps were drawn by Lowry et al. (1989) of the Seismological Observatory, based on information contained in questionnaires distributed widely in the immediate vicinity of the earthquake and also less densely over the North Island extending from Whangarei to Masterton. Press and other media reports were also used when sufficiently specific.

EDGECUMBE EARTHQUAKE
1987 MARCH 02

Magnitude:	6.1 M_L 6.6 M_s
Epicentre:	37.89° S 176.80° E
Depth:	10 km (restricted)

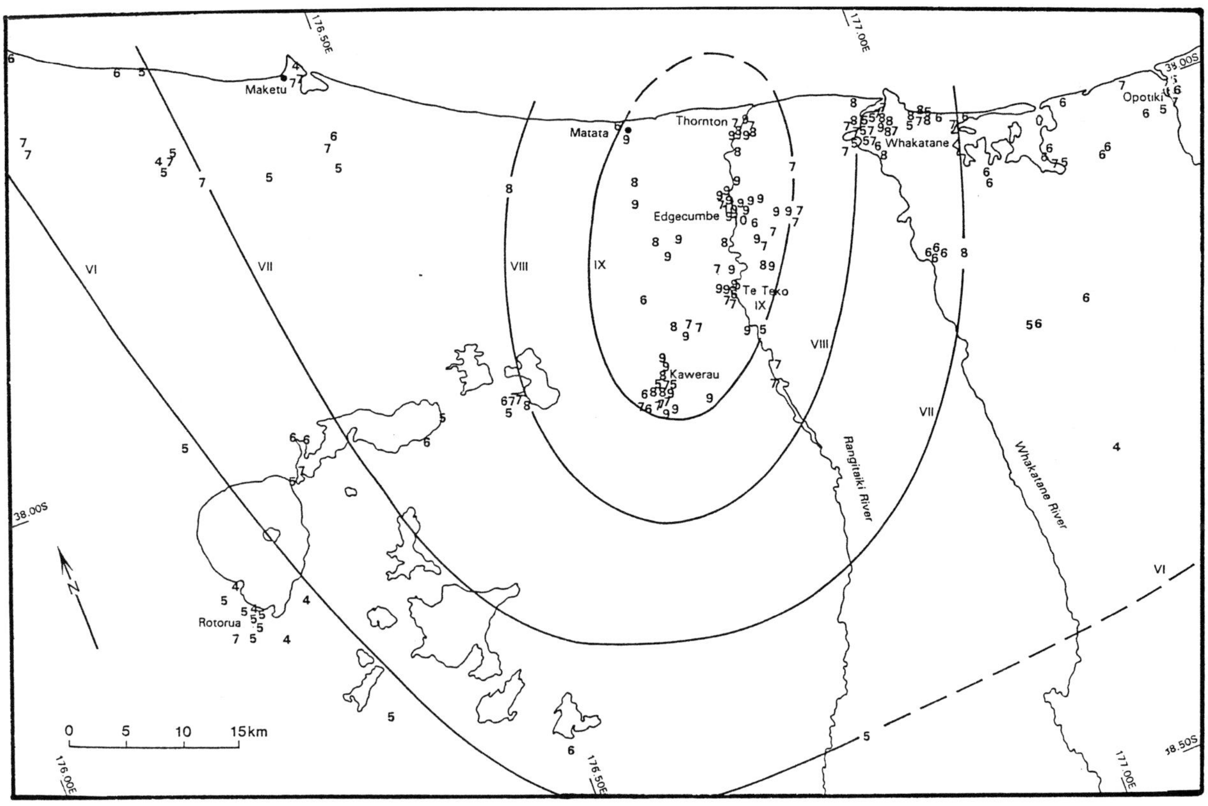

after Lowry et al. (1989)

118. ISOSEISMAL MAP OF THE PEGASUS BAY EARTHQUAKE - 1987 MARCH 08

DATE:	1987 MARCH 08
TIME:	19:17:59.0 UT
MAGNITUDE:	5.2 M_L
EPICENTRE:	43.22°S 173.20°E (2)
DEPTH:	30 km (restricted)

Six days after the disastrous earthquake at Edgecumbe, on 1987 March 08 a damaging earthquake of magnitude 5.2 M_L occurred at Pegasus Bay in the South Island.

'It was felt strongly in Christchurch, only 50 km away. There is a report of a cracked pavement at Brighton. A number of houses in North Canterbury suffered chimney damage and had items knocked off shelves. The earthquake was felt as far away as Greymouth and Wellington.' (Seismological Observatory bulletin E 171)

The isoseismal map is reproduced from the Seismological Observatory bulletin E 171.

PEGASUS BAY EARTHQUAKE
1987 MARCH 08

Magnitude:	5.2 M_L
Epicentre:	43.22° S 173.20° E
Depth:	30 km (restricted)

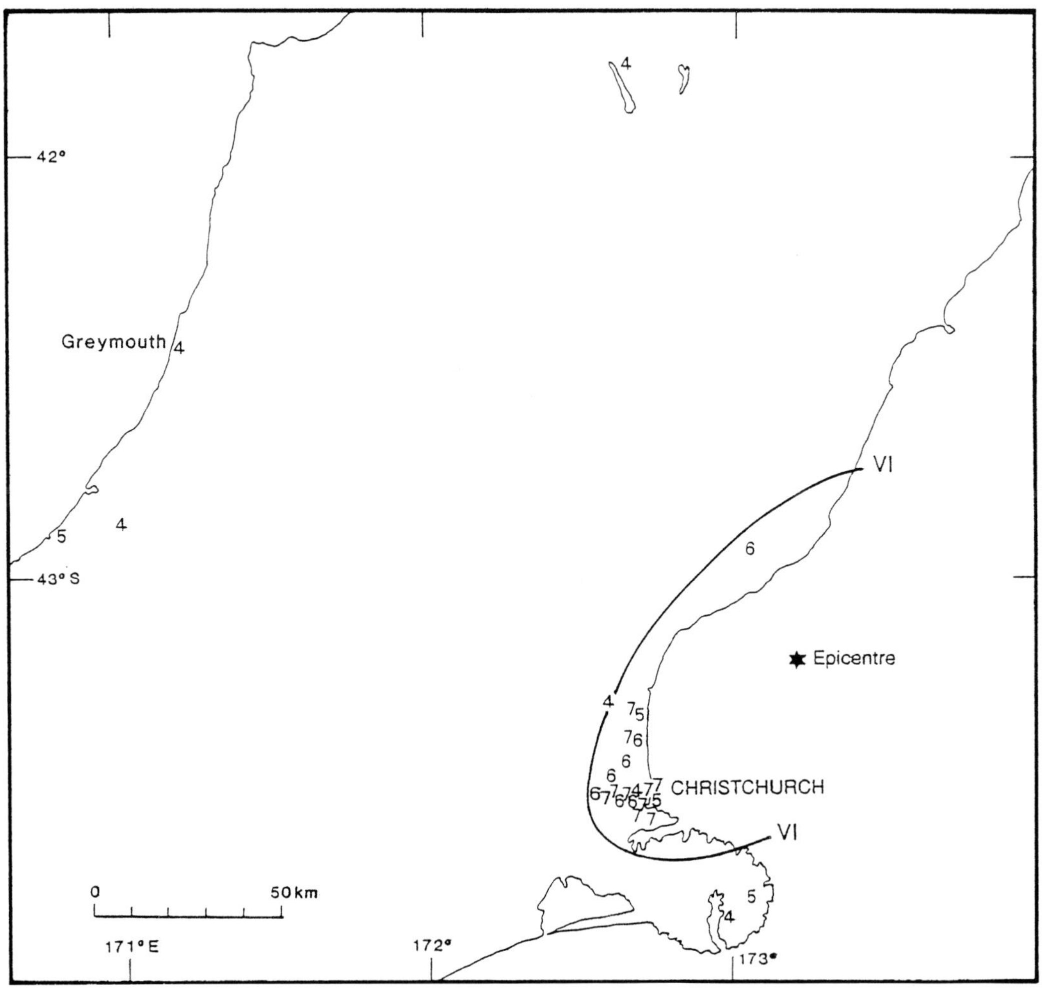

119. ISOSEISMAL MAP OF THE TE ANAU EARTHQUAKE - 1988 JUNE 03

DATE:	1988 JUNE 03
TIME:	23:27:34.7 UT
MAGNITUDE:	6.1 M_L; 6.7 M_s (Dowrick & Smith 1990); 6.7 M_w (Anderson et al. 1993)
EPICENTRE:	45.10°S 167.17°E (2)
DEPTH:	57 km

'At 23 27 hr on 1988 June 3 UTC (11 27 hr on June 4 NZST) the South Island of New Zealand was shaken by a magnitude $M_L = 6.1$ ($M_s = 6.7$) earthquake .... Reported intensities of ground shaking were greatest in the Milford Sound-Te Anau-Manapouri area. Numerous minor landslides and rockfalls occurred in the ranges of Fiordland, one of which blocked the Te Anau-Milford Sound road. A seiche about 1 m high was also reported from the North Fiord of Lake Te Anau. Because the earthquake was one of the largest events to occur in the Fiordland Seismic Region in recent years and because it produced some of the highest intensities ever reported from the region, three portable seismographs were dispatched to the epicentral area.' (Reyners et al. 1991)

'The Te Anau earthquake of June 3rd produced a report of MM9 from Manapouri.... The area in which this shock was felt extended from the southern shore of the South Island to Westport and there was even a report of it being felt in Wellington. Although damaging intensities were experienced only in thinly populated areas, electricity supplies to Christchurch and Invercargill were interrupted when a switch at Manapouri Power Station was tripped by the earthquake.' (Seismological Observatory bulletin E 172)

The isoseismal map has been reproduced from Reyners et al. (1991). The parameters of the earthquake are also from Reyners et al (1991) and differ from those in the Seismological Observatory bulletin E 172.

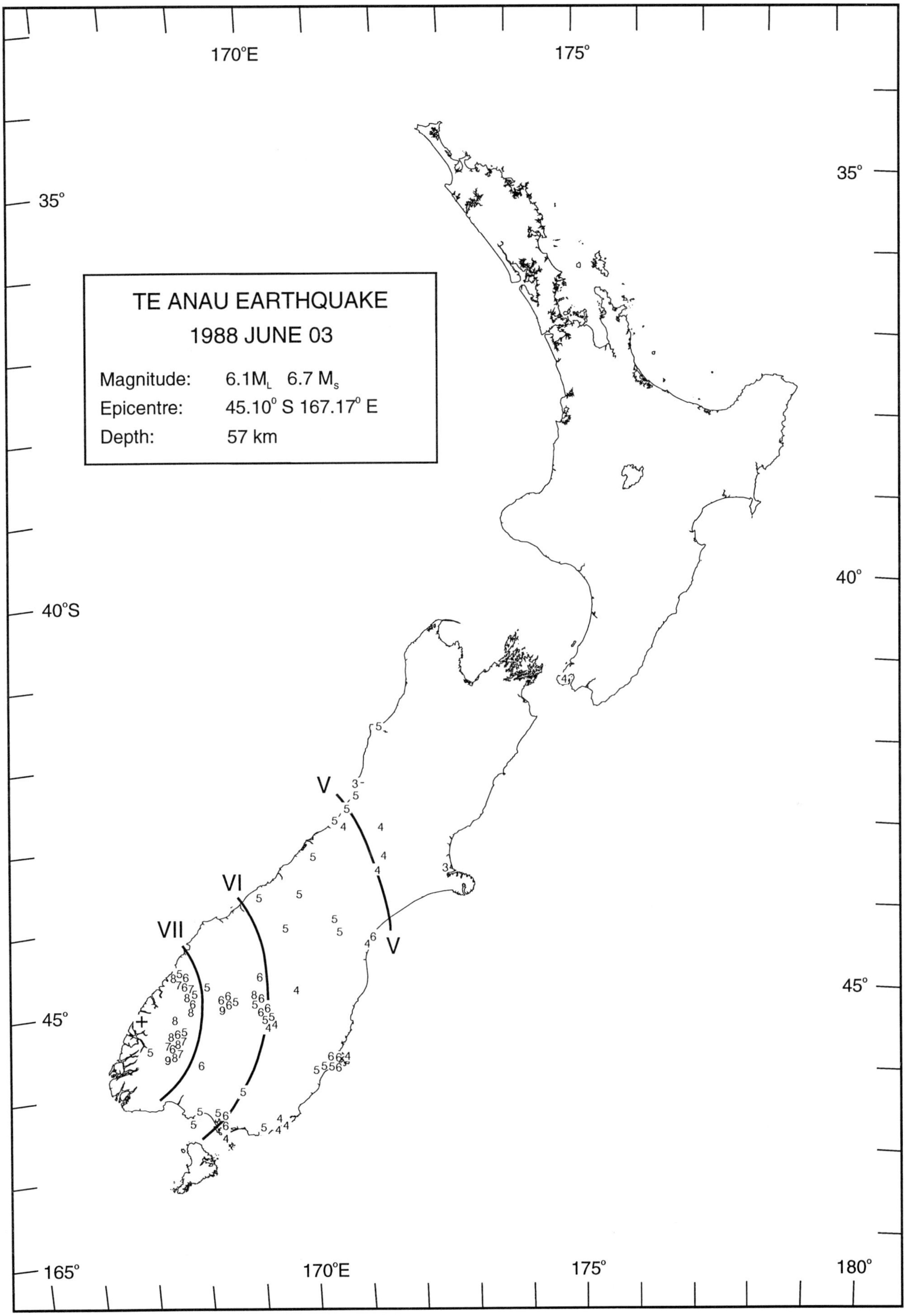
TE ANAU EARTHQUAKE
1988 JUNE 03
Magnitude: 6.1M_L 6.7 M_s
Epicentre: 45.10° S 167.17° E
Depth: 57 km
170°E
175°
35°
40°S
40°
45°
165°
180°
V
VI
VII

120. ISOSEISMAL MAP OF THE DOUBTFUL SOUND EARTHQUAKE - 1989 MAY 31

DATE:	1989 MAY 31
TIME:	05:54:23.1 UT
MAGNITUDE:	6.1 M_L; 6.2 M_s (ISC); 6.4 M_w (Anderson et al. 1993)
EPICENTRE:	45.33°S 166.87°E (2)
DEPTH:	23 km (restricted)

At 05:54 UT on 1989 May 31 the second earthquake to strike the Te Anau region within the year occurred some 40 km south west of the earlier event and was widely felt from Stewart Island to Christchurch and Westport. The maximum intensity of MM6 was experienced in the vicinity of Te Anau and Manapouri and at Mount Aspiring Station.

The isoseismal map is reproduced from the Seismological Observatory bulletin E 173.

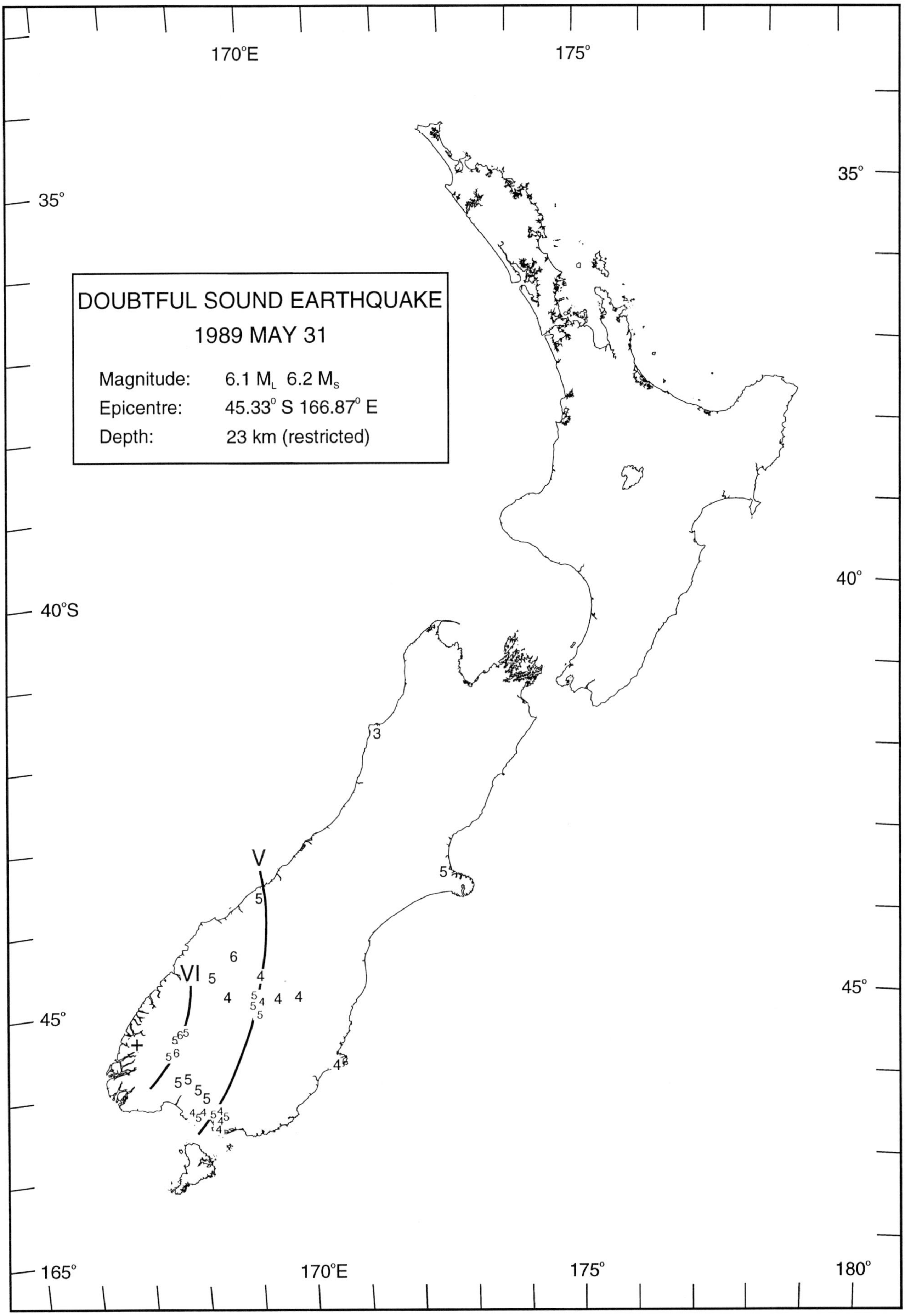
DOUBTFUL SOUND EARTHQUAKE
1989 MAY 31
Magnitude: 6.1 M_L 6.2 M_S
Epicentre: 45.33° S 166.87° E
Depth: 23 km (restricted)
170°E
175°
35°
40°S
40°
45°
165°
180°
V
VI

121. ISOSEISMAL MAP OF THE LAKE TENNYSON EARTHQUAKE - 1990 FEBRUARY 10

DATE:	1990 FEBRUARY 10
TIME:	03:27:42.0 UT
MAGNITUDE:	5.8 M_L ; 5.9 M_S ISC
EPICENTRE:	42.25°S 172.65°E (2)
DEPTH:	13 km

'Event 90/776 (M_L 5.8) occurred near Lake Tennyson in North Canterbury on February 10. It was followed within 30 minutes by aftershocks of M_L 5.5 and 5.3. A field survey using portable digital equipment recorded over 4000 aftershocks during the following week. The main shock shattered mountain ridges in the vicinity of the epicentre, and in the surrounding forest caused healthy beech trees to snap off at a surprisingly constant 1.5 metres above the ground. (Seismological Observatory bulletin E 174)

The earthquake was felt from Wanganui in the North Island to Akaroa in the South Island.

The isoseismal map is reproduced from the Seismological Observatory bulletin E 174.

LAKE TENNYSON EARTHQUAKE
1990 FEBRUARY 10

Magnitude:	5.8 M_L 5.9 M_s
Epicentre:	42.25° S 172.65° E
Depth:	13 km

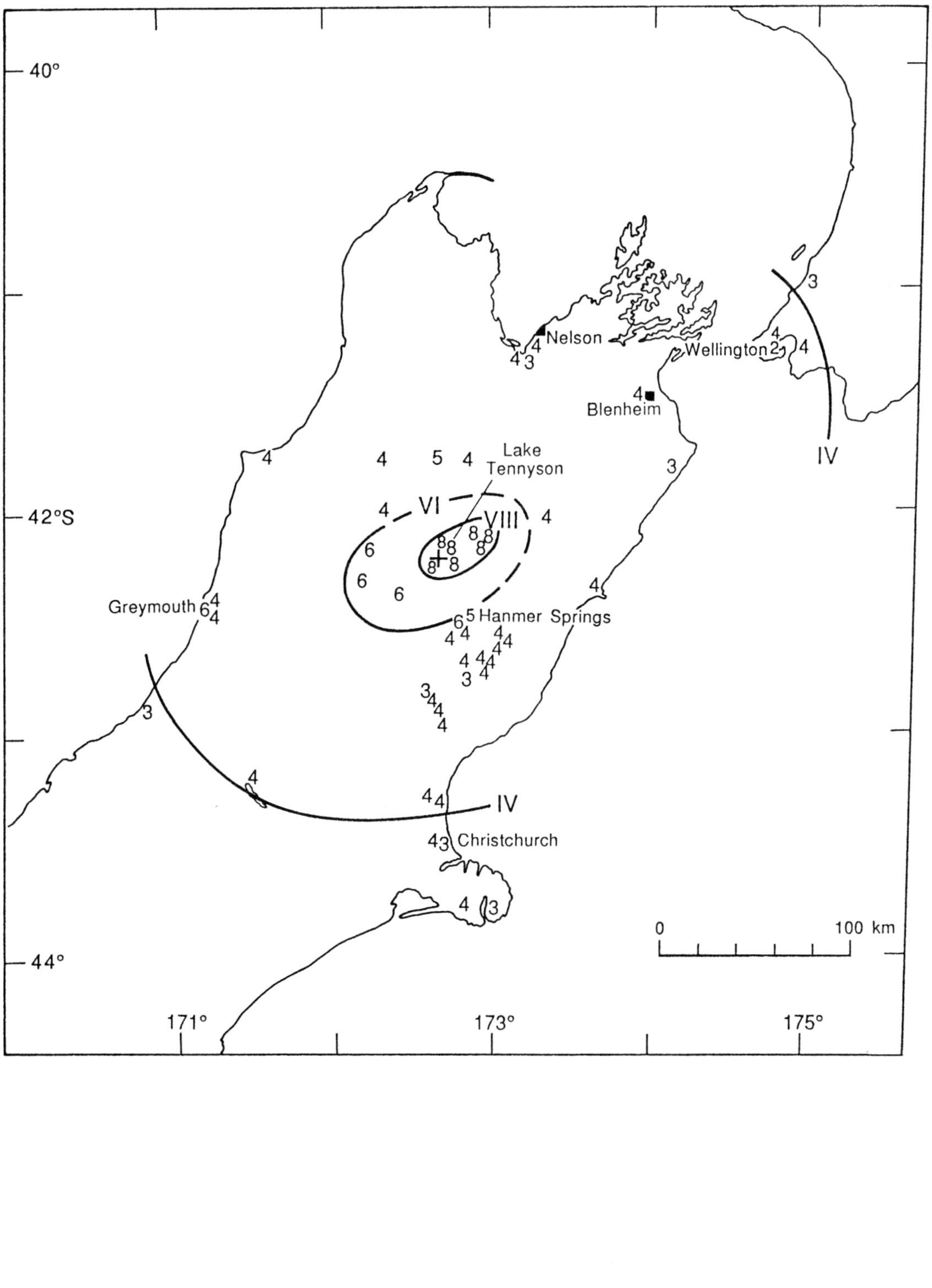

122. ISOSEISMAL MAP OF THE WEBER I EARTHQUAKE - 1990 FEBRUARY 19

DATE:	1990 FEBRUARY 19
TIME:	05:34:37.7 UT
MAGNITUDE:	6.1 M_L (Robinson 1994); 6.3 M_s ISC; 5.9 m_b PDE; 6.2 M_w Harvard Moment Tensor
EPICENTRE:	40.36°S 176.36°E (Robinson 1994)
DEPTH:	24 km

'Nine days later [after the Lake Tennyson earthquake], event 90/3443 (M_L 5.9) [M_l 6.1 revised, Robinson (1994)] occurred near the settlement of Weber in southern Hawke's Bay. It was followed on May 13 by event 90/6657 (M_L 6.2)].

'Aftershocks of these two events were not as prolific as for the Tennyson event in February, but there was one on August 15 (90/9539) of M_L 5.6 [M_L 5.5 revised, Robinson (1994)] at a depth of 42 km [30.85 km Robinson (1994)].' (Seismological Observatory bulletin E 174)

The earthquake was felt from Waihi in the North Island to Maruia in the northern South Island. The maximum intensity MM7 occurred at Te Uri, within the epicentral region, and at Aramoana east of Dannevirke.

'[The earthquake] occurred within the upper part of the subducting Pacific plate on a steeply NW dipping normal fault, as defined by relocated aftershocks. The mechanism of this event reflects down-dip tension within the subducting plate due to "slab-pull", as is common in New Zealand.' (Robinson 1992)

The isoseismal map is taken from the Seismological Observatory bulletin E 174. The hypocentral parameters are from Robinson (1994), being the result of a special study on the Weber events. A temporary array of portable seismographs was deployed after the earthquake in February and again after the earthquake in May.

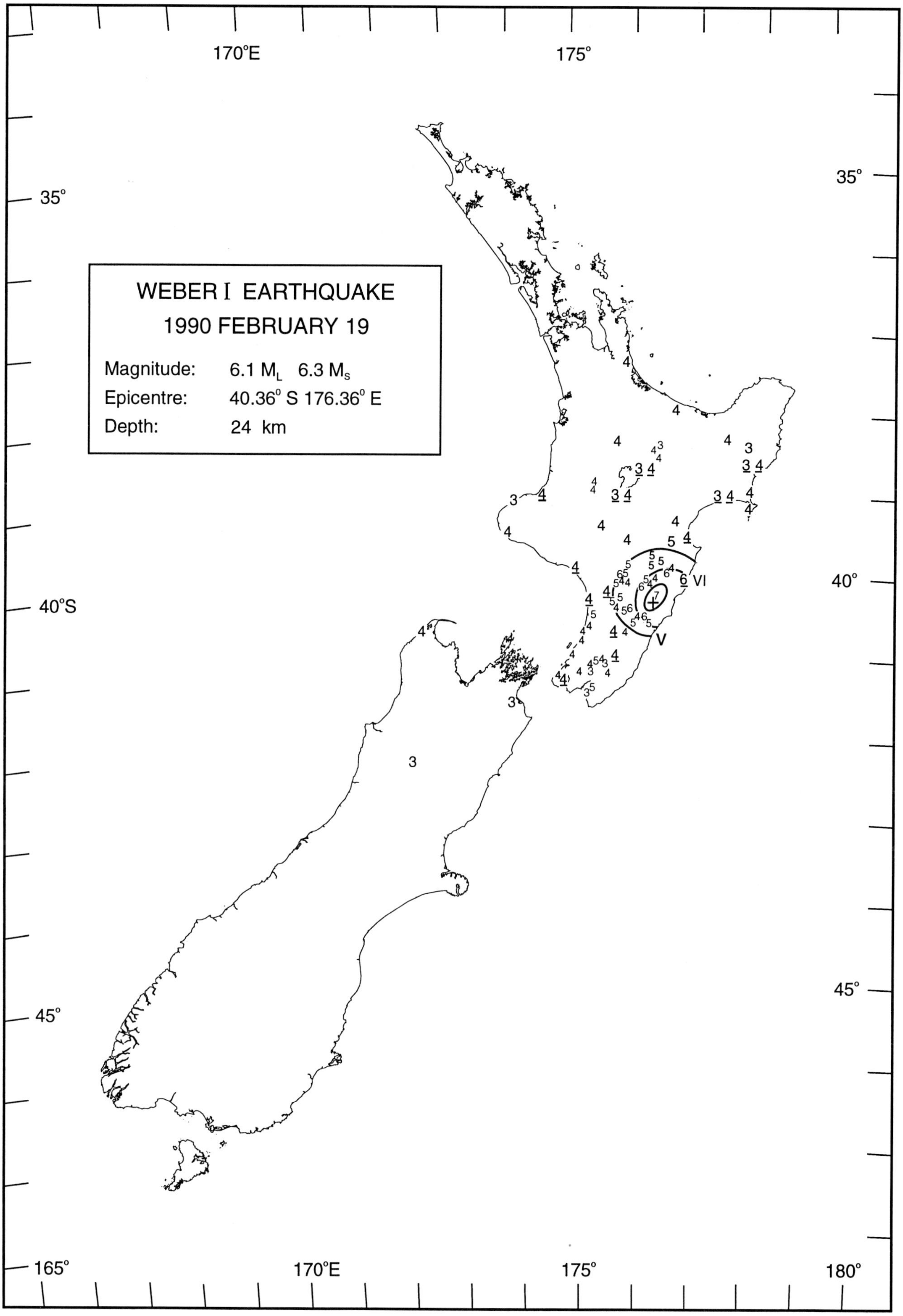

WEBER I EARTHQUAKE
1990 FEBRUARY 19
Magnitude: 6.1 M_L 6.3 M_s
Epicentre: 40.36° S 176.36° E
Depth: 24 km
170°E
175°
35°
40°S
40°
45°
165°
180°
VI
V

123a. ISOSEISMAL MAP OF THE WEBER II EARTHQUAKE - 1990 MAY 13

DATE:	1990 MAY 13
TIME:	04:23:09.3 UT
MAGNITUDE:	6.2 M_L (Robinson 1994); 6.2 M_s ISC; 6.0 m_b PDE; 6.4 M_w Havard Moment Tensor
EPICENTRE:	40.28°S 176.30°E (Robinson 1994)
DEPTH:	12 km

'[The magnitude M_L 5.9 earthquake at Weber in February] was followed on May 13 by event 90/6657 (M_L 6.2). Intensity MM9 was reported from near the epicentre. Damage in Dannevirke, 20 km away and the nearest town of any size, was surprisingly light. One might have expected damage, especially to old brick buildings of which there are many in the town. Detailed reports are available from engineers who visited immediately after the earthquakes.

'Aftershocks of these two events were not as prolific as for the Tennyson event in February, but there was one on August 15 (90/9539) of M_L 5.6 [M_L 5.5 revised, Robinson (1994)] at a depth of 42 km [30.85 km revised, Robinson (1994)].

'After the May 13 earthquake, an experiment was conducted to assess the microzoning effects in the Wellington region, about 200 km away, using a newspaper questionnaire. From nearly 3000 responses, evidence has been found for markedly higher intensities on the areas of poorly consolidated sediments, and lower intensities in the hill suburbs. Intensities were all moderate, up to MM6, so extrapolation to higher intensities must be done with caution, but indications are that those areas that experienced enhanced intensities in that earthquake might also expect to sustain greater damage in stronger earthquakes.' (Seismological Observatory bulletin E 174)

The earthquake was felt from Hamilton in the North Island to Takamatua Bay near Akaroa in the South Island, with one instance of MM9 near the town of Weber and several MM8 within 20 km of the epicentre. Berryman & Beanland (1990) note that no surface faulting was observed for this or the February event.

'The second event, May 13 1990, M_w 6.4, occurred on an imbricate thrust fault in the overlying Australian plate, dipping shallowly NW, with a component of dextral strike-slip motion. The slip vector is close to the direction of plate convergence.' (Robinson 1992)

The isoseismal map, not published previously, includes all reports, New Zealand wide. The detailed map of the epicentral area appears on the following pages. The hypocentral parameters are from Robinson (1994), being the result of a special study of the Weber events. A temporary array of portable seismographs was deployed after the earthquake in February and again after the earthquake in May.

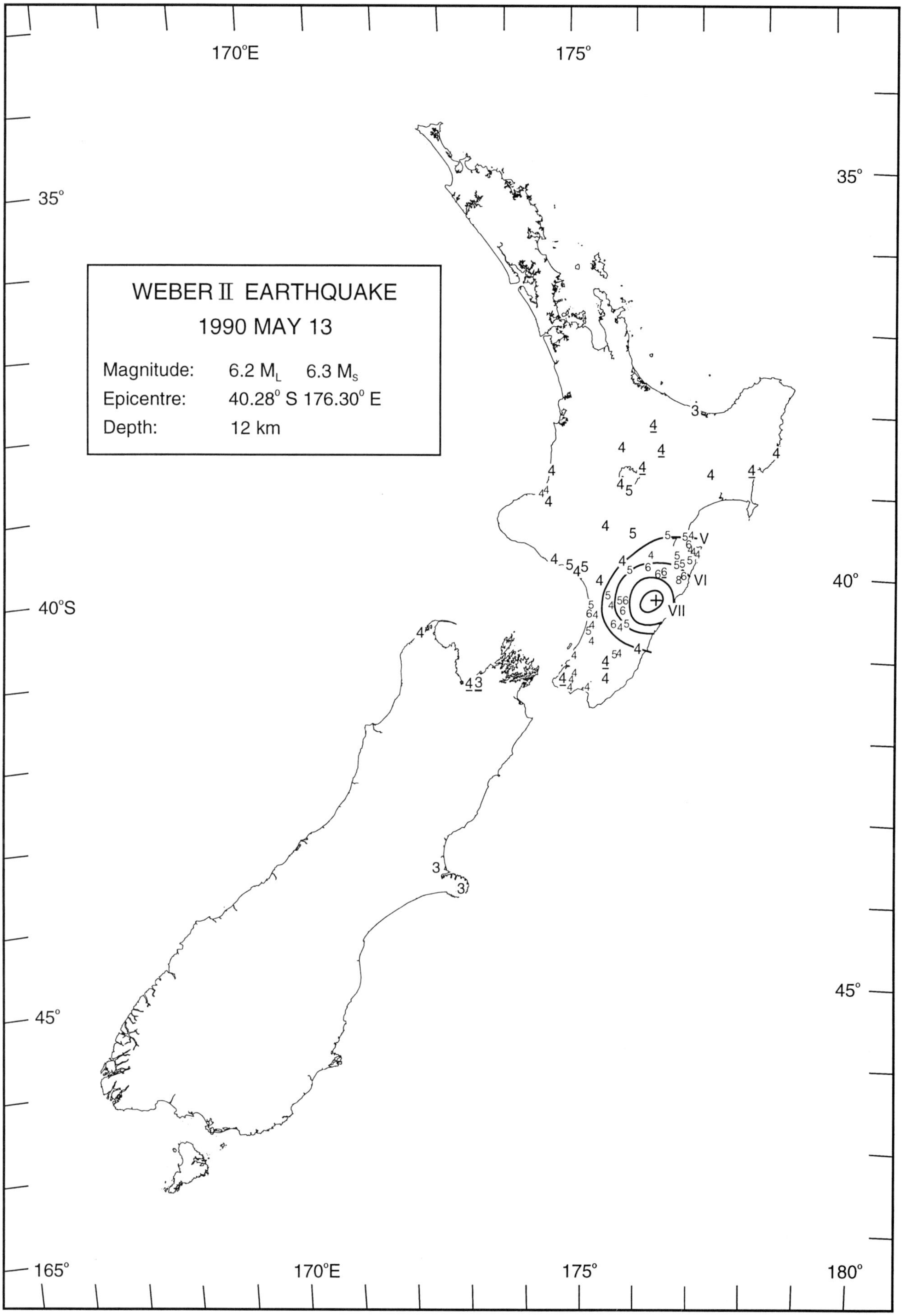
WEBER II EARTHQUAKE
1990 MAY 13
Magnitude: 6.2 M_L 6.3 M_S
Epicentre: 40.28° S 176.30° E
Depth: 12 km
170°E
175°
35°
40°S
40°
45°
165°
180°
V
VI
VII

123b. ISOSEISMAL MAP OF THE WEBER II EARTHQUAKE - 1990 MAY 13

DATE:	1990 MAY 13
TIME:	04:23:09.3 UT
MAGNITUDE:	6.2 M_L (Robinson 1994); 6.2 M_s ISC; 6.0 m_b PDE; 6.4 M_w Havard Moment Tensor
EPICENTRE:	40.28°S 176.30°E (Robinson 1994)
DEPTH:	12 km

A series of four large earthquakes with epicenters very close to one another occurred in 1990-1992 in the Weber region of the southeast coast of the North Island, New Zealand. The region is one of oblique plate convergence and subduction, the plate interface being at about 20 km depth. The first event, February 19, 1990, M_L 6.1, M_w 6.2, occurred within the upper part of the subducting Pacific plate on a steeply northwest dipping normal fault, as defined by relocated aftershocks. The mechanism of this event reflects downdip tension within the subducting plate due to "slab pull" forces, as is common in New Zealand. The second event, May 13, 1990, M_L 6.2, M_w 6.4, occurred on an imbricate fault in the overlying Australian plate, dippling shallowly northwest, with components of both thrusting and dextral strike-slip motion. The slip vector is similar to the direction of plate convergence. It is unlikely that the spatial and temporal closeness of these events is coincidence (a similar pair occurred to the south in 1942), but the coupling mechanism is not clear, perhaps being related to a seismic slip downdip on the subduction interface. Two subsequent events, both of M_L 5.5, occurred on August 15, 1990, and March 2, 1992, and extended the aftershock zone of the deeper first main shock to the northeast, matching the extent of the shallower second main shock. (Robinson 1994)

The isoseismal map is taken from the Seismological Observatory bulletin E 174. The hypocentral parameters are from Robinson (1994), being the result of a special study of the Weber events. A temporary array of portable seismographs was deployed after the earthquake in February and again after the earthquake in May.

WEBER II EARTHQUAKE
1990 MAY 13

Magnitude:	6.2 M_L 6.3 M_S
Epicentre:	40.28° S 176.30° E
Depth:	12 km

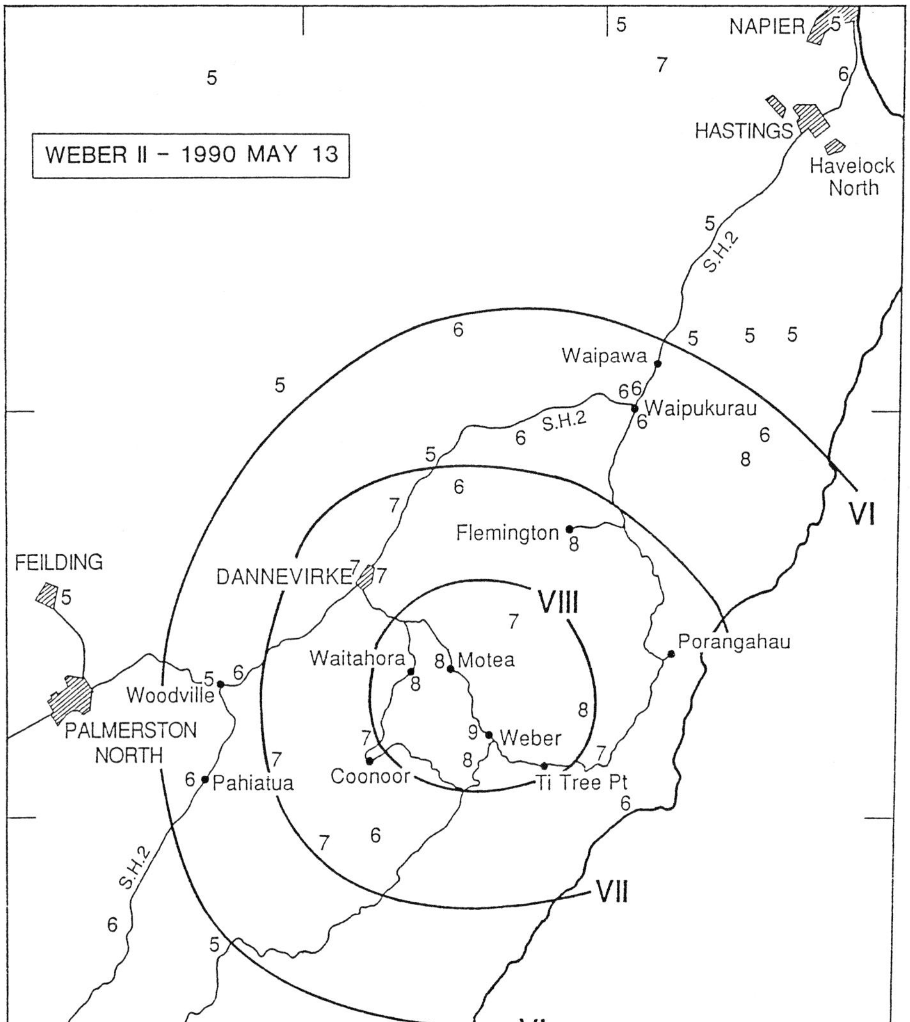

APPENDIX : SELECTED BIBLIOGRAPHY

1. Marlborough earthquake 1848 October 15

Dowrick, D. J. 1991. A revision of attenuation relationships for Modified Mercalli intensity in New Zealand earthquakes. *Bulletin of the New Zealand National Society for Earthquake Engineering 24*: 210-224.

Eiby, G. A. 1980. The Marlborough earthquakes of 1848. *DSIR bulletin* 225.

Eiby, G. A. 1973. A descriptive catalogue of New Zealand earthquakes. Part 2 - shocks felt from 1846 to 1854. *New Zealand journal of geology and geophysics 16*: 857-907.

Hayes, R. C. 1936. The seismicity of New Zealand cities and towns. *New Zealand journal of science and technology 18*: 608-612.

References to earlier works appear in the above publications.

2. Wairarapa earthquake 1855 January 23

Adkin, G. L. 1921. Porirua Harbour: a study of its shoreline and other physiographic features. *Transactions of the New Zealand Institute 53*: 144-156.

Darby, J. D., Beanland, S. 1992. Possible source models for the 1855 Wairarapa earthquake, New Zealand. *Journal of geophysical research 89 (B9)*: 12375-12389.

Dellow, G. D. 1988. Earthquake generated landslides in the Wellington-Lower Hutt area. MSc Thesis. University of Canterbury Christchurch, New Zealand.

Eiby, G. A. 1988. Documenting New Zealand earthquakes. *In*: Lee, W. H. K., Meyers, H., Shimazaki, K. *(eds)* "Historical Seismograms and Earthquakes of the World", London: Academic Press.

Eiby, G. A. 1989. 1855 - Observation and inference. *Bulletin of the New Zealand National Society for Earthquake Engineering 22*: 10-11.

Eiby, G. [A.], 1990. Changes to Porirua Harbour in about 1855: historical tradition and geological evidence. *Journal of the Royal Society of New Zealand 20*: 233-248.

Fairless, G. 1984. Liquefaction case histories in New Zealand. *Research report no. 84-18*. Dept. of Civil Engineering, University of Canterbury, Christchurch, New Zealand.

Fairless, G. J., Berrill, J. B. 1984. Liquefaction during historic earthquakes in New Zealand. *Bulletin of the New Zealand National Society for Earthquake Engineering 17*: 280-291.

Grapes, R. H. 1988. Geology and revegetation of an 1855 landslide, Ruamahanga River, Kopuaranga, Wairarapa. *Tuatara 30*: 77-83.

Grapes, R. H. 1989. The Wairarapa fault: displacements and paleoseismicity. *Bulletin of the New Zealand National Society for Earthquake Engineering 22*: 13-14.

Grapes, R. H. 1990. The 1855 earthquake and the Wairarapa. Feature article in *Wairarapa Times Age*, 23 January 1990.

Grapes, R. H. 1991. Aggradation surfaces and implications for displacement rates along the Wairarapa Fault, southern North Island, New Zealand. *CATENA 18*: 453-469.

Grapes, R. H. 1992a. When the big one came. *News VUW 1*: 11-13.

Grapes, R. H. 1992b. When the big one hit Wellington. *Victoria News 5(7)*: 1-4.

Grapes, R. H. 1992c. Wellington on a roll. Feature article in *The Dominion*, 18 September 1992.

Grapes, R. H., Hardy, E. F., Wellman, H. W. 1984. The Wellington, Mohaka and Wairarapa Faults. *Publication No. 9*, Geology Dept., Victoria University of Wellington, New Zealand.

Grapes, R. H., Wellman, H. 1988. The Wairarapa Fault. *Publication no. 4*. Research School of Earth Sciences, Geology Board of Studies, Victoria University of Wellington, New Zealand.

Hayes, R. C. 1936. The seismicity of New Zealand cities and towns. *New Zealand journal of science and technology 18*: 608-612.

Lyell, C. 1856. Sur les éffects du tremblement de terre du 23 Janvier, 1855, à la Nouvelle Zélande. *Bulletin de la Société Géologique de France série 2*, 13: 661-667.

Lyell, C. 1868. *Principles of Geology*, 10th ed. Vol. 2: 82-89. London: John Murray.

Mills, C., Carter, C. R. 1855. Report of the Commission Appointed by His Honour the Superintendent to Inquire into the Amount of Damage Sustained by the City of Wellington and Suburbs, from the Earthquake of the 23rd January 1855 and also to Report on the Material and Mode of Building, Best Calculated to Resist the Effects of the Earthquakes Peculiar to New Zealand. *New Zealand Government Gazette (Province of Wellington) 2(14)*: 116-126.

Ongley, M. 1943. The surface trace of the 1855 earthquake. *Transactions of the Royal Society of New Zealand 73*: 84-89.

Robbins, R. G. 1958. Direct effect of the 1855 earthquake on the vegetation of the Orongorongo Valley, Wellington. *Transactions of the Royal Society of New Zealand 85*: 205-212.

Stevens, G. R. 1973. Late Holocene marine features adjacent to Port Nicholson, Wellington, New Zealand. *New Zealand journal of geology and geophysics 16*: 455-484.

Stevens, G. R. 1989. Historical aspects of the 1855 earthquake. *Bulletin of the New Zealand National Society for Earthquake Engineering 22*: 30-32.

Sykes, L. R. 1989. Great earthquakes of 1855 and 1931 in New Zealand: evidence for seismic slip along down going plate boundary and implications for seismic potential of Cascadia Subduction Zone (Abstract). *EOS Transactions of the American Geophysical Union 70(43)*: 1331.

Roberts, E. 1855. Appendix F. Mem. on the earthquake in the islands of New Zealand, 23 January 1855. p. 471-2. *In*: Taylor, R. "Te Ika a Maui, or New Zealand and its inhabitants." 1st ed. London: Wertheim and MacIntosh 490 p.
Wellman, H. W. 1967. Tilted marine beach ridges at Cape Turakirae, New Zealand. *Journal of Geosciences, Osaka City University 10*: 123-129.
Wellman, H. W. 1970. Tilted marine beach ridges at Cape Turakirae, New Zealand. *Tuatara 17*: 82-92. (A longer account than the 1967 paper).

3. Cape Farewell earthquake 1868 October 18

Anderson, H., Beanland, S., Blick, G., Darby, D., Downes, G., Haines, J., Jackson, J., Robinson, R., Webb, T. 1994. The 1968 May 23 Inangahua, New Zealand, earthquake: an integrated geological, geodetic, and seismological source model. *New Zealand journal of geology and geophysics 37*: 59-86.
Eiby, G. A. 1968. An annotated list of New Zealand earthquakes, 1460-1965. *New Zealand journal of geology and geophysics 11*: 630-647.
Smith, W. D., Berryman, K. R. 1986. Earthquake hazard in New Zealand: inferences from seismology and geology. *Royal Society of New Zealand bulletin. 24*: 223-243.

4a. Oamaru earthquake I 1876 February 25
4b. Oamaru earthquake II 1876 February 25
4c. Oamaru earthquake III 1876 April 11

Reyners, M. in prep. The earthquakes of 1876, and the seismicity of Oamaru.

5a, 5b. North Canterbury earthquake 1888 August 31

Cowan, H. A. 1991. The North Canterbury earthquake of September 1, 1888. *Journal of the Royal Society of New Zealand 21*: 1-12
References to other earlier papers on this earthquake are listed in the above paper, including the well known work of A. McKay.

6. Waikato Heads earthquake 1891 June 23

Hogben, G. 1891. Notes on the earthquake of 24th June, 1891. *Transactions of the New Zealand Institute 24*: 574-577. Isoseismal map is on p. 580.

7. Nelson earthquake 1893 February 11

Anderson, H., Beanland, S., Blick, G., Darby, D., Downes, G., Haines, J., Jackson, J., Robinson, R., Webb, T. 1994. The 1968 May 23 Inangahua, New Zealand, earthquake: an integrated geological, geodetic, and seismological source model. *New Zealand journal of geology and geophysics 37*: 59-86.
Hogben, G. 1893. The Nelson earthquake of the 12th of February, 1893. *Transactions and proceedings of the New Zealand Institute 26*: 347-354.

8. Cape Turnagain earthquake 1904 August 08

Downes, G. in prep. The 1904 Cape Turnagain earthquake.

9. East Cape earthquake I 1914 October 06
10. East Cape earthquake II 1914 October 28
11. Bay of Plenty earthquake 1914 November 22

Morgan, P.G. 1920. Earthquakes in the Gisborne-East Cape district, 1914. *New Zealand Geological Survey bulletin 21 (new series)*: 81-83.
Reyners, M. in prep. The East Cape earthquakes of 1914.

12. Hawke's Bay earthquake 1921 June 28

Bullen, K. E. 1937. The Hawke's Bay earthquake of 1921, June 29. *New Zealand journal of science and technology 19*: 199-205.

13a, 13b. Motunau earthquake 1922 December 25

Skey, H. F. 1925. Report on the North Canterbury earthquake of 25th December, 1922. *Records of the Survey of New Zealand 1*: 69-70.
Jobberns, G. 1925. Report on the Motunau earthquake phenomena, 25th December, 1922. *Records of the Survey of New Zealand 1*: 70.

14. Arthur's Pass earthquake 1929 March 09

Speight, R. 1933. The Arthur's Pass earthquake of 9th March, 1929. *New Zealand journal of science and technology 15*: 173-182.
Yang, J.S. 1992. Landslide mapping and major earthquakes on the Kakapo Fault, South Island, New Zealand. *Journal of the Royal Society of New Zealand 22*: 205-212.

15. Buller earthquake 1929 June 16
(also referred to as the Murchison earthquake)

Adams, C. E. 1929-30. Report of the Dominion Astronomer and Seismologist for the year 1929. *Dominion Observatory [New Zealand] bulletin R 15.*
Adams, J. 1981. Earthquake dammed lakes in New Zealand. *Geology 9*: 215-219.
Anderson, H., Beanland, S., Blick, G., Darby, D., Downes, G., Haines, J., Jackson, J., Robinson, R., Webb, T. 1994. The 1968 May 23 Inangahua, New Zealand, earthquake: an integrated geological, geodetic, and seismological source model. *New Zealand journal of geology and geophysics 37*: 59-86.
Bastings, L. 1933. Some seismological aspects of the Buller earthquake, 1929, June 16-17. *New Zealand journal of science and technology 15*: 128-142.
Bastings, L. 1936. Some seismological aspects of the Buller earthquake. Part II - The P waves in the distant hemisphere. *New Zealand journal of science and technology 18*: 396-409.
Bastings, L. 1937. Some seismological aspects of the Buller earthquake. Part III - Singly and multiply reflected P waves. *New Zealand journal of science and technology 19*: 180-188.

Bastings, L. 1937. Some seismological aspects of the Buller earthquake. Part IV - Singly and multiply reflected S waves. *New Zealand journal of science and technology 19*: 256-261.

Bastings, L. 1938. Some seismological aspects of the Buller earthquake (concluded). *New Zealand journal of science and technology 19*: 388-400.

Berryman, K. R. 1980. Late Quaternary movement on White Creek Fault, South Island, New Zealand. *New Zealand journal of geology and geophysics 23*: 93-101.

Dowrick, D. J. 1994. Damage and intensities in the magnitude 7.8 1929 Murchison, New Zealand, earthquake. *Bulletin of the New Zealand National Society for Earthquake Engineering* 27: 190-204.

Ferrar, H. T., Grange, L. I. 1929. Geological reconnaissance in the Murchison earthquake area. *New Zealand journal of science and technology 11*: 185-191.

Ferrar, H. T. 1930. The Murchison (New Zealand) earthquake, 17th June, 1929. *Geological magazine 67 (789)*: 132-134.

Ford, C. R., Butcher, H., Mitchell, A. S. 1929. The South Island earthquake of June 17th, 1929. *Report of the New Zealand Institute of Architects Investigation Committee. Wellington, New Zealand.*

Fyfe, H. E. 1929. Movement on White Creek Fault, New Zealand, during the Murchison Earthquake of 17th June, 1929. *New Zealand journal of science and technology 11*: 192-197.

Henderson, J. 1937. The West Nelson earthquakes of 1929 (with notes on the geological structure of West Nelson). *New Zealand journal of science and technology 19*: 65-144.

Suggate, R. P. 1988. Quaternary deposition and deformation in the Buller and tributary valleys. The Maruia Falls. *New Zealand Geological Survey record 25*: 45-51.

Suggate, R. P. 1990. Vertical displacement associated with the White Creek fault accompanying the 1929 Murchison (New Zealand) earthquake. *New Zealand Geological Survey report EDS 131.*

16. Hawke's Bay earthquake 1931 February 02

Adams, C. E. 1931-32. Report of the Dominion Astronomer and Seismologist for the year ended 31st December, 1931. *Dominion Observatory [New Zealand] bulletin R 17.*

Adams, C. E., Barnett, M. A. F., Hayes, R. C. 1933. The Hawke's Bay earthquake of 3rd February, 1931. Seismological report of the Hawke's Bay earthquake of 3rd February, 1931. *New Zealand journal of science and technology 15*: 93-107.

Brodie, A., Harris, A. G. 1933. The Hawke's Bay earthquake of 3rd February, 1931. Damage to buildings. *New Zealand journal of science and technology 15*: 108-115.

Bullen, K. E. 1938. An analysis of the Hawke's Bay earthquakes during February, 1931. *New Zealand journal of science and technology 19*: 497-519.

Callaghan, F. R. 1933. The Hawke's Bay earthquake of 3rd February, 1931. General description. *New Zealand journal of science and technology 15*: 3-38.

Dahm, C. G. 1933. Epicenter of the Hawke Bay (New Zealand) earthquake and travel times of condensational waves. *Bulletin of the Seismological Society of America 23*: 139-157.

Dowrick, D. J. In prep. Damage and intensities in the 1931 Hawke's Bay, New Zealand, earthquake.

Gibowicz, S. J. 1973. Variation of the frequency-magnitude relation during the 1931 Hawke's Bay, 1934 Pahiatua, and 1942 Wairarapa aftershock sequences. *New Zealand journal of geology and geophysics 16*: 1009-1045.

Haines, A. J., Darby, D. J. 1987. Preliminary dislocation models for the 1931 Napier and 1932 Wairoa earthquakes. *New Zealand Geological Survey report EDS 114.*

Henderson, J. 1933. The Hawke's Bay earthquake of 3rd February, 1931. The geological aspects of the Hawke's Bay earthquakes. *New Zealand journal of science and technology 15*: 38-76.

Hull, A. 1990. Tectonics of the 1931 Hawke's Bay earthquake. *New Zealand journal of geology and geophysics 33*: 309-320.

Marshall, P. 1933. The Hawke's Bay earthquake of 3rd February, 1931. Effects of earthquake on coast-line near Napier. *New Zealand journal of science and technology 15*: 79-92.

Robinson, I. R., Benjamin, H. L. 1932. Effects of earthquakes on electrical supply systems. *Proceedings of the New Zealand Society of Civil Engineers 19*: 256-289.

Strong, S. W. S. 1933. The Hawke's Bay earthquake of 3rd February, 1931. The Sponge Bay uplift, Gisborne and the Hangaroa mud blowout. *New Zealand journal of science and technology 15*: 76-78.

Walshe, H. E. 1937. The Wairoa earthquake of 16th September, 1932. 2. Earth movements in Hawke's Bay District disclosed by triangulation. *New Zealand journal of science and technology 18*: 852-854.

17. Aftershock of the Hawke's Bay earthquake 1931 February 13

Adams, C. E. 1932. Report of the Dominion Astronomer and Seismologist for the year ended 31st December, 1931. *Dominion Observatory* [New Zealand] *bulletin R 17.*

Dominion Observatory [New Zealand] bulletin E 27. Seismological reports for February 1931.

18. Southland earthquake 1931 September 15

Adams, C. E. 1932. Report of the Dominion Astronomer and Seismologist for the year ended 31st December, 1931. *Dominion Observatory [New Zealand] bulletin R 17.*

19. Bay of Plenty earthquake 1931 September 21

Adams, C. E. 1932. Report of the Dominion Astronomer and Seismologist for the year ended 31st December, 1931. *Dominion Observatory [New Zealand] bulletin R 17.*

20. Bay of Plenty earthquake 1932 March 05

Adams, C. E. 1932-33. Report of the Dominion Astronomer and Seismologist for the year ended 31st December, 1932. *Dominion Observatory [New Zealand] bulletin R 18.*

21. Hawke's Bay earthquake 1932 May 05

Adams, C. E. 1932-33. Report of the Dominion Astronomer and Seismologist for the year ended 31st December, 1932. *Dominion Observatory [New Zealand] bulletin R 18.*

Dominion Observatory [New Zealand] bulletin E 33. Seismological reports for April, May, June 1932.

22. Taranaki Bight earthquake 1932 July 20

Dominion Observatory [New Zealand] bulletin E 33. Seismological reports for July, August, September, 1932.

23. Wairoa earthquake 1932 September 15

Bullen, K. E. 1938. The Wairoa earthquake of 1932 September 15. *New Zealand journal of science and technology B20*: 31-43.

Haines, A. J., Darby, D. J. 1987. Preliminary dislocation models for the 1931 Napier and 1932 Wairoa earthquakes. *New Zealand Geological Survey report EDS 114.*

Hayes, R. C. 1937. The Wairoa earthquake of 16th September, 1932. 4. The seismological aspects of the Wairoa earthquake. *New Zealand journal of science and technology 18*: 858-865.

Henderson J. 1937. The Wairoa earthquake of 16th September, 1932. 3. Notes on the Wairoa earthquake. *New Zealand journal of science and technology 18*: 854-857.

Ongley, M. 1937. The Wairoa earthquake of 16th September, 1932. 1. Field Observations. *New Zealand journal of science and technology 18*: 845-851.

Smith, R. T. 1933-34. Wairoa River Bridge. *Proceedings of the New Zealand Society of Civil Engineers 20*: 140-177.

Walshe, H. E. 1937. The Wairoa earthquake of 16th September, 1932. 2. Earth movements in Hawke's Bay District disclosed by triangulation. *New Zealand journal of science and technology 18*: 852-854.

24a, 24b. Pahiatua earthquake 1934 March 05

Bullen, K. E. 1936. On near earthquakes in the vicinity of New Zealand. *New Zealand journal of science and technology 18*: 493-507.

Bullen, K. E. 1938. On the epicentre of the 1934 Pahiatua earthquake. *New Zealand journal of science and technology B20*: 62-66.

Dominion Observatory [New Zealand] bulletin S 27. Earthquakes in New Zealand (including earthquake summaries for the year 1934).

Gibowicz, S. J. 1973. Variation of the frequency-magnitude relation during the 1931 Hawke's Bay, 1934 Pahiatua, and 1942 Wairarapa aftershock sequences. *New Zealand journal of geology and geophysics 16*: 1009-1045.

Hayes, R. C. 1935. Seismic waves and crustal structure in the New Zealand region. *New Zealand journal of science and technology 17*: 519-528.

Hayes, R. C. 1937. The Pahiatua earthquake of 1934 March 5. *New Zealand journal of science and technology 19*: 382-388.

25. Northern Hawke's Bay earthquake 1934 March 15

Bullen, K. E. 1936. On near earthquakes in the vicinity of New Zealand. *New Zealand journal of science and technology 18*: 493-507.

Hayes, R. C. 1935. Seismic waves and crustal structure in the New Zealand region. *New Zealand journal of science and technology 17*: 519-528.

26. Southern Hawke's Bay earthquake 1938 December 15

Dominion Observatory [New Zealand] bulletin S 56 Earthquakes in New Zealand (including earthquake summaries for the year 1938).

27. Charles Sound earthquake 1938 December 16

Dominion Observatory [New Zealand] bulletin S 56. Earthquakes in New Zealand (including earthquake summaries for the year 1938).

28. Wairarapa I earthquake 1942 June 24

Aked, W. E. 1945. Damage to buildings in the City of Wellington by earthquake, 1942. *Proceedings of the New Zealand Institution of Engineers 31*: 132-169. A discussion follows the article on p. 169-183.

Gibowicz, S. J. 1973. Variation of the frequency-magnitude relation during the 1931 Hawke's Bay, 1934 Pahiatua, and 1942 Wairarapa aftershock sequences. *New Zealand journal of geology and geophysics 16*: 1009-1045.

Hayes, R. C. 1942. The Wairarapa earthquake of 24th June, 1942. A preliminary report on some of the seismological aspects of the earthquake. *Dominion Observatory [New Zealand] bulletin S 66.*

Luke, K. E. 1943. City of Wellington, N.Z. Report on damage by earthquakes, 1942. (unpublished) *City Engineer's Report to Wellington City Council.*

Northey, R. D. 1974. Earthquake damage survey: distribution of insurance claims in Wellington arising from earthquakes in 1942 and 1968. *In:* Grant-Taylor T.L. *et al.* 1974. Microzoning for earthquake effects in Wellington, N.Z. *DSIR bulletin 213.*

Ongley, M. 1943. Wairarapa earthquake of 24th June, 1942, together with a map showing surface traces of faults recently active. *New Zealand journal of science and technology B25*: 67-78.

Webb, T. H. 1989. The 1942 Wairarapa earthquakes. *Bulletin of the New Zealand National Society for Earthquake Engineering 22*: 33.

29. Wairarapa II earthquake 1942 August 01

Aked, W. E. 1945. Damage to buildings in the City of Wellington by earthquake, 1942. *Proceedings of the New Zealand Institution of Engineers 31*: 132-169. A discussion follows the article on p. 169-183.

Berryman, K. 1977. Dating recent tectonic events in parts of New Zealand with dendrochronology. (unpublished) Wellington, BSc (Hons) paper. Victoria University of Wellington, Wellington, New Zealand.

Gibowicz, S. J. 1973. Variation of the frequency-magnitude relation during the 1931 Hawke's Bay, 1934 Pahiatua, and 1942 Wairarapa aftershock sequences. *New Zealand journal of geology and geophysics 16*: 1009-1045.

Hayes, R. C. 1942. The Wairarapa earthquake of 24th June, 1942. A preliminary report on some of the seismological aspects of the earthquake. *Dominion Observatory [New Zealand] bulletin S 66.*

Luke, K. E. 1943. City of Wellington, N.Z. Report on damage by earthquakes, 1942. *(unpublished) City Engineer's Report to Wellington City Council.*

Neef, G. 1976. Probable faulting on a minor, reverse bedding fault adjacent to the Alfredton Fault, north Wairarapa, during the 1 August 1942 earthquake (Note*). New Zealand journal of geology and geophysics 19*: 737-742.

Ongley, M. 1943. Wairarapa earthquake of 24th June, 1942, together with a map showing surface traces of faults recently active. *New Zealand journal of science and technology B25*: 67-78.

Webb, T. H. 1989. The 1942 Wairarapa earthquakes. *Bulletin of the New Zealand National Society for Earthquake Engineering 22*: 33.

30. Puysegur Bank earthquake 1945 September 01

Eiby, G. A. 1983. New Zealand Seismological Report 1945-47. *Seismological Observatory [New Zealand] bulletin E 166.*

31. Lake Coleridge earthquake 1946 June 26

Eiby, G. A. 1983. New Zealand Seismological Report 1945-47. *Seismological Observatory [New Zealand] bulletin E 166.*

Eiby, G. A. 1990. The Lake Coleridge earthquakes of 1946. *Bulletin of the New Zealand National Society for earthquake engineering 23*: 150-158.

32. Waiau earthquake 1948 May 22

Eiby, G. A. 1953. The Waiau earthquakes of May, 1948. *Proceedings of the 7th Pacific Science Congress 2*: 622-628.

Eiby, G. A. 1982. New Zealand Seismological Report 1948-49-50. *Seismological Observatory [New Zealand] bulletin E 164.*

Hayes, R. C. 1949. Earthquakes in New Zealand during the year 1948. *New Zealand journal of science and technology B31*: 37-40.

33. Cook Strait earthquake I 1950 January 7

34. Cook Strait earthquake II 1950 January 12 20:48
Cook Strait earthquake III 1950 January 12 20:49

Eiby, G. A. 1982. New Zealand Seismological Report 1948-49-50. *Seismological Observatory [New Zealand] bulletin E 164.*

Hayes, R. C. 1951. Seismological Observatory. *25th Annual Report of the Department of Scientific and Industrial Research.*

Hayes, R. C. 1952. The Cook Strait earthquakes: 1950 Jan-Feb. *New Zealand journal of science and technology B33: 309-318.*

Seismological Observatory [New Zealand] bulletin S 96. Earthquakes in New Zealand during the year 1950.

35. Huiarau Range earthquake 1950 February 28

Eiby, G. A. 1982. New Zealand Seismological Report 1948-49-50. *Seismological Observatory [New Zealand] bulletin E 164.*

Hayes, R. C. 1951. Seismological Observatory. *25th Annual Report of the Department of Scientific and Industrial Research.*

Hayes, R. C. 1952. Earthquake origins in New Zealand during the year 1950. *New Zealand journal of science and technology B33*: 306-308.

Seismological Observatory [New Zealand] bulletin S 96. Earthquakes in New Zealand during the year 1950.

36. Stephens Island earthquake 1950 March 13

Eiby, G. A. 1982. New Zealand Seismological Report 1948-49-50. *Seismological Observatory [New Zealand] bulletin E 164.*

Hayes, R. C. 1951. Seismological Observatory. *25th Annual Report of the Department of Scientific and Industrial Research.*

Seismological Observatory [New Zealand] bulletin S 96. Earthquakes in New Zealand during the year 1950.

37. North Island earthquake 1950 June 17

Eiby, G. A. 1982. New Zealand Seismological Report 1948-49-50. *Seismological Observatory [New Zealand] bulletin E 164.*

Hayes, R. C. 1951. Seismological Observatory. *25th Annual Report of the Department of Scientific and Industrial Research.*

Seismological Observatory *[New Zealand] bulletin* S 96. Earthquakes in New Zealand during the year 1950.

38. Cheviot earthquake 1951 January 10
39. Aramoana earthquake 1951 February 10
40. Bay of Plenty earthquake 1951 March 28

Seismological Observatory [New Zealand] bulletin E 123. Seismological Report for January, February, March, 1951.

Seismological Observatory [New Zealand] bulletin S 97. Earthquakes in New Zealand during the year 1951.

41. Cape Runaway earthquake 1951 April 23
42. Western Hawke's Bay earthquake 1951 June 24

Seismological Observatory [New Zealand] bulletin E 124. Seismological Report for April, May, June, 1951.

Seismological Observatory [New Zealand] bulletin S 97. Earthquakes in New Zealand during the year 1951.

43. Fiordland earthquake 1951 July 07
44. Northwest Nelson earthquake 1951 October 03

Seismological Observatory [New Zealand] bulletin E 125. Seismological Report for July, August, September, 1951.

Seismological Observatory [New Zealand] bulletin S 97. Earthquakes in New Zealand during the year 1951.

45. Southern Hawke's Bay earthquake 1952 August 28

Seismological Observatory [New Zeaand] bulletin E 129. Seismological Report for July, August, September, 1952.

Seismological Observatory [New Zealand] bulletin S 98. Earthquakes in New Zealand during the year 1952.

46. Foxton earthquake 1953 April 11
47. Tokaanu earthquake 1953 July 04
48. Bay of Plenty earthquake 1953 September 29
49. Northern Taranaki Bight earthquake 1953 October 18

Seismological Observatory [New Zealand] bulletin S 99. Earthquakes in New Zealand during the year 1953.

50. Fiordland earthquake 1954 May 04

Seismological Observatory [New Zealand] bulletin S 106. Earthquakes in New Zealand during the year 1954.

51. Waipuru earthquake 1955 February 08

Seismological Observatory [New Zealand] bulletin E 136. New Zealand Seismological Report 1955.

52. Bay of Plenty earthquake 1956 January 30
53. Tokaanu earthquake 1956 March 02

Seismological Observatory [New Zealand] bulletin E 137. New Zealand Seismological Report 1956.

54. East Cape earthquake 1956 December 28

Seismological Observatory [New Zealand] bulletin E 137. New Zealand Seismological Report 1956.

International Seismological Summary for 1956. Kew Observatory, Richmond, Surrey, 1963.

55. Southern Kermedec Ridge earthquake 1957 February 09
56. Tongariro earthquake 1957 February 22
57. North Island earthquake 1957 March 13
58. Castlepoint earthquake 1957 August 21
59. South Taranaki Bight earthquake 1957 September 26

Seismological Observatory [New Zealand] bulletin E 138. New Zealand Seismological Report 1957.

60a, 60b. Ashley Clinton earthquake 1958 January 31
61. Bay of Plenty earthquake 1958 December 10

Seismological Observatory [New Zealand] bulletin E 139. New Zealand Seismological Report 1958.

62. Marlborough Sounds earthquake 1959 May 22

Seismological Observatory [New Zealand] bulletin E 140. New Zealand Seismological Report 1959.

63. East Cape earthquake 1960 February 03
64. Acheron River earthquake 1960 February 21

Seismological Observatory [New Zealand] bulletin E 141. New Zealand Seismological Report 1960.

65. Fiordland earthquake 1960 May 24

Hamilton, R. M. 1966. The Fiordland earthquake sequence of 1960, and seismic velocities beneath New Zealand. *New Zealand journal of geology and geophysics 9*: 224-238.

Seismological Observatory [New Zealand] bulletin E 141. New Zealand Seismological Report 1960.

66. North Island earthquake 1961 February 03
67. Dannevirke earthquake 1961 May 14
68. Barrier Range earthquake 1961 July 04
69. Bay of Plenty earthquake 1961 July 26
70. Hikurangi Trench earthquake 1961 December 27

Seismological Observatory [New Zealand] bulletin E 142. New Zealand Seismological Report 1961.

71. Aria earthquake 1962 January 23

Seismological Observatory [New Zealand] bulletin E 143. New Zealand Seismological Report 1962.

72. Westport earthquake 1962 May 10

Adams, R.D., Le Fort, J. H., 1963. The Westport earthquake, May 1962. *New Zealand journal of geology and geophysics* 6: 487-509.

Seismological Observatory [New Zealand] bulletin E 143. New Zealand Seismological Report 1962.

73. Marlborough earthquake 1962 July 29

Seismological Observatory [New Zealand] bulletin E 143. New Zealand Seismological Report 1962.

74. Bruce Bay earthquake 1962 October 15

Anderson H., Webb, T., Jackson, J. 1993. Focal mechanisms of large earthquakes in the South Island of New Zealand: implications for the accommodation of Pacific-Australia plate motion. *Geophysical journal international 115*: 1032-1054.

Seismological Observatory [New Zealand] bulletin E 143. New Zealand Seismological Report 1962.

75. Huiarau Range earthquake 1963 April 12

76. South Taranaki earthquake 1963 July 14

Seismological Observatory [New Zealand] bulletin E 144. New Zealand Seismological Report 1963.

77. Mangonui earthquake 1963 November 16

78. Peria earthquakes 1963 December 22

Eiby, G. A. 1964a. The Northland earthquakes of 1963 November-December and the seismicity of Northland. *New Zealand journal of geology and geophysics 7:* 745-765.

Eiby, G. A. 1964b. Earthquakes in Northland. *New Zealand engineering 19*: 125-129.

Seismological Observatory [New Zealand] bulletin E 144. New Zealand Seismological Report 1963.

79. Big Bay earthquake 1964 March 08

Anderson H., Webb, T., Jackson, J. 1993. Focal mechanisms of large earthquakes in the South Island of New Zealand: implications for the accommodation of Pacific-Australia plate motion. *Geophysical journal international 115:* 1032-1054.

Seismological Observatory [New Zealand] bulletin E 145. New Zealand Seismological Report 1964.

80. Chatham Rise earthquake 1965 April 11

Anderson H., Webb, T., Jackson, J. 1993. Focal mechanisms of large earthquakes in the South Island of New Zealand: implications for the accommodation of Pacific-Australia plate motion.l *Geophysical journal international 115:* 1032-1054.

Seismological Observatory [New Zealand] bulletin E 146. New Zealand Seismological Report 1965.

81. Lake Te Anau earthquake 1965 May 20

Hamilton, R. M., Evison, F. F. 1967. Earthquakes at intermediate depths in south-west New Zealand. *New Zealand journal of geology and geophysics 10:* 1319-1329.

Seismological Observatory [New Zealand] bulletin E 146. New Zealand Seismological Report 1965.

82. Bay of Plenty earthquake 1965 June 15

83. Bay of Plenty earthquake 1965 December 08

Seismological Observatory [New Zealand] bulletin E 146. New Zealand Seismological Report 1965.

84a, 84b. Gisborne earthquake 1966 March 04

Hamilton, R. M., Lensen, G. J., Skinner, R. I., Hall, O. J., Andrews, A. L., Strachen, C.M., Glogau, O. M. 1969. Gisborne earthquake, New Zealand, March 1966. *DSIR bulletin 194.*

Seismological Observatory [New Zealand] bulletin E 148. New Zealand Seismological Report 1966.

Webb, T. H., Wesnousky, S. G., Helmberger, D. V. 1985. A body wave analysis of the Gisborne, New Zealand, earthquake. *Tectonophysics 113:* 271-182.

85. Seddon earthquake 1966 April 23

Adams, R. D., Lensen, G. J., Strachan, C. M., Laws, H.R. 1970. Seddon earthquake, New Zealand, April 1966. *DSIR bulletin 199.*

Anderson H., Webb, T., Jackson, J. 1993. Focal mechanisms of large earthquakes in the South Island of New Zealand: implications for the accommodation of Pacific-Australia plate motion. *Geophysical journal international 115:* 1032-1054.

Seismological Observatory [New Zealand] bulletin E 148. New Zealand Seismological Report 1966.

86. Palmerston North earthquake 1967 January 16

87. Owahanga earthquake 1967 March 24

88. Te Horo earthquake 1967 September 21

89. Whangaehu earthquake 1967 December 20

Seismological Observatory [New Zealand] bulletin E 149. New Zealand Seismological Report 1967.

90a, 90b. Inangahua earthquake 1968 May 23

Adams, R. D., Eiby, G. A., Lowry, M. A. 1969. A preliminary seismological report. *Bulletin of the New Zealand Society for Earthquake Engineering 2*: 9-18.

Adams, R. D., Lowry, M. A., Ware, D. E. 1971. . New Zealand Seismological Report Inangahua earthquakes 1968. *Seismological Observatory [New Zealand] bulletin E 147.* [In this publication, there is a bibliography containing references to all publications relating to the Inangahua earthquake and its effects, as known at August 1970.]

Anderson, H., Beanland, S., Blick, G., Darby, D., Downes, G., Haines, J., Jackson, J., Robinson, R., Webb, T. 1994. The 1968 May 23 Inangahua, New Zealand, earthquake: an integrated geological, geodetic and seismological source model. *New Zealand journal of geology and geophysics 37:* 59-86. This paper contains a comprehensive list of papers relating to the Inangahua earthquake.

Anderson H., Webb, T., Jackson, J. 1993. Focal mechanisms of large earthquakes in the South Island of New Zealand: implications for the accommodation of Pacific-Australia plate motion. *Geophysical journal international 115*: 1032-1054.

Bulletin of the New Zealand Society for Earthquake Engineering 2: no. 1 (1969) was devoted entirely to the Inangahua earthquakes.

91. Solander earthquake 1968 September 25

Anderson H., Webb, T., Jackson, J. 1993. Focal mechanisms of large earthquakes in the South Island of New Zealand: implications for the accommodation of Pacific-Australia plate motion. *Geophysical journal international 115:* 1032-1054.

Seismological Observatory [New Zealand] bulletin E 151. New Zealand Seismological Report 1968.

92. Palliser Bay earthquake 1968 November 01

Grant-Taylor, T. L., Adams, R. D., Hatherton, T., Milne, J. D. G., Northey, R. D., Stephenson, W. R. 1974. Microzoning for earthquake effects in Wellington, New Zealand. *DSIR bulletin 213.*

Seismological Observatory [New Zealand] bulletin E 151. New Zealand Seismological Report 1968.

93. Fiordland earthquake 1969 January 02

94. North Island earthquake 1969 May 23

Seismological Observatory [New Zealand] bulletin E 153. New Zealand Seismological Report 1969.

95. Bay of Plenty earthquake 1970 July 27

Seismological Observatory [New Zealand] bulletin E 157. New Zealand Seismological Report 1969.

96. Maruia Springs earthquake 1971 August 13

Anderson, H., Webb, T., Jackson, J. 1993. Focal mechanisms of large earthquakes in the South Island of New Zealand: implications for the accommodation of Pacific-Australia Plate motion. *Geophysical journal international 115:* 1032-1054.

Seismological Observatory [New Zealand] bulletin E 150. New Zealand Seismological Report 1971.

97. Hawke's Bay earthquake 1971 October 31

Seismological Observatory [New Zealand] bulletin E 150. New Zealand Seismological Report 1971.

98. Te Aroha earthquake 1972 January 08

Adams, R. D., Muir, M. G., Kean, R. J. 1972. Te Aroha earthquake, 9th January, 1972. *Bulletin of the New Zealand National Society for Earthquake Engineering 5:* 54-58

Seismological Observatory [New Zealand] bulletin E 152. New Zealand Seismological Report 1972.

99. North Island earthquake 1973 January 05

Seismological Observatory [New Zealand] bulletin E 154. New Zealand Seismological Report 1973.

100. Dunedin earthquake 1974 April 09

Adams, R. D., Kean, R. J. 1974. The Dunedin earthquake, 9 April 1974. Part 1: Seismological studies. *Bulletin of the New Zealand National Society for Earthquake Engineering 7:* 115-122.

Bishop, D. G. 1974. The Dunedin earthquake, 9 April 1974. Part 2: Local effects. *Bulletin of the New Zealand National Society for Earthquake Engineering 7:* 123-129.

Seismological Observatory [New Zealand] bulletin E 155. New Zealand Seismological Report 1974.

101. Milford Sound earthquake 1974 September 20

Seismological Observatory [New Zealand] bulletin E 155. New Zealand Seismological Report 1974.

102. Opunake earthquake 1974 November 05

Robinson, R., Calhaem, I. M., Thomson, A. A. 1976. The Opunake, New Zealand, earthquake of 5 November 1974. *New Zealand journal of geology and geophysics 19*: 335-345.

Seismological Observatory [New Zealand] bulletin E 155. New Zealand Seismological Report 1974.

103. Kapiti Island earthquake 1975 January 04

Adams, R. D. Ware, D. E. 1977. Subcrustal earthquakes beneath New Zealand: locations determined with a laterally inhomogeneous velocity model. *New Zealand journal of geology and geophysics 20:* 59-83.

Seismological Observatory bulletin E 156. New Zealand Seismological Report 1975.

104. Hen and Chickens Island earthquake 1975 February 11

105. Oringi earthquake 1975 June 10

Seismological Observatory [New Zealand] bulletin E 156. New Zealand Seismological Report 1975.

106. Milford Sound earthquake 1976 May 04

Anderson, H., Webb, T., Jackson, J. 1993. Focal mechanisms of large earthquakes in the South Island of New Zealand: implications for the accommodation of Pacific-Australia Plate motion. *Geophysical journal international 115:* 1032-1054.

Eiby, G. A. 1978. The Milford Sound earthquake of 1976 May 4. *Bulletin of the New Zealand National Society for Earthquake Engineering 11:* 191 - 192

Seismological Observatory [New Zealand] bulletin E 158. New Zealand Seismological Report 1976.

107. Te Puke earthquake 1976 October 27

108. Golden Bay earthquake 1976 December 03

Seismological Observatory [New Zealand] bulletin E 158. New Zealand Seismological Report 1976.

109. Korakanui earthquake 1976 December 05

Eiby, G. A. 1977. Anomalous intensities due to the Korakonui earthquake, 5 December 1976. *Bulletin of the New Zealand National Society for Earthquake Engineering 10:* 167-169.

Seismological Observatory [New Zealand] bulletin E 158. New Zealand Seismological Report 1976.

110. Cape Campbell earthquake 1977 January 18

Seismological Observatory [New Zealand] bulletin E 159. New Zealand Seismological Report 1977.

111. Matata earthquake 1977 May 31

Richardson, W. P. 1989. The Matata earthquake of 1977 May 31: a recent event near Edgecumbe, Bay of Plenty, New Zealand. *New Zealand journal of geology and geophysics 32:* 17-30.

Seismological Observatory [New Zealand] bulletin E 159. New Zealand Seismological Report 1977.

112. Puysegur Bank earthquake 1979 October 12

Anderson, H., Webb, T., Jackson, J. 1993. Focal mechanisms of large earthquakes in the South Island of New Zealand: implications for the accommodation of Pacific-Australia Plate motion. *Geophysical journal international 115:* 1032-1054.

Seismological Observatory [New Zealand] bulletin E 161. New Zealand Seismological Report 1979.

Webb, T. H., Lowry, M. A. 1982. The Puysegur Bank earthquake of 1979 October 12. *New Zealand journal of geology and geophysics 25:* 383-395.

113. Hawke's Bay earthquake 1982 September 02

Anderson, H., Webb, T., Jackson, J. 1993. Focal mechanisms of large earthquakes in the South Island of New Zealand: implications for the accommodation of Pacific-Australia Plate motion. *Geophysical journal international 115:* 1032-1054.

Reyners, M. 1984. The Hawke's Bay, New Zealand, earthquake of 1982 September 02 and deformation in the interior of the subducted Pacific plate. *Tectonophysics 106*: 259-273.

Seismological Observatory [New Zealand] bulletin E 165. New Zealand Seismological Report 1982.

114. Waiotapu earthquake 1983 December 14

Seismological Observatory [New Zealand] bulletin E 167. New Zealand Seismological Report 1983.

Smith, E. G. C., Scott, B. J., Latter, J. H. 1984. The Waiotapu earthquake of 1983 December 14. *Bulletin of the New Zealand National Society for Earthquake Engineering 17:* 272-279.

115. Motu River earthquake 1984 March 08

Reyners, M., Hodder, S. 1985. The Motu River earthquake of 8 March 1984. *Bulletin of the New Zealand National Society for Earthquake Engineering 18:* 41-54.

Seismological Observatory [New Zealand] bulletin E 168. New Zealand Seismological Report 1984.

116. Macauley River earthquake 1984 June 24

Anderson, H., Webb, T., Jackson, J. 1993. Focal mechanisms of large earthquakes in the South Island of New Zealand: implications for the accommodation of Pacific-Australia Plate motion. *Geophysical journal international 115:* 1032-1054.

Seismological Observatory [New Zealand] bulletin E 168. New Zealand Seismological Report 1984.

117a, 117b. Edgecumbe earthquake 1987 March 02

A collection of 21 papers on the Edgecumbe earthquake sequence including Lowry, M. A. *et al.* are contained in a special issue of the *New Zealand journal of geology and geophysics 32:* 1-190. Other related articles are listed in the various papers in this volume.

Lowry, M. A., Ede, S. C., Harris, J. S. 1989. Assessment of seismic intensities resulting from the 1987 Edgecumbe earthquake, New Zealand, and implications for modernising the intensity scale. *New Zealand journal of geology and geophysics 32:* 145-152.

Seismological Observatory [New Zealand] bulletin E 171. New Zealand Seismological Report 1987.

118. Pegasus Bay earthquake 1987 March 08

Seismological Observatory [New Zealand] bulletin E 171. New Zealand Seismological Report 1987.

119. Te Anau earthquake 1988 June 03

Anderson, H., Webb, T., Jackson, J. 1993. Focal mechanisms of large earthquakes in the South Island of New Zealand: implications for the accommodation of Pacific-Australia Plate motion. *Geophysical journal international 115:* 1032-1054.

Seismological Observatory [New Zealand] bulletin E 172. New Zealand Seismological Report 1988.

Reyners, M., Gledhill, K., Waters, D. 1991. Tearing of the subducted Australian Plate during the Te Anau, New Zealand, earthquake of 1988 June 3. *Geophysical journal international 104:* 105-115.

120. Doubtful Sound earthquake 1989 May 31

Anderson, H., Webb, T., Jackson, J. 1993. Focal mechanisms of large earthquakes in the South Island of New Zealand: implications for the accommodation of Pacific-Australia Plate motion. *Geophysical journal international 115:* 1032-1054.

Seismological Observatory [New Zealand] bulletin E 173. New Zealand Seismological Report 1989.

121. Lake Tennyson earthquake 1990 February 10

Seismological Observatory [New Zealand] bulletin E 174. New Zealand Seismological Report 1990.

122. Weber I earthquake 1990 February 19

Robinson, R. 1992. Analysis of the Weber (Dannevirke) earthquakes of 2 February and 13 May 1990, and their aftershocks. *Institute of Geological & Nuclear Sciences Contract Report 1992/64* prepared for Earthquake and War Damage Commission.

Robinson, R. 1994. Shallow subduction tectonics and fault interaction: Weber, New Zealand, earthquake sequence of 1990-1992. *Journal of geophysical research 99B:* 9663-9679.

Seismological Observatory [New Zealand] bulletin E 174. New Zealand Seismological Report 1990.

123a, 123b. Weber II earthquake 1990 May 13

Berryman, K., Beanland, S. 1990. Reconnaissance of the epicentral area of May 13, 1990 "Weber 2" earthquake, Dannevirke district. *New Zealand Geological Survey EDS Immediate Report 90/5.*

Johnstone, P. G., Potangaroa, R., (edited by Moss, P. J.) 1993. Reconnaissance report on the Weber earthquake - 13 May 1990. *Bulletin of the New Zealand National Society for Earthquake Engineering 26*: 222-239.

Robinson, R. 1992. Analysis of the Weber (Dannevirke) earthquakes of 2 February and 13 May 1990, and their aftershocks. *Institute of Geological & Nuclear Sciences Contract Report 1992/64* prepared for Earthquake and War Damage Commission.

Robinson, R. 1994. Shallow subduction tectonics and fault interaction: Weber, New Zealand, earthquake sequence of 1990-1992. *Journal of geophysical research 99B:* 9663-9679.

Seismological Observatory [New Zealand] bulletin E 174. New Zealand Seismological Report 1990.